METHODS IN CELL BIOLOGY

VOLUME 34

Vectorial Transport of Proteins into and across Membranes

Series Editor

LESLIE WILSON

Department of Biological Sciences
University of California, Santa Barbara
Santa Barbara, California

ASCB

METHODS IN CELL BIOLOGY

Prepared under the Auspices of the American Society for Cell Biology

VOLUME 34
Vectorial Transport of Proteins into and across Membranes

Edited by

ALAN M. TARTAKOFF

INSTITUTE OF PATHOLOGY
CASE WESTERN RESERVE UNIVERSITY SCHOOL OF MEDICINE
CLEVELAND, OHIO

ACADEMIC PRESS, INC.
Harcourt Brace Jovanovich, Publishers

San Diego   New York   Boston
London   Sydney   Tokyo   Toronto

Copyright © 1991 BY ACADEMIC PRESS, INC.
All Rights Reserved.
No part of this publication may be reproduced or transmitted in any form or
by any means, electronic or mechanical, including photocopy, recording, or
any information storage and retrieval system, without permission in writing
from the publisher.

Academic Press, Inc.
San Diego, California 92101

United Kingdom Edition published by
ACADEMIC PRESS LIMITED
24-28 Oval Road, London NW1 7DX

Library of Congress Catalog Card Number: 64-14220

ISBN   0-12-564134-6   (alk. paper)

PRINTED IN THE UNITED STATES OF AMERICA
91 92 93 94   9 8 7 6 5 4 3 2 1

For Paola Ymayo,
Daniela Helen Elisabeth,
Joseph Michael,
and Laura

CONTENTS

CONTRIBUTORS

Numbers in parentheses indicate the pages on which the authors' contributions begin.

YOSHINORI AKIYAMA, Institute for Virus Research, Kyoto University, Kyoto 606, Japan (189)

KEVIN P. BAKER, Department of Biochemistry, Biocenter, University of Basel, CH-4056 Basel, Switzerland (377)

JAMES P. BELTZER, Department of Biochemistry, Biocenter, University of Basel, CH-4056 Basel, Switzerland (287)

GÜNTER BLOBEL, The Laboratory of Cell Biology, Howard Hughes Medical Institute, Rockefeller University, New York, New York 10021 (263)

ANDERS BRANDT,[1] Department of Physiology, Biocenter, University of Basel, CH-4056 Basel, Switzerland (369)

LORNA BRUNDAGE, Department of Biological Chemistry, Molecular Biology Institute, University of California, Los Angeles, Los Angeles, California 90024 (147)

GERALD COHEN,[2] Department of Cell Biology and Anatomy, Mount Sinai School of Medicine, New York, New York 10029 (303)

PAULA COLLINS, Department of Biochemistry and Molecular Biology, University of Massachusetts Medical School, Worcester, Massachusetts 01655 (223)

ROSS E. DALBEY, Department of Chemistry, The Ohio State University, Columbus, Ohio 43210 (39)

ARNOLD J. M. DRIESSEN,[3] Department of Biological Chemistry, Molecular Biology Institute, University of California, Los Angeles, Los Angeles, California 90024 (147)

STEPHEN D. FULLER, The Cell Biology Program and The Biological Structures and Biocomputing Program, European Molecular Biology Laboratory, D-6900 Heidelberg, Germany (1)

REID GILMORE, Department of Biochemistry and Molecular Biology, University of Massachusetts Medical School, Worcester, Massachusetts 01655 (223)

LYNNE GILSON, Department of Microbiology and Molecular Genetics, Harvard Medical School, Boston, Massachusetts 02115 (205)

BENJAMIN S. GLICK, Biocenter, University of Basel, CH-4056 Basel, Switzerland (389)

DIRK GÖRLICH, Central Institute of Molecular Biology, D-1115 Berlin-Buch, Germany (241)

JOSEPH P. HENDRICK, Department of Biological Chemistry, Molecular Biology Institute, University of California, Los Angeles, Los Angeles, California 90024 (147)

VICTORIA HINES, Department of Biochemistry, Biocenter, University of Basel, CH-4056 Basel, Switzerland (377)

[1]*Present address:* Department of Physiology, Carlsberg Laboratory, DK-2500 Copenhagen, Denmark.
[2]*Present address:* Sandoz Forschungsinstitut, G.m.b.H., A-1235 Vienna, Austria.

[3]*Permanent address:* Department of Microbiology, University of Groningen, 9751 NN Haren, The Netherlands.

MAURICE HOFNUNG, Unité de Programmation Moléculaire et Toxicologie Génétique, CNRS UA 1444, Institut Pasteur, 75015 Paris, France (77)

TSUNEO IMANAKA, Department of Microbiology and Molecular Pathology, Teikyo University, Sagamiko, Kanagawa 199-01, Japan (303)

KOREAKI ITO, Institute for Virus Research, Kyoto University, Kyoto 606, Japan (189)

THOMAS JASCUR, Department of Biochemistry, Biocenter, University of Basel, CH-4056 Basel, Switzerland (359)

JULIE JOHNSON, Department of Biochemistry and Molecular Biology, University of Massachusetts Medical School, Worcester, Massachusetts 01655 (223)

KENNETH KEEGSTRA, Department of Botany, University of Wisconsin, Madison, Wisconsin 53706 (327)

KENNAN KELLARIS, Department of Biochemistry and Molecular Biology, University of Massachusetts Medical School, Worcester, Massachusetts 01655 (223)

ROBERTO KOLTER, Department of Microbiology and Molecular Genetics, Harvard Medical School, Boston, Massachusetts 02115 (205)

UTE C. KRIEG, Department of Biochemistry, Biocenter, University of Basel, CH-4056 Basel, Switzerland (409, 419)

TEYMURAS V. KURZCHALIA, Central Institute of Molecular Biology, D-1115 Berlin-Buch, Germany (241)

PAUL B. LAZAROW, Department of Cell Biology and Anatomy, Mount Sinai School of Medicine, New York, New York 10029 (303)

HSOU-MIN LI, Department of Botany, University of Wisconsin, Madison, Wisconsin 53706 (327)

JIAN P. LIAN, Department of Cell and Molecular Biology, Boston Biomedical Research Institute, Boston, Massachusetts 02114 (167)

COLIN MANOIL, Department of Genetics, University of Washington, Seattle, Washington 98195 (61)

SHIN-ICHI MATSUYAMA, Institute of Applied Microbiology, The University of Tokyo, Yayoi, Tokyo 113, Japan (107)

GIOVANNI MIGLIACCIO, The Laboratory of Cell Biology, Howard Hughes Medical Institute, Rockefeller University, New York, New York 10021 (263)

SHOJI MIZUSHIMA, Institute of Applied Microbiology, The University of Tokyo, Yayoi, Tokyo 113, Japan (107)

CHRISTOPHER NICCHITTA, The Laboratory of Cell Biology, Howard Hughes Medical Institute, Rockefeller University, New York, New York 10021 (263)

SHARYN E. PERRY, Department of Botany, University of Wisconsin, Madison, Wisconsin 53706 (327)

NIKOLAUS PFANNER, Institut für Physiologische Chemie, Universität München, D-8000 München 2, Germany (345)

PETER RAPIEJKO, Department of Biochemistry and Molecular Biology, University of Massachusetts Medical School, Worcester, Massachusetts 01655 (223)

TOM A. RAPOPORT, Central Institute of Molecular Biology, D-1115 Berlin-Buch, Germany (241)

JOACHIM RASSOW, Institut für Physiologische Chemie, Universität München, D-8000 München 2, Germany (345)

PHILIPP E. SCHERER, Department of Biochemistry, Biocenter, University of Basel, CH-4056 Basel, Switzerland (409, 419)

ELMAR SCHIEBEL, Department of Biological Chemistry, Molecular Biology Institute, University of California, Los Angeles, Los Angeles, California 90024 (147)

ANDRÉ SCHNEIDER, Biocenter, University of Basel, CH-4056 Basel, Switzerland (401)

RACHEL C. SKVIRSKY,[4] Department of Microbiology and Molecular Genetics, Harvard Medical School, Boston, Massachusetts 02115 (205)

GILLIAN SMALL, Department of Anatomy and Cell Biology, University of Florida at Gainesville, Gainesville, Florida 32610 (303)

THOMAS SÖLLNER, Institut für Physiologische Chemie, Universität München, D-8000 München 2, Germany (345)

MARTIN SPIESS, Department of Biochemistry, Biocenter, University of Basel, CH-4056 Basel, Switzerland (287)

PHANG C. TAI, Department of Cell and Molecular Biology, Boston Biomedical Research Institute, Boston, Massachusetts 02114 (167)

ROLF THIERINGER,[5] Department of Cell Biology and Anatomy, Mount Sinai School of Medicine, New York, New York 10029 (303)

GUOLING TIAN, Department of Microbiology and Molecular Genetics, Harvard Medical School, Boston, Massachusetts 02115 (167)

HAJIME TOKUDA, Institute of Applied Microbiology, The University of Tokyo, Yayoi, Tokyo 113, Japan (107)

DAVID VAUX, The Cell Biology Program and The Biological Structures and Biocomputing Program, European Molecular Biology Laboratory, D-6900 Heidelberg, Germany (1)

HANS PETER WESSELS, Department of Biochemistry, Biocenter, University of Basel, CH-4056 Basel, Switzerland (287)

WILLIAM WICKNER, Department of Biological Chemistry, Molecular Biology Institute, University of California, Los Angeles, Los Angeles, California 90024 (147)

MARTIN WIEDMANN, Central Institute of Molecular Biology, D-1115 Berlin-Buch, Germany (241)

HAODA XU, Department of Microbiology and Molecular Genetics, Harvard Medical School, Boston, Massachusetts 02115 (167)

JACK N. YU, Department of Cell and Molecular Biology, Boston Biomedical Research Institute, Boston, Massachusetts 02114 (167)

[4]*Present address:* Biology Department, University of Massachusetts at Boston, Harbor Campus, Boston, Massachusetts 02125.

[5]*Present address:* Department of Biochemistry, Merck, Sharpe & Dohme Pharmaceuticals, Rathway, New Jersey 07065.

PREFACE

The presence of membranes around and within cells has made necessary the evolution of efficient mechanisms that cause selective vectorial translocation of macromolecules across lipid bilayers. The speed and accuracy with which translocation occurs are impressive. This thermodynamically improbable topological feat has fascinated cell biologists, who have used biochemical, electrophysiologic, genetic, microscopic, and molecular biological approaches for its analysis.

Biogenetic translocation of proteins into and across membranes in higher eukaryotic cells is known to occur at four sites: the rough endoplasmic reticulum (RER), mitochondria, chloroplasts, and peroxisomes. Due to extensive vesicular communication between the RER and downstream compartments along the secretory and endocytic paths, the ultimate location of membrane proteins inserted into the RER may be the RER, Golgi membranes, endosomes, lysosomes, or the plasma membrane. The observation that three other sites of membrane insertion exist suggests that there is absolutely no vesicular communication among these four compartments; were communication to exist, one would expect it to have been exploited by evolution for delivery of membrane proteins from one class of membranes to another. Nothing is known of those constraints which apparently prohibit membrane fusion between these compartments.

In addition to the four sites of membrane insertion mentioned, a fifth membrane in eukaryotic cells is equipped with a conspicuous translocation apparatus for newly synthesized proteins: the double membrane which surrounds the nucleus. In this case, all indications are that only soluble proteins and ribonucleoproteins, as opposed to membrane proteins, are translocated through the pores. Apart from the biosynthetic activity of the ribosomes attached to the outer nuclear membrane—which appears equivalent to the rest of the RER—there is no evidence that the nuclear membrane functions as an independent site for insertion of a distinct class of membrane proteins. Nevertheless, very few proteins of the inner nuclear membrane have been studied; therefore, the questions of their biosynthetic origin and possible, unidentified biosynthetic functions in the inner membrane remain unanswered.

One of the major successes of cell biology has been the achievement of a unified model, which in general terms, can explain translocation of newly synthesized proteins across any of the four compositionally and functionally distinct membranes mentioned above. Thus, much of the protein translocation in eukaryotes exhibits the following three characteristics: (1) the protein undergoing translocation has certain key covalent and conformational features ("signals"); (2) the protein forms a complex with soluble factors which help maintain it in an "unfolded" conformation; and (3) this complex facilitates targeting (in certain cases, by direct interaction) to membrane components which accomplish translocation.

The specific features which pertain at each of these three steps depend on the destination of the protein. Thus, the covalent "signals" which direct a protein into the RER are substantially different from those which would lead a protein into mitochondria or peroxisomes. With the exception of hsp70, which participates in import into both mitochondria and the RER, the soluble factors also appear to be distinct. The sites of translocation across each of the membranes in question, although only partly described in molecular terms, also appear distinct.

There are close analogies between this eukaryotic scenario and what is known of protein export to the outer membrane and periplasm of gram-negative bacteria, where the prospect for genetic analysis is especially great. Moreover, certain prokaryotic and eukaryotic signals and factors required for translocation are extensively interchangeable.

A further complexity has emerged in several cases where a primary translocation event is followed by relocation of the product. Hence, pairs of mutually compatible and unambiguous targeting signals must coexist within individual polypeptides. This is the case for targeting first to and then within chloroplasts and mitochondria, and also for the secretion by gram-negative bacteria of proteins such as cholera toxin, which appear first to pass across the inner membrane into the periplasmic space and then traverse the outer membrane.

Apart from translocation, which can be explained according to the basic three-step translocation model, a number of "alternate" translocation mechanisms have been detected. Each is independent of "conventional" signal sequences analogous to those which function in conjunction with the bacterial *SEC* gene products or the higher eukaryotic SRP-dependent path. These include:

1. The secretion by yeast (Kuchler *et al.*, 1989) and, possibly, higher eukaryotic cells (Rubartelli *et al.*, 1990; Young *et al.*, 1988), of proteins which lack obvious signal sequences and appear not to be delivered to the cisternal space of the RER. In yeast, these events make use of membrane proteins which show homology to the multidrug transporter.
2. The delivery of cytosolic antigen fragments to the secretory path, apparently at the level of the RER, prior to their expression at the cell surface (Cresswell, 1990; Yewdell and Beenick, 1990).
3. The *SEC* gene independent translocation of selected proteins, such as the small M13 coat protein of the inner membrane of virus-infected *Escherichia coli* (Wickner, 1988).

4. The secretion to the extracellular medium of hemolysin and certain colicins by *E. coli*. These events do not involve conventional signal sequences. Secretion may pass via the zones of adhesion between the inner and outer membranes and involve proteins related to the multidrug transporter (Gray *et al.*, 1989; Hartlein *et al.*, 1983; see also R. Skvirsky *et al.*, Chapter 9, this volume).

5. The secretion to the extracellular medium of other colicins which also lack conventional signal sequences. In these cases, secretion involves activation of host cell phospholipases by bacteriocin release proteins (Luria and Suit, 1987; Lazdunski, 1990; Pugsley and Schwartz, 1985).

There are also several striking examples of protein passage across membranes from the ecto- to endodomain (i.e., outside to inside). Among these are: (1) the entry of many protein toxins into animal cells, often requiring endocytosis into an acidic compartment (Olsnes and Sandvig, 1985); and (2) the entry of colicins into bacteria (Luria and Suit, 1987; Lazdunski, 1990; Pugsley and Schwartz, 1985).

With regard to nucleic acid transport across membranes, apart from bacterial transformation, the only events which have been studied at length concern transport of RNAs across the nuclear membrane. All evidence is consistent with transport via nuclear pores; however, since protein transport does occur across so many varieties of membranes, it is certainly premature to exclude the possibility of nucleic acid transport across membranes which lack obvious pores. Several covalent features of RNA and specific interactions are essential for RNA transport across the nuclear membrane:

1. For 5s RNA, association with a ribosomal protein (L5) or transcription factor (TFIIIA) precedes export from the nucleus (Guddat *et al.*, 1990).

2. Several base changes and interruption of 3′ and 5′ processing interrupt tRNA export (Tobian *et al.*, 1985); however, it is not known which proteins, if any, participate.

3. snRNAs exit from the nucleus, undergo further 5′ cap methylation in the cytoplasm, and then return to the nucleus in association with snRNP proteins which they have collected from the cytoplasm. The return leg of this cycle depends on acquisition of a trimethyl guanosine cap structure (Parry *et al.*, 1989; Hamm *et al.*, 1990).

4. mRNAs with monomethyl guanosine caps are exported, possibly by some of the large number of proteins with which they are associated in the nucleus (Dreyfuss, 1986; Hamm and Mattaj, 1990).

In each of these cases, it is likely that RNA–protein complexes, as opposed to RNA alone, are exported. Remarkably little is known of the transport machinery.

Furthermore, there are a number of nearly unexplored examples of DNA and RNA transport. For example: (1) the passage of viral genomes into the host cell nucleus; (2) the apparent transfer of entire chromosomes between nuclei in heterokaryons (Nilsson-Tillgren *et al.*, 1981; Dutcher, 1981); and (3) the import of nuclearly encoded RNA and protein–DNA conjugates into mitochondria (Chang and Clayton, 1989; Vestweber and Schatz, 1989).

For the analysis of protein translocation enumerated above, cell-free systems have been essential. Moreover, major genetic contributions have been important in their own right and have provided key material for the cell-free analysis of bacteria and yeast. The present volume is designed to provide detailed methodological information for such cell-free analysis. My hope is that it will both lead to elaboration of those lines of research which are already underway and contribute to analogous investigations of some of the less understood translocation phenomena involving both proteins and nucleic acids.

There have been only a few methodological volumes dedicated to protein transport across membranes. Volume 96 (1983) of *Methods in Enzymology* has several chapters on import into the RER; Volume 97 (1983) has chapters on prokaryotes, mitochondria, and chloroplasts; and Volume 194 (1990) covers import into the RER in yeast.

Special thanks to Marie Ward for her help in preparing this volume.

ALAN M. TARTAKOFF

REFERENCES

Chang, D., and Clayton, D. (1989). *Cell (Cambridge, Mass.)* **56**, 131–139.
Cresswell, P. (1990). *Nature (London)* **343**, 593–594.
Dreyfuss, G. (1986). *Annu. Rev. Cell Biol.* **2**, 459–498.
Dutcher, S. (1981). *Mol. Cell Biol.* **1**, 245–253.
Gray, L., Baker, K., Keny, B., Mackman, N., Haigh, R., and Holland, I. B. (1989). *J. Cell Sci. Suppl.* **11**, 45–57.
Guddat, U., Bakken, A., and Pieler, T. (1990). *Cell (Cambridge, Mass.)* **60**, 619–628.
Hamm, J., and Mattaj, I. (1990). *Cell (Cambridge, Mass.)* **63**, 109–118.
Hamm, J., Darzynkiewicz, E., Tahara, S., and Mattaj, I. (1990). *Cell (Cambridge, Mass.)* **62**, 569–577.
Hartlein, M., Schliessl, Wagner, W., Rdest, U., Kreft, J., and Goebel, W. (1983). *J. Cell Biochem.* **22**, 87–97.
Kuchler, K., Sterne, R., and Thorner, J. (1989). *EMBO J.* **13**, 3973–3984.
Lazdunski, C. (1990). *In* "Dynamics and Biogenesis of Membranes" (J. Op den Kamp, ed.), pp. 269–289. Springer-Verlag, New York.
Luria, S., and Suit, J. (1987). *In* "*E. coli* and *S. typhimurium*" (F. Neihardt, ed.), pp. 1615–1624. American Society for Microbiology, Washington, D.C.
Nilsson-Tillgern, T., Gjermansen, C., Kielland-Brandt, M., Petersen, J., and Holmberg, S. (1981). *Carlsberg Res. Commun.* **46**, 65–76.
Olsnes, S., and Sandvig, K. (1985). *In* "Endocytosis" (I. Patsen and M. Willingham, eds.), pp. 195–234. Plenum, New York.

Parry, H., Scherly, D., and Mattaj, I. (1989). *Trends Biochem. Sci.* **14**, 15–19.

Pugsley, A., and Schwartz, M. (1985). *FEMS Microbiol. Rev.* **32**, 3–38.

Rubartelli, A., Cozzolino, F., Talio, M., and Sitia, R. (1990). *EMBO J.* **9**, 1503–1510.

Tobian, J., Drinkard, L., and Zasloff, M. (1985). *Cell (Cambridge, Mass.)* **43**, 415–422.

Vestweber, D., and Schatz, G. (1989). *Nature (London)* **338**, 170–172.

Wickner, W. (1988). *Biochemistry* **27**, 1081–1086.

Yewdell, J., and Beenick, J. (1990). *Cell (Cambridge, Mass.)* **62**, 203–206.

Young, P., Hazuda, D., and Simon, P. (1988). *J. Cell Biol.* **107**, 447–456.

Chapter 1

The Use of Antiidiotype Antibodies for the Characterization of Protein–Protein Interactions

DAVID VAUX AND STEPHEN D. FULLER

The Cell Biology Program and
The Biological Structures and Biocomputing Program
European Molecular Biology Laboratory
D-6900 Heidelberg, Germany

I. Introduction

Many questions in biology can be reduced to a problem of protein–protein interactions. These proteins may be enzyme and substrate, receptor and ligand, cell surface, and extracellular matrix or targeting signals on

1

two organelles within a cell. A wide variety of biochemical methods are available for the analysis of interactions between proteins, and many biological processes have been characterized using them. However, some interactions are not easily addressed using biochemical techniques. These include transient or conformation-dependent interactions, such as the Lys-Asp-Glu-Leu (KDEL) receptor system, and interactions that are individually weak but reach a useful level when multiplied in repetitive polymeric structures, such as the alphavirus budding signal. In these situations it is sometimes possible to obtain results by making use of the ability of the immune system to generate networks of immunoglobulins (network antibodies), which can then be used as specific high-affinity probes for both partners in a protein–protein interaction. In this chapter we outline the theoretical basis for the analysis of protein–protein interactions using network antibodies, consider the advantages and limitations of the approach, and then discuss two experimental systems that illustrate many of these points.

A.   Network Theory

An immunoglobulin (Ig) molecule consists of a basic heterotetrameric unit containing two identical heavy chains and two identical light chains linked by disulfide bonds. Each chain is encoded by a separate gene and is organized into domains; a chain consists of one variable region and one or more constant domains. The antigen-combining site is assembled from the variable domains of both heavy and light chains, with the predominant contacts occurring between antigen and three hypervariable regions within the variable domain, known as the complementarity determining regions (CDRs). A central concept in immunology is that the antigen-combining site of an Ig is a novel structure that may itself be immunogenic (Jerne, 1974). The epitopes of an Ig-variable region are described as idiotopes; the array of idiotopes on an immunoglobulin constitutes its idiotype and antibodies elicited against them are described as antiidiotype antibodies. The humoral response to a self immunoglobulin contains only antiidiotype antibodies that fall (in the simplest analysis) into two classes—those recognizing idiotopes that lie wholly within the antigen-combining site of the first immunoglobulin, and those recognizing idiotopes that lie wholly outside this region. The latter give rise to "framework" antiidiotypes, whereas the former give rise to "internal image" antiidiotypes, so called because the antigen-combining site of the antiidiotype antibody is related spatially to the original antigen. This network of interconnected antibodies has many implications for our understanding of diversity and the control of the immune system (Burdette and Schwarz, 1987; Coutinho, 1989; Jerne, 1974, 1985).

B. Structure of Antiidiotype Antibodies

Antiidiotype antibodies are immunogloblins with same structural organization as any other immunoglobulin. There are several solved crystal structures for antiidiotype Igs and for Ig–antigen complexes (e.g., Amit *et al.*, 1986), but only a single example exists of an idiotype–antiidiotype complex in which the idiotype–antigen complex is also solved (Bentley *et al.*, 1990). In the single example known of antigen–idiotype complex and idiotype–antiidiotype complex, the contact residues important in the binding of the idiotype to the antigen are significantly different from the contact residues important for the interaction of the idiotype with the antiidiotype. At first sight, this result calls into question the idea of an internal image in an antiidiotype antibody. However, this complex involves an antilysozyme antibody that is known to have a large area of contact with a relatively flat face of the antigen, and it is possible that this is not a good example from which to generalize. The authors conclude that "it is likely that the practical use of this approach will decrease with the increasing complexity of the target antigens, from oligopeptides to fully folded, multi-subunit proteins" (Bentley *et al.*, 1990), and this is possible. Nonetheless, we still do not know whether the interaction between an antiidiotype antibody and its antigen, the idiotype Ig, differs in any general way from the interaction of an antibody with a non-Ig antigen. In the many cases in which short oligopeptide signals have been identified, this technology may be ideal. This is a very important point, because we are forced to assume that the same constraints that govern the size and accessibility of non-Ig epitopes are extendable to idiotopes. This is an explicit assumption underlying the use of internal image antiidiotype antibodies for probing protein–protein interactions.

By definition, idiotype Ig is recognized by an internal image antiidiotype only when the antiidiotype Ig antigen-combining site mimics the starting antigen used to elicit the idiotype antibody. Thus, if the starting antigen is a "ligand," one can seek to identify the "receptor" by virtue of its affinity for the antiidiotype antibody, and this is the basis for the experimental approach described here (Fig. 1). Several important consequences flow from the fact that an antiidiotype antibody can only mimic the original antigenic epitope within the constraints of an immunoglobulin framework.

A relevant question concerns the way in which an antiidiotype Ig assembles an antigen-combining site that mimics the original antigen. The potential for an epitope to give rise to a useful internal image antiidiotype will

ROUND	IMMUNOGEN	RESPONSE	SPECIFICITY
1	ligand	idiotype	ligand
2	idiotype	antiidiotype	receptor *(and idiotype)*
3	antiidiotype	anti-antiidiotype	ligand *(and antiidiotype)*

FIG. 1. The relationship between the members of an antibody network is shown with the corresponding components of a receptor–ligand system.

depend upon the ability of the CDRs to reproduce the shape of the original epitope. This is an important constraint of the technique. One may imagine that an epitope ranges in complexity from a simple structure that may be copied using a variety of primary amino acid sequences, to a structure so complex that the only way in which it may be recreated is by the repetition of exactly the same amino acid sequence. In a characterization of an antiidiotype antibody raised against a neutralizing monoclonal antibody to type 3 reovirus, Greene and colleagues sequenced the variable region of an internal image antiidiotype Ig and found strong homology between the sequence of the heavy chain CDR II and the sequence of the epitope recognized by the neutralizing monoclonal antibody (Bruck *et al.*, 1986; Williams *et al.*, 1988, 1989). When other antiidiotype CDR sequences are compared with the sequence of the original antigen, it will be possible to assess whether this is an unusual occurrence. Until then, we have no information on how frequently the antiidiotype CDRs must converge upon the sequence of the original antigen. Thus, one would predict that the CDRs of internal image antiidiotype antibodies will range from perfect sequence homology with the underlying epitope, to no apparent sequence homology, although the spatial structure is still being copied.

Another structural assumption that is made when protein–protein interactions are modeled with antiidiotype antibodies is that the constraint imposed upon the antiidiotype CDRs by the Ig framework will not prevent the whole molecule from being a useful mimic of the starting antigen. Thus, even if the ligand structure is mimicked exactly by the internal image antiidiotype antibody, this will not be useful if the ligand-binding site on the receptor lies within a cleft smaller than the dimensions of the variable region domain of the immunoglobulin.

C.   Antiidiotype Antibodies as Probes of Protein–Protein Interactions

Sege and Peterson were the first to show that internal image antiidiotype antibodies could be exploited for the analysis of protein–protein interactions (Sege and Peterson, 1983). If an antibody recognizes a specific ligand, then a subset of its internal image antiidiotypes will mimic the ligand and bind specifically to its receptor. Thus, it is possible to produce antibodies that define previously unidentified receptors starting from a knowledge of the ligand alone, using the terms ligand and receptor in their most general sense. This approach has been used to study chloroplast import receptors (Pain *et al.*, 1988) and the import machinery of mitochondria (Murakami *et al.*, 1990; Pain *et al.*, 1990), as well as the interaction of intermediate filaments with the nuclear envelope (Djabali *et al.*, 1991). A wide-ranging survey of the use of this technology is found in "Methods in Enzymology," (Volume 178). Internal image antiidiotype antibodies that usefully mimic ligand will be referred to simply as antiidiotypes in the remainder of this chapter.

The expanding field receptor mimicry by antiidiotype reagents has benefited from a variety of strategies for antibody production, depending upon the available knowledge of the starting epitope (for general reviews, see Gaulton and Greene, 1986; Farid and Lo, 1985). Most of the early studies involved the use of genetically disparate rabbits for the idiotype and antiidiotype immunizations. This approach suffers from the problem that serologically detectable differences on immunoglobulins (allotypes) will result in an antiallotype response as well as an antiidiotype response. This difficulty can be overcome by two rounds of affinity selection, or by using genetically identical (syngeneic) animals for both the first- and second-round immunizations. If mice are used, the approach also offers the advantage that antiidiotype responses may be immortalized by hybridoma fusion to generate monoclonal antiidiotype reagents. The examples given in this chapter concentrate exclusively on the use of monoclonal antibody methods.

STRATEGY FOR ANTIIDIOTYPE PRODUCTION

The general strategy for antiidiotype production rests on the available knowledge of the system under study. The methods fall into two groups: if it is possible to select a single relevant idiotype antibody, then the purified immunoglobulin is used as the antigen; if no assay is available to identify a specific idiotype, then a mixture of idiotypes is used for immunization. In the former case, specificity controls are relatively obvious and the selection of relevant antiidiotypes may be straightforward. In the latter case, all of the selection is performed on the putative antiidiotypes and the simple specificity checks involving idiotype competition are not available. Thus, the screening method is of crucial importance and special controls (see below) become essential.

D.   Sequential Immunization Using a Defined Idiotype

If the exact ligand is known, it should be possible to identify a relevant idiotype simply because it is able to recognize the ligand. It is important to recognize that the method will only work if the *exact* ligand structure is immunogenic in the system used—it is not sufficient that the same protein, or even the same region of the protein, be immunogenic. The ideal idiotype is a monoclonal antibody that has been shown efficiently to inhibit the interaction between the ligand and receptor. Failing this, it is possible to use antibodies against small peptides known to be or to contain the ligand as idiotypes, even if it is not possible to show inhibition in a functional assay. This sequential approach is experimentally straightforward and success depends on the exact ligand being sufficiently immunogenic to produce an appropriate idiotype, which is itself immunogenic enough to result in a usable internal image antiidiotype.

Although this approach was pioneered using outbred rabbits for immunization, recent studies have often made use of inbred animals strains to avoid allotype differences. We have found that the best antiidiotype response in mice is usually obtained by immunization of the syngeneic mouse strain with fixed idiotype-secreting hybridoma cells. It is also possible to generate antiidiotype responses by growing hybridomas as ascitic tumors in unirradiated syngeneic animals, although this is not optimal because the animals rarely survive long enough for isotype switching to occur and the titers of the IgM produced are not usually high.

The antiidiotype response may show an unusual or unexpected time course in comparison with the immune response to an antigen that is not a syngeneic immunoglobulin. For example, it is often observed that the initial response at 2–3 weeks after immunization gives the best antiidiotype reactivity in the serum of the recipient animal (Marriott *et al.,* 1987). Subsequent boosts with idiotype immunoglobulin may result in the reduction or even disappearance of this response. Insufficient experience has yet accumulated to say whether or not this means that the animals should be taken for fusion as soon as they show an antiidiotype response. In our hands, however, the fusion is normally carried out after a primary immunization and a single boost.

E. Generation of Monoclonal Antiidiotype Antibodies by Paired *in Vitro* Immunization

In the absence of an assay for the selection of a single relevant idiotype, the best idiotype pool would consist of similar amounts of all antibodies against all of the epitopes present in the antigen, regardless of their relative immunogenicity. Such a mixture is not easily obtained by animal immunization because multiple doses of antigen are used and the resulting polyclonal response is strongly shaped by the relative strengths of the epitopes and the contribution of T cell modulation. A closer approach to this ideal may be obtained from primary *in vitro* immunization, in which all possible B cell responses occur and the amplification due to the repeated stimulation of an immunodominant clone is not seen. This also has the advantage of rapidity; most protocols for *in vitro* immunization specify 5 days between starting the immunization and the fusion (for a discussion of *in vitro* immunization, see Reading, 1982; Vaux, 1990b).

It would be possible to attempt to collect all of the idiotypes resulting from the *in vitro* immunization by myeloma fusion and hybridoma selection, but this introduces another level at which the response may become restricted. A technically simple and theoretically attractive alternative is to take the medium conditioned by the *in vitro* response, assume that it contains small amounts of all of the idiotype immunoglobulins secreted as a result of the initial immunization, and use this mixture as the antigen for a second round of *in vitro* immunization (Vaux, 1990a; Vaux *et al.*, 1988). In this way, all of the idiotypes are presented to the immune system for the generation of antiidiotypes, which can then be rescued from the second *in vitro* immunization by myeloma fusion and hybridoma formation. Provided that the reconstituted immune systems immunized *in vitro* are genetically identical no response to the constant region framework of the idiotype antibodies will be seen.

The paired *in vitro* immunization approach has the advantages of speed and simplicity and results in the production of monoclonal antibodies rather than highly variable polyclonal antiidiotype antisera. It also has the theoretical advantage that the immune response both to the starting antigen and subsequently to the idiotype mixture is as complete as possible because antigen-specific T suppression is circumvented (Schrier and Lefkovits, 1979). It is also possible that the predominance of pentavalent IgM in the idiotype mixture may enhance the antiidiotype response.

The disadvantages of paired *in vitro* immunization flow from the fact that the idiotype response to the initial antigen is used as an unfractionated mixture for the second-round immunization. Therefore, the starting idiotype is not isolated and cannot be used to validate the putative anti-idiotype by competition studies. There are two ways in which this may be overcome. First, the cells from the first *in vitro* immunization may be subject to myeloma fusion to produce a panel of idiotype monoclonal antibodies. If a member of this panel competes for antigen binding with the putative antiidiotype, this identifies a pair of antibodies that deserve further characterization. Alternatively, the putative antiidiotype antibody may be selected by some indirect method based on assumptions about the protein–protein interaction under study, and then used as antigen for a third round of immunization (Fig. 1). If the putative antiidiotype indeed bears an "internal image" of the starting antigen, a subset of the antibodies that result in the third round will not only bind to the immunizing putative antiidiotype, but will also recreate the specificity of the idiotype and bind to the starting antigen. In both of the examples considered in this chapter, this latter approach was used successfully. In one case, the third-round response was rescued by myeloma fusion and monoclonal anti-antiidiotype antibodies were obtained (Vaux *et al.*, 1988). These antibodies are now functionally equivalent to the relevant idiotype that was present in, but not isolated from, the mixture used as antigen to generate the antiidiotype response.

THEORETICAL LIMITATIONS TO THE ANTIIDIOTYPE APPROACH

It is important to recognize that the approach will be unable to work for some protein–protein interactions. If the murine immune system is indeed "complete" in terms of network theory, then any idiotype immunoglobulin will generate antiidiotypes (Perelson, 1989), but there may be other constraints that prevent these from being useful mimics of the original antigen (see above). More fundamentally, there remains the possibility that the region of the protein surface relevant for interaction with the receptor may not be accessible to an immunoglobulin or sufficiently immunogenic in the immune system used. Thus, the existence of antisera or monoclonal antibodies that inhibit the interaction offers great encouragement that an antiidiotype approach may help to identify the receptor. However, even this is not a perfect predictor because it is possible that such antibodies interfere by steric hindrance rather than by recognizing the ligand site exactly.

In the rest of this chapter we will consider two experimental systems in which antiidiotype antibodies have provided probes for protein–protein interactions that had proved difficult to analyze by conventional biochemical means.

II. Identification of the KDEL Receptor

A. Receptor-Mediated Retention of Soluble Resident Proteins of the Endoplasmic Reticulum

A great deal of work over the past several years has focused on understanding the selective transport of proteins from the endoplasmic reticulum (ER) to the later stages of the secretory pathway. Initially this work focused on attempts to identify a sorting signal within transported proteins, which would mark them for transport and distinguish them from resident proteins of the ER. Resident proteins would then remain in the ER because they lack this signal (Hurtley and Helenius, 1989; Pelham, 1989; Rose and Doms, 1988). Early results suggested that such a signal must exist because transport of a number of secretory and plasma membrane proteins was dramatically slowed down or stopped by mutations that removed or altered regions of the protein. These results are now seen as reflections of changes in the overall structure of the modified proteins that lead to improper folding.

It is now generally accepted that proteins must fold properly and oligomerize in order to pass through the secretory pathway. A particularly clear example of this is the synthesis of immunoglobulin chains. Both correct disulfide bond formation and oligomerization are required of immunoglobulins before they exit the ER. The formation of the disulfide bonds, both intramolecular and intermolecular, requires catalysis from the ER resident protein, protein disulfide isomerase (PDI) (Freedman, 1984). This soluble protein not only has a role in catalyzing disulfide exchange to reach the equilibrium disulfide bond configuration, but has been implicated in a number of other enzymatic activities (such as proline hydroxylase) that are required for attaining the correct folding for some proteins. PDI has been shown to be cross-linked to nascent immunoglobulin chains *in vivo* (Roth and Pierce, 1987). The oligomerization of heavy and light chains has been shown to involve another soluble ER resident protein, immunoglobulin heavy-chain binding protein (BiP) (Bole *et al.,* 1986; Haas and Wabl, 1983). This protein binds to partially folded immunoglobulin chains and can be released upon hydrolysis of ATP. Although BiP is not a heat-shock protein in mammals, sequence homology indicates that it is a member of the hsp70 family of heat-shock proteins (Haas and Meo, 1988; Munro and Pelham, 1986), which are believed to mediate folding and unfolding of proteins in other compartments of the cell. In summary, the immunoglobulin example shows that transport from the ER requires proper folding and that the folding process is catalyzed in cells by resident proteins of the ER.

For other proteins, proper folding and oligomerization may require the assistance of other enzymatic activities, including oligosaccharide addition, isomerization of prolines by *cis-* and *trans-*proline isomerase, hydroxylation of prolines or lysines by the respective hydroxylases, and even the addition of a glycolipid anchor. The need for proper folding and oligomerization prior to transport has now been shown for a large number of membrane and secretory protein (Hurtley and Helenius, 1989). The lack of success of the search for defined signals for transport from the ER coupled with an experiment using peptides to demonstrate fluid-phase transport from the ER to the Golgi apparatus (Wieland *et al.,* 1987) has motivated the present consensus view that transport to the latter stages of the secretory pathway is the default for proteins that are free to move in the ER (Pelham, 1989). No transport signal is believed to be necessary.

The lack of a requirement for a transport signal and the presence of soluble resident proteins within the lumen of the ER is a paradox. The apparent contradiction is exacerbated by the fact that the proteins that reside in the ER lumen are present at very high concentrations. PDI, for example, is present at approximately 0.6 mM in the ER lumen of hepatocytes [calculated from Freedman (1984) using the morphometric data of Weibel *et al.,* (1969)], whereas BiP is almost as abundant. A pivotal finding in the field was made by Munro and Pelham, who sequenced BiP (Munro and Pelham, 1986) and noticed that its carboxy-terminal four amino acids Lys-Asp-Glu-Leu (KDEL) were identical to those of the soluble ER resident protein PDI. These authors later showed that deletion or modification of this sequence caused secretion of BiP and that the carboxy-terminal KDEL signal could confer ER retention on the normally secreted protein lysozyme (Munro and Pelham, 1987). This work established KDEL as a necessary and sufficient signal for retention of soluble proteins in the ER lumen.

The functioning of this signal is unique because it is constrained by several features of ER retention. First, the abundance of proteins such as PDI is so great that the number of copies of the KDEL signals in the ER would outnumber any of the identified membrane proteins in the ER. Hence stoichiometric binding to an ER membrane protein would not be a plausible mechanism for retention. Second, studies of the mobility of BiP in *Xenopus* oocytes (Ceriotti and Colman, 1988) showed that the protein diffuses much more rapidly than a membrane protein but somewhat more slowly than a secreted protein. The rate of diffusion through the ER of BiP did not change when its KDEL signal was truncated. This experiment suggests that signal recognition does not occur in the ER at all, although a weak interaction of the body of BiP with the walls of the ER does appear to be present. In place of restricted movement in the ER, Munro and Pelham

(1987) presented an elegant recycling model for retention. In this model, resident ER proteins pass with other proteins from the ER into a later (salvage) compartment and are retrieved by a process of interacting with a receptor for the KDEL signal and recycling to the ER (Fig. 2). Experimental evidence of the passage of ER proteins to later compartments of the secretory pathway has been obtained by expressing secretory proteins with retention signals. When cathepsin D is modified to contain a KDEL tail, the modified protein does not pass to the later stages of the secretory pathway and is primarily retained in the ER. However, some of this protein acquires Glc-N-Ac-1-phosphate, a post-ER carbohydrate modification (Pelham, 1988). Hence, retained proteins see post-ER compartments. A similar experiment has been done with fusion proteins in the yeast *Saccharomyces cerevisiae*, in which the retention signal is known to be HDEL (Dean and Pelham, 1990). Together these experiments support a

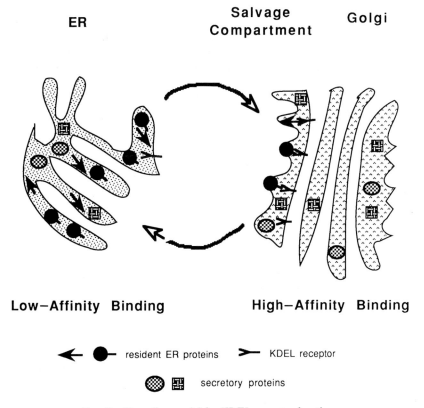

ER   **Salvage Compartment**   **Golgi**

Low—Affinity Binding   **High—Affinity Binding**

←●— resident ER proteins   ➤— KDEL receptor

⊛ ▦ secretory proteins

FIG. 2. Recycling model for KDEL receptor function.

mechanism for retention in which the receptor acts as a guard that returns escaping ER proteins to their proper location rather than a chain that ties them down continuously.

1. Problems with Conventional Approaches

Conventional approaches to the identification of the KDEL receptor are frustrated by several unique features of the recycling model for retention (Kelly, 1990; Vaux *et al.*, 1990). First, the fact that the receptor functions through recycling allows a few copies of receptor to control the localization of large numbers of ER proteins, but it also allows the actual number of copies of the receptor to be very small. Second, the kinetics of recycling will determine the steady-state distribution of the receptor and hence there is no *a priori* knowledge of its localization in the cell. Third, the recycling model requires that the receptor have a strong affinity for ligand in the salvage compartment, where it binds, and a much weaker one in the ER, where it is released. A simple way to accomplish this would be to follow the model of the well-characterized cell-surface receptors that function in endocytosis. Some of these receptors have an affinity for ligand that is dependent upon the environment of the compartment. This allows movement of ligand from the cell surface to the endosome. Typical changes in affinity are approximately 10^2. Fourth, the function of several of the proteins being retained by the system is the mediation of protein folding. For proteins such as BiP, this function requires that they interact with denatured proteins. Hence, binding of such a protein to another protein need not occur via the specific interaction with the KDEL tail. These features of the proposed mechanism create the formidable task of identifying a minor protein with an unknown localization whose specific affinity for ligand is critically influenced by unknown environmental factors.

These factors have caused difficulties in attempts to identify the receptor. Affinity chromatography is an obvious approach because the ligands are well defined, and nonrecognized analogs such as KDELGL are available. Unfortunately, we know very little about the environment of the ER and even less about the environment of our putative salvage compartment. Conditions for affinity chromatography cannot be optimized in an absence of knowledge about the environment in which binding is optimal. Worse, the ability of ER proteins to interact reversibly with other proteins as part of their role in folding gives rise to a very high background binding. In our hands, simple affinity chromatography with a column containing a coupled KDEL tails shows binding of a large number of proteins, making the specific receptor binding difficult to distinguish. An alternative approach that avoids these problems with direct binding studies makes use of the power of yeast genetics to identify proteins involved in assayable cellular functions (see below). This approach is complicated by the likelihood that, as required by the recycling model, environmental factors affect the binding of the ligand to the receptor. A mutation that altered the salvage compartment environment to resemble that of the ER would cause secretion of ER proteins because they would not be recognized efficiently, whereas one that changed the ER environment to resemble that of the salvage compartment would inhibit the release of ligand in the ER, saturating the limited amount of receptor in the cell and also resulting in resident protein secretion. Evidence that changes in compartment environment can give rise to protein secretion is provided by the work of Booth and Koch (1989), who showed that changing Ca^{2+} levels leads to the secretion of ER proteins. Of course, any putative receptor must be shown to have the appropriate binding characteristics for KDEL-terminated proteins and to cause the expected changes in retention of ER proteins upon mutation or exchange between organisms; however, another method for identifying a candidate receptor would be a useful alternative to these approaches.

2. ADVANTAGES OF THE ANTIIDIOTYPE APPROACH

The antiidiotype approach to the identification of the KDEL receptor can avoid several pitfalls of the other approaches. The requirement for this approach to be successful is that the first-round antibodies will mimic the recognition features of the receptor toward the ligand and that the second-round, antiidiotypic antibody will form a prototypic ligand for the receptor that will allow the relevant interactions for ligand recognition. The antiidiotype approach avoids the difficulties that affect the affinity approach because only the KDEL interaction is modeled by the antibody rather than the other features of a typical KDEL protein. In particular, by freezing the recognized features of the ligand in the antiidiotype, one might avoid the effects of environment on the affinity of receptor for ligand. For this purpose, the fact that an antiidiotype need not be a perfect molecular mimic of the ligand is an advantage. Further, the problems of interaction of the body of the protein with a putative receptor can be better controlled because the region of the ER protein that mediates folding has been eliminated from the system. The approach is complementary to a genetic one because only the KDEL-binding aspect of the receptor–ligand interaction is examined by the use of antiidiotypic antibodies rather than all cellular changes that give rise to secretion of ER proteins. Two very critical requirements for the validity of this approach must be met. First, the first-round antibodies must be capable of mimicking the receptor's recognition of the ligand. This can be tested by using the knowledge given by mutagenesis studies about the specificity of binding of the receptor to KDEL sequences. Second, the antiidiotype must embody the features of the ligand that are recognized by the receptor. This second requirement is addressed by experiments using reactivity with a panel of first-round antibodies to test for the presence of a public idiotope and by examination of the third-round anti-antiidiotypic antibodies. Finally, although our *a priori* knowledge about the receptor is limited, there are features such as localization to the early portions of the secretory pathway, the expected transmembrane nature of the protein, the response of the protein level in the cell to ligand concentration, and the binding of authentic ligands to the receptor that could be used to support the identification of a putative receptor.

B.   Identification of the Receptor by Antiidiotype Antibodies

1.   FIRST-ROUND ANTIBODIES

The table of KDEL-related sequences (Table I) indicates our approach to the characterization of the specificity of the first-round antibodies. It is known that all the KDEL-terminated sequences displayed should interact

TABLE I

SEQUENCES OF SYNTHETIC PEPTIDES USED

Name	Sequence	Origin
KAVK	NH_2-KAVKDEL-CO_2H	Short PDI
KDDD	NH_2-KDDDQKAVKDEL-CO_2H	Long PDI
KETE	NH_2-KETEKESTEKDEL-CO_2H	grp 94
KEED	NH_2-KEEDTSEKDEL-CO_2H	Long BiP
KSEK	NH_2-KSEKDEL-CO_2H	Short BiP
KEEE	NH_2-KEEESPGQAKDEL-CO_2H	Calreticulin
KDELGL	NH_2-KEEDTSEKDELGL-CO_2H	Nonrecognized tail
KX_5[a]	NH_2-KXXXXXKDEL-CO_2H	Mixed tail
HDEL	NH_2-KDDDGDYFEHDEL-CO_2H	*Saccharomyces cerevisiae* BiP

[a] The peptide KX_5KDEL consists of a mixture of analogs containing any one of alanine, aspartic acid, histidine, glutamine, leucine, tyrosine, or lysine at each of the positions X. KDEL-terminated peptides are identified throughout by their unique amino-terminal residues.

specifically with the receptor but that the KDELGL-terminated sequence should not interact. The HDEL-terminated sequence is known to be a ligand for the yeast receptor but not for the mammalian receptor. Antipeptide antibodies raised in rabbits using sequences such as KDDD give rise to antisera that recognize the specific KDEL-terminated peptide but not the KDELGL-terminated peptide. In early bleeds of the rabbits, recognition is observed of a large number of KDEL-terminated peptides; however, this broad specificity narrows in later bleeds to generate antibodies that only recognize the specific KDEL-terminated protein corresponding to the immunizing sequence (Fuller *et al.*, 1991). The most striking case of this is seen with an antiserum generated against a randomized peptide sequence KX_5 terminated with a KDEL tail. The amino acids contributing to the randomized tail were selected so that the previously described retention signal peptides would not be present in the mixture. Early bleeds of these rabbits recognize more than 25 identifiable protein spots on a two-dimensional gel of hybridoma lysate (Fig. 3). These spots are also found in lysates derived from microsomal preparations and are seen in the

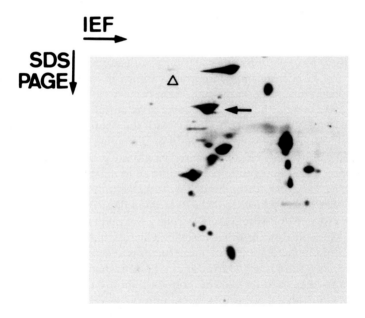

FIG. 3. Western blot from a two-dimensional gel of mouse hybridoma lysate with a rabbit polyclonal raised against keyhole limpet hemocyanin-coupled KX_5. The arrow marks the positions of PDI and the triangle marks the location of the strongest carboxypeptidase-insensitive reactivity.

soluble protein fraction of a TX114 separated protein lysate (Fuller *et al.*, 1991). This result indicates that the KDEL retention system controls a much larger number of proteins than have already been identified.

An important feature of the recognition of the KDEL tails by the first-round antibodies is that this recognition should be dependent on the complete KDEL sequence. A protocol was developed for carboxypeptidase treatment under mild conditions, which should destroy the KDEL sequence. An important feature of the carboxypeptidase treatment was that the digestions needed to be performed at low pH (pH 5.5) to be efficient. Control experiments showed that this pH requirement is not an effect on the activity of the carboxypeptidase A but rather on the susceptibility of the tail to carboxypeptidase digestion. This may indicate that an environmental effect on tail exposure could modulate its interaction with the receptor. The carboxypeptidase treatment can be performed in solution or on proteins transferred electrophoretically to nitrocellulose. The result on the complex pattern given by KX_5 is particularly dramatic. The majority of the spots no longer react with KX_5 after digestion. Reprobing the digested blot with an antibody against the whole of PDI reveals that at least this ER protein is still present after digestion. The specificity of this treatment is highlighted by the fact that the activity of carboxypeptidase A is reduced for acidic residues, suggesting that only the carboxy-terminal leucine will be removed in mild carboxypeptidase digestion of a KDEL protein.

In contrast to many workers' experience with antipeptide antibodies, the KDEL- and HDEL-terminated peptides shown in Table I have all given rise to useful antisera upon immunization of mice or rabbits. "Useful," in this context, means that the antibodies give an ER staining pattern by indirect immunofluorescence and react with the appropriate proteins after Western blotting. Immunoprecipitation of the native protein requires that the reactive antibodies be concentrated by affinity purification and that the immunoprecipitation conditions be optimized to maximize exposure of the KDEL tail. This very fortunate behavior of the anti-KDEL peptide antibodies may reflect the fact that the corresponding epitope is a signal that must be exposed for recognition during intracellular transport. For this case, and presumably for other cases in which signal recognition is being studied, the characterization of the structures recognized by antipeptide antibodies as "unfoldons" (Laver *et al.*, 1990) is inappropriate.

The antibodies used for the first round of antiidiotype production were generated and tested using the experience gained from the polyclonal rabbit sera. The first round of *in vitro* immunization was begun with the KETE peptide, which had been shown in several immunizations in rabbits to give rise to an antiserum that reacted with a very broad range of KDEL proteins but that had little carboxypeptidase-insensitive reactivity. We expected a similar broad range of reactivities to be produced in the first round of *in vitro* immunization, making this peptide an optimal starting point for this approach to an antiidiotype. We also employed *in vivo* immunization using fixed hybridoma cells from a monoclonal antibody that had been generated against the KDDD peptide and that had been shown to lose reactivity with PDI after digestion with carboxypeptidase A.

2. GENERATION AND SCREENING OF THE ANTIIDIOTYPE ANTIBODIES

Once the starting immunogens had been selected, two separate approaches were used for the generation of the anti-idiotypic antibodies: *in vivo* and *in vitro* immunization (Fig. 4). The first took advantage of an available hybridoma line, 1D3, which was generated by *in vivo* immunization of BALB/c mice with unconjugated peptide KDDD and fusion with Ag8 myeloma cells. The antiidiotypic monoclonal antibody 9D6 was produced by fusing Ag8 myeloma cells to spleen cells from a BALB/c mouse

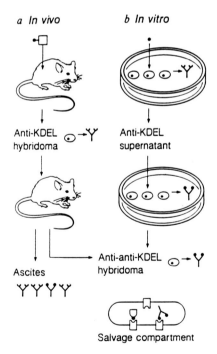

FIG. 4. Schematic of the production of the antiidiotype antibodies against the KDEL receptor. Reproduced from Kelly (1990) by permission from *Nature* (*London*) Vol. 345 pp. 480–481, Copyright © 1990 Macmillan Magazines Ltd.

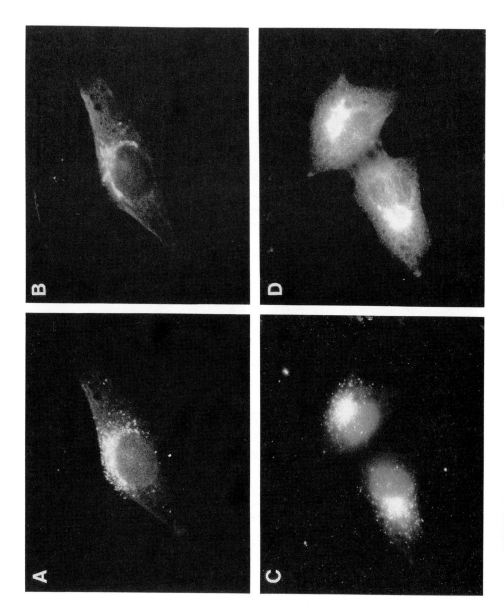

immunized with washed 3% (w/v) paraformaldehyde-fixed 1D3 hybrid-oma cells. The advantages of this *in vivo* approach are that it utilizes very standard technology and that the first-round antibody is available as a monoclonal antibody for use in later work. Monoclonal antibody 5D3 was produced by paired *in vitro* immunization using 100 μg of the unconjugated HPLC-purified KETE peptide as starting antigen. The advantages of this approach include speed (4 weeks as opposed to 6 months in the *in vivo* scheme), the ability to use very small amounts of antigen, and the minimization of selection during the generation of the antibodies because it utilizes a primary response. Potential disadvantages of the *in vitro* approach include the more elaborate and demanding methodology, which is not standard in many labs, and the fact that the primary response will produce an IgM.

Both *in vivo* and *in vitro* immunizations involve fusions and require screening a large collection of hybridoma supernatants for the expression of the antiidiotype. From the description of the expected characteristics of the putative receptor above, it is clear that the assumptions made in developing a screen need to be examined very carefully. A first and very useful screen was the examination of the immunofluorescence patterns given by the hybridoma supernatants. We screened for antibodies that gave staining patterns consistent with localization to the salvage compartment. Although searching for such patterns involves an assumption about the morphology of a compartment that is poorly characterized, it is a reliable way of excluding hybridomas. For example, we excluded antibodies that gave staining patterns characteristic of cytoskeletal elements or nuclear components as well as those that produced no detectable staining and concentrated on hybridomas whose staining patterns were reticular, vesicular, or Golgi-like (Fig. 5). A second screen was based on a completely different set of assumptions about the receptor and about the antiidiotype antibodies that would be expected to recognize it. We assumed that the true antiidiotype would be a sufficiently good mimic of the KDEL ligand that it would be recognized by antibodies directed against the tail. For practical reasons, the screening of mouse hybridomas for reactivity with antitail antibodies is most easily performed using antitail antibodies raised in rabbits. The screen is conveniently done by impregnating nitrocellulose sheets in ammonium sulfate-purified rabbit serum, incubating with hybridoma supernatents in a 96-well dot-blotting manifold, and visualizing

---

Fig. 5. Characteristic staining patterns in NRK cells for the antiidiotype 5D3 (panels A and C) and a Golgi marker (B) and an ER marker, PDI (D).

A

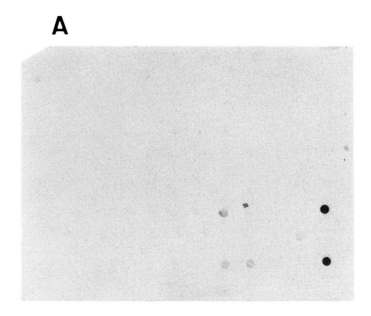

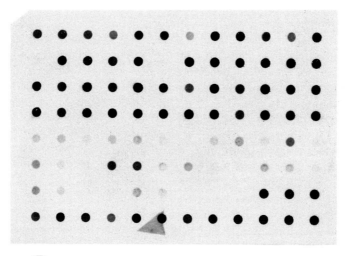

B

FIG. 6.   Screening of hybridoma supernatants for reactivity with rabbit anti-KAVK. Panel A shows the initial screening of 48 wells in duplicate. Two positive clones were selected in this screen. The positive duplicates at the lower right are 9D6 supernatants. Panel B compares the reactivity of separate clones of 9D6 (top four rows) with that of an irrelevant hybridoma (bottom four rows).

the reactivity with antimouse second antibody (Fig. 6). The screen is relatively rapid, easily interpreted, and allowed us to select several clones that had the appropriate binding specificity.

3. CHARACTERISTICS OF THE ANTIIDIOTYPE ANTIBODIES

The results of the two immunizations were two antibodies, 9D6, an IgG generated by *in vivo* immunization with anti-KDDD hybridoma cells, and 5D3, an IgM generated by paired *in vitro* immunization with KETE (Vaux *et al.*, 1990). The two antibodies gave nearly identical staining patterns by immunofluorescence and both showed reactivity with the anti-KDEL antibodies produced in rabbits. Immunoblotting of cell lysates showed that both antibodies (Fig. 7) reacted strongly with a band at 72 kDa, which was present in a wide range of species. Cross-reactivity was demonstrated by precipitation with one antibody followed by blotting with

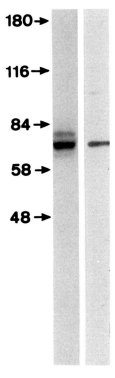

FIG. 7. Western blotting of the two independantly derived antiidiotypic antibodies 5D3 (anti-anti-KETE, right) and 9D6 (anti-anti-KDDD, left).

the other and confirmed that the two independently derived antiidiotypes were identifying the identical protein. This 72-kDa protein became our candidate for the KDEL receptor. One supporting piece of evidence for the identity of this protein is that intracellular complexes between the antibody and the antigen can be isolated from the hybridoma cells. This demonstrates that the binding site for the antiidiotype and hence the putative binding site for the KDEL ligand faces the lumen, as it must to function as a receptor. The distribution of reactivity of these antibodies among species is revealing. A strongly reactive band is seen in all mammalian species tested and in plants in which KDEL and HDEL are both recognized. It is not seen in blots of *S. cerevisiae*, in which HDEL but not KDEL is recognized as the retention signal. This specificity matches that seen in the first-round antibodies that react with KDEL-terminated peptides but not HDEL-terminated ones. The antiidiotype antibodies do not recognize KDEL peptides (Fig. 8), ruling out the possibility that we had inadvertently selected a first-round response in our screening.

As mentioned previously, the use of monoclonal antibodies in the generation of antiidiotypes has many advantages. In the KDEL receptor work, the use of monoclonal antibodies was critical at several stages. First, the ability to eliminate the anti-KDEL response from the antiidiotype was necessary for our screening approach. The presence of a small amount of anti-ER resident protein staining would have swamped the signal from the

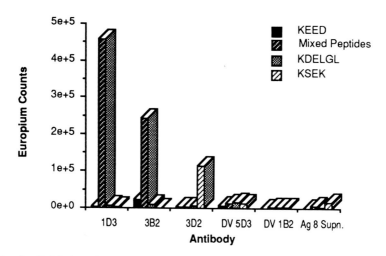

Fig. 8. Solid-phase immunoassay of idiotype and antiidiotype antibodies on peptides performed as in Vaux *et al.* (1990). Mixed peptides comprise KEED, KDDD, KSEK, and KAVK. 1D3 and 3B2 are anti-KDDD. 3D2 is anti-KSEK. 5D3 and 1B2 are antiidiotypes generated by paired *in vitro* immunization. Ag 8 Supn is the supernatant from the nonsecreting myeloma line used in the fusion that generated 1D3 and 3B2.

salvage compartment because the ER proteins are present in such high amounts. Second, the generation of an antibody that reacts with a lumenally oriented site in the secretory pathway creates problems for the hybridoma. The cells clone inefficiently and grow poorly. We have frequently had to restart growth from early passages of the hybridomas because later passages lost the ability to secrete the antiidiotype. In a polyclonal system, this poor growth and lack of stability would cause selection against the cells that were producing the antibody of interest. Certainly one would expect a large variability in the reactivity of different bleeds of the same animal if a response were generated at all. Only through the monoclonal approach can one work with a defined reagent. A third advantage of the monoclonal approach is that the response of the hybridomas to producing the antiidiotype could be studied. This is particularly interesting because one could expect that the antiidiotype should compete for the ligand-binding site on the receptor and hence interfere with the functioning of the retention system. Although one possible cell response would be the up-regulation of resident ER proteins so that a small leak could be tolerated as occurs in the retention-defective mutants in yeast (Hardwick *et al.*, 1990), the cells respond by up-regulating the receptor manyfold so that the level of receptor expression is above that of the antibody. This dramatic up-regulation suggests that the protein with which the antiidiotype reacts is an important protein for cell function and indicates that the level of KDEL tails may regulate the level of receptor expression. Finally, a major advantage of the monoclonal approach is that the sequences of the antiidiotypes can be obtained by cDNA cloning from the hybridomas. This is now relatively straightforward using polymerase chain reaction (PCR) primers based on the strongly conserved regions of the immunoglobulin message and allows an independent approach to the characterization of the ligand-binding site. It is worth noting that several of the recent examples of identifications of receptors by antiidiotypes that have come under recent criticism have in common the fact that polyclonal antibodies were used (Meyer, 1990).

4. Characteristics of the Third-Round Antibodies

A third round of immunization is particularly revealing in the case of the KDEL tail system (Vaux *et al.*, 1990). The first-round antibodies were generated against two different peptides (KDDD and KETE) and the original anti-KDDD monoclonal 1D3 showed negligible reactivity with any of the other peptides except KAVK, which is a proper subset of its sequence. If the second-round antibody is a good mimic of the characteristics of the KDEL tail that allow it to be recognized by the receptor, one would expect that a third-round antibody should be specific for the ligand.

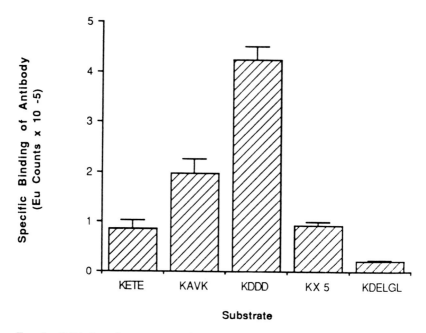

FIG. 9.   Solid-phase immunoassay of third-round antisera (anti-5D3) on peptides performed as in Vaux *et al.* (1990).

This third-round reactivity with the first-round immunogen is one demonstration that the second-round antibody is a true internal image antiidiotype. Antisera were produced in mice by immunizing with fixed 9D6 or
5D3 cells and were screened for reactivity with the set of peptides described in the Table I. The expected reactivity is observed; the third-round
antibodies do react with their respective first-round immunogens. Interestingly, the third-round antibodies show a broader specificity than was seen
in the first round (Fig. 9). This broader specificity is a better simulation of
the specificity of the receptor because it includes all the known KDEL
signals but excludes the signals known to be nonfunctional in mammals,
such as KDELGL and HDEL (Vaux *et al.*, 1990). This result indicates
that the second-round antibody was a very good mimic for the ligand
recognized as the receptor.

5.   EVIDENCE THAT THE 72-kDa PROTEIN IS A PHYSIOLOGICAL KDEL RECEPTOR

Identification of the 72-kDa protein as a putative receptor is, of course,
only the initial step toward establishing that this protein functions as the

physiological receptor for retention of ER proteins *in vivo*. The third-round antibody response shows that the specificity is appropriate for receptor function. Independent evidence can be obtained by a biochemical or genetic demonstration of receptor function. The localization of the antigen provides one such piece of evidence. Not only is the localization of the 72-kDa protein in normal cells consistent with the localization of the receptor to a region intermediate between the ER and Golgi (Fig. 5), but a variety of treatments that are known to affect other components of the intermediate compartment in defined ways have the same effect on the 72-kDa distribution. These treatments include 15°C treatment and behavior during treatment with and recovery from Brefeldin A. Another piece of support for the role of the protein is its orientation; the 72-kDa protein is a transmembrane protein whose ligand-binding site is lumenal and which is capable of binding the antiidiotype during its transit through the secretory pathway. Further support is given by direct binding studies with a soluble fragment of the 72-kDa protein that confirm that this protein binds to KDEL-terminated peptides but not to the unrecognized peptide, KDELGL. As predicted from the model, the binding of the isolated 72-kDa protein to ligand is dependent on environment. The binding to peptides is relatively weak but that to isolated PDI is stronger and shows a K_D of ~ 20 μM. This affinity is appropriate to the task of retaining proteins that are present at near-millimolar concentrations. It is also sufficiently weak that modulation of affinity by the same factors seen in environment-dependent cell surface receptors, such as the transferrin receptor, would allow the release of ligand into the ER. Hence, the 72-kDa protein has the characteristics predicted by the model of Munro and Pelham (1987) for the KDEL receptor.

The observation that the level of receptor was up-regulated in the hybridoma cell suggested that looking for the regulation of the 72-kDa protein as a function of the concentration of KDEL tails in the ER could provide evidence for the relevance of this protein to ER retention. To modulate the level of KDEL tails in the ER, we incubated cells with a peptide containing a consensus glycosylation signal and a KDEL carboxy terminus that had been rendered hydrophobic by esterification (Fuller *et al.*, in preparation). The esterified peptide crossed cell membranes and entered the ER, where it was trapped by glycosylation and deesterification. The resulting accumulation of this KDEL peptide in the secretory pathway resulted in the secretion of a fraction of the ER resident protein PDI and in a marked and rapid increase in the level of the 72-kDa KDEL-binding protein. This demonstrated that the KDEL retention system is saturable in animal cells and that the level of the 72-kDa protein responds to the concentration of KDEL tails *in vivo*, supporting its role as a receptor for retention.

Pelham and co-workers have identified two mutations, *erd1* and *erd2*, in the yeast *S. cerevisiae*, which cause secretion of HDEL proteins (Hardwick *et al.*, 1990; Lewis *et al.*, 1990; Semenza *et al.*, 1990). The *erd1* mutation has been localized to the late Golgi and appears to have its effect on retention in an indirect way (Hardwick *et al.*, 1990). Exchange of the *erd2* gene between the yeasts *S. cerevisiae* and *Kluyveromyces lactis* causes a change in the specificity of the *S. cerevisiae* retention so that it is more like that of *K. lactis* (Lewis *et al.*, 1990). This experiment has been interpreted as a demonstration that *erd2* is the physiological HDEL receptor in yeast. A human homologue of *erd2* has been cloned by PCR and has a molecular mass near 27 kDa (Lewis and Pelham, 1990). This and a variety of other criteria demonstrate that *erd2* is not closely related to the 72-kDa protein. The human gene was not capable of mediating the retention of KDEL in *S. cerevisiae* and no binding data of the *erd2*-encoded protein to a KDEL ligand are available. The case for the *erd2* product as the receptor is complicated by the fact that *erd1* and several *sec* mutants show the *erd*-encoded phenotype. Hence several distinct mutations can lead to ER protein secretion (Semenza *et al.*, 1990). Such a result might be expected from the complexity of the recycling and retention system described above. Further, the localization of this protein as determined by examining mammalian cells that overexpress an ERD2 fusion protein containing a *myc*-encoded epitope is broader than that expected for the intermediate compartment (Lewis and Pelham, 1990). The fact that *ERD2* appears to encode a seven-helix membrane-spanning protein (Semenza *et al.*, 1990) may indicate that it is a channel that controls the environment of either the ER or the salvage compartment and hence has strong effects on ER retention. The 72-kDa protein could then be the binding component of the system whose activity is modulated by an ERD2-controlled environment. Another possibility is that mammals have more that one system for retaining ER proteins. A great deal of work remains to be done to understand the physiological roles of these proteins (Pelham, 1990).

III.   Budding of Alphaviruses

A.   Background

The alphaviruses are a well-studied family of enveloped viruses with a single positive-strand RNA genome (Schlesinger and Schlesinger, 1986). The cell biology of alphavirus infection has been widely studied, and the processing of the viral structural polyprotein to deliver the spike glycoproteins to the plasma membrane of the infected cell has been described in detail (Simons and Warren, 1984). Virus budding occurs at the plasma membrane of the infected cell and results in the enclosure of the nucleo-capsid within a layer of viral glycoproteins embedded in a host cell-derived lipid bilayer (Fig. 10) (Simons and Garoff, 1980). Host cell membrane proteins are almost completely excluded from this forming envelope, and considerable interest has focused on the sorting mechanism responsible for this effect. A specific interaction between the cytoplasmic domains of the spike glycoprotein complex and the surface of the nucleocapsid was proposed by Simons and Garoff (1980), supported by the results of cross-linking experiments (Garoff and Simons, 1974). Further experimental support came from octyl-β-d-glucopyranoside extraction experiments (Helenius and Kartenbeck, 1980). Reconstitution of spike glycoprotein binding to isolated nucleocapsids was only achieved at low efficiency, insufficient to further characterize the interaction. This may be because the individual interaction between a spike glycoprotein trimer and the nucleo-capsid is of low affinity, and can only direct specific viral budding because it is reiterated many times during this process. This view is supported by a three-dimensional reconstruction of Sindbis virus, in which the interactions of the cytoplasmic tail with the capsid are directly visualized (Fuller, 1987).

B.   An Antiidiotype Antibody Approach

Any direct interaction between the spike glycoproteins and the nucleo-capsid must occur in the cytoplasm, so it is to the cytoplasmic domains of the spike glycoproteins that one must first look for nucleocapsid recognition signals (Fig. 11). In the alphaviruses, the spike glycoprotein complex consists of a homotrimer of a heterooligomer of two [(E1,E2)$_3$] or three [(E1,E2,E3)$_3$] members. Only the E2 spike glycoprotein has a significant cytoplasmic domain, so this is the most likely location for a signal. In Semliki Forest virus (SFV), the E2 cytoplasmic domain consists of 31 amino acid residues. If a specific signal exists and constitutes an immunogenic epitope, then it should be possible to elicit an antibody response to the signal. This antibody would in turn be the appropriate idiotype from which to generate an internal image antiidiotype that would mimic the signal and bind to the nucleocapsid. Thus, a simple version of this approach would begin with the preparation of monoclonal antipeptide antibodies to a 31-residue synthetic peptide corresponding to the entire cytoplasmic domain of E2 (E2c). However, it is not possible to know in advance the antigenic complexity of such a peptide, and it is possible that a

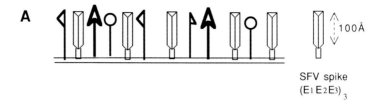

SFV spike
$(E_1 E_2 E_3)_3$

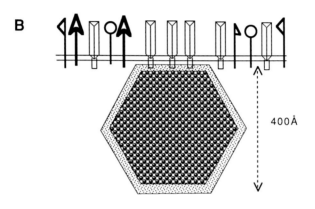

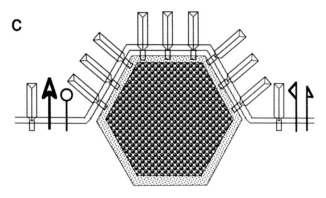

Fig. 10. Schematic view of the budding of the SF virion. Budding is believed to be initiated when the concentration of spikes at the plasma membrane reaches a threshold value that allows lateral association of spikes to exclude host cell proteins (A → B). The patches of spikes form binding sites for the preassembled capsid (B), which interacts with the cytoplasmic tails of the E2 proteins. It is this critical interaction that is modeled by the antiidiotype approach. The later stages of budding proceed by lateral spike–spike and by capsid–spike interactions to cause the envelopment of the capsid (C). The final virion is 700 Å in diameter. The nucleocapsid contains 180 copies of the capsid protein arranged with $T = 3$ triangulation number enclosing the negative strand of RNA. The spike complexes are present as 80 trimers $(E1E2E3)_3$ in a $T = 4$ packing, which is complementary to that of the capsid.

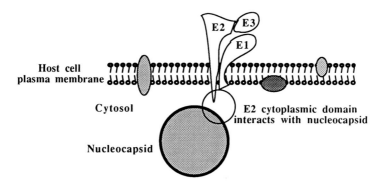

FIG. 11. The final stage of assembly of SFV involves nucleocapsid budding through the plasma membrane of the infected host cell. Inclusion of viral spike glycoproteins and exclusion of host cell proteins is ensured by a specific nucleocapsid interaction.

number of epitopes will elicit an antibody response, only one of which is relevant. Figure 12 shows that E2c is immunologically complex; in the BALB/c mouse it elicits responses to at least four different epitopes (Fig. 12B). The response of rabbits to this peptide appears simpler, and only one strong epitope is seen (Fig. 12A). Without an assay to tell us which, if any, of these structures constitutes the signal in E2c, we are not justified in using a single monoclonal antipeptide antibody as the idiotype. There are two possible solutions: either prepare antiidiotypes using a mixture of all available antipeptide monoclonal antibodies as antigen, or use a functional assay to identify the relevant idiotype. Neither of these approaches is trivial; the first may fail if the initial peptide contains an irrelevant but immunodominant epitope, and the second implies the existence of an assay that could already be used to study the interaction without using antibodies at all.

The cytoplasmic domain of SFV E2 has been used as the initial antigen for a paired *in vitro* immunization (Vaux *et al.,* 1988). The unconjugated 31-residue synthetic peptide was found to be strongly immunogenic in BALB/c mice, giving rise to antibodies that recognized the peptide and the whole E2 molecule both by immunoprecipitation and immunofluorescence on permeabilized infected cells. After paired *in vitro* immunization, hybridomas was selected for their ability to label infected but not uninfected cells. From 110 wells containing hybridomas, 5 wells were obtained that showed this ability; all gave similar staining patterns in infected cells. Antibody from a stable cloned cell line from one of these wells, F13, was also shown to immunoprecipitate nucleocapsids from homogenates of infected cells.

A

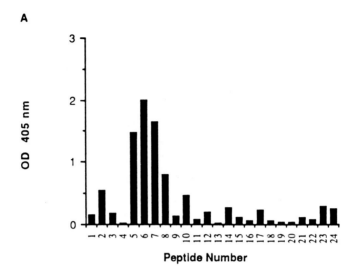

B

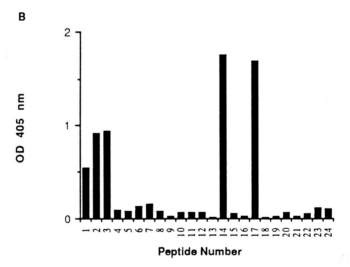

FIG. 12. Epitope mapping of polyclonal responses in the rabbit (A) and mouse (B) to a 31-residue synthetic peptide corresponding to the entire cytoplasmic domain of SFV E2. Mapping is carried out using overlapping 8-residue peptides on plastic rods; peptide 1 is residues 1–8, peptide 2 is residues 2–9, and so on. Measured values are the optical density of a colored product formed by alkaline phosphatase conjugated to the detecting antibody.

The demonstration that F13 does recognize nucleocapsids is consistent with it being an internal image antiidiotype, but of course does not prove it. This point is of crucial importance. If a relevant idiotype had been identified initially, then an essential control would be to show that the idiotype competes with antigen for antiidiotype binding. Unfortunately, this control is not available after paired *in vitro* immunization because no intermediate specific idiotype has been isolated.

An alternative method for confirming that the putative antiidiotype does constitute an internal image of the original antigen is to use the immunoglobulin for the generation of a third-round, or anti-antiidiotype, response. If the tested antibody is a true internal image antiidiotype antibody, the third-round response will recreate the original idiotype specificity and recognize the starting antigen. Moreover, this third-round response may be captured by hybridoma production, giving rise to a monoclonal antibody that is effectively identical to the relevant idiotype from the first immunization. Thus, paired *in vitro* immunization followed by third-round antibody production offers a way to make monoclonal antibodies against the "ligand" and "receptor" of a protein–protein interaction without pure receptor or a functional assay.

The F13 putative internal image antiidiotype was subjected to this analysis (Fig. 13). Monoclonal antibodies from the third-round response recognized the immunizing F13 immunoglobulin, and a subset of them also recognized the initial E2 cytoplasmic domain peptide (Fig. 14). A stably cloned hybridoma from the third-round immunization, 3G10, secretes a third-round antibody that competitively inhibited the binding of F13 to infected cells (Fig. 15). Furthermore, the binding of the F13 antiidiotype antibody to fixed, permeabilized SFV-infected cells could be inhibited with the peptide corresponding to the entire E2 cytoplasmic domain (D. Vaux, unpublished observations; Metsikko and Garoff, 1990). These results strongly support the identification of F13 as a true internal image antiidiotype antibody, capable of binding specifically to the E2 cytoplasmic domain receptor site on the viral nucleocapsid.

The F13 antibody has proved to be a useful reagent for examining the formation and behavior of this receptor site. The first important observation was that F13-reactive nucleocapsids were not restricted to SFV-infected cells, but were also found with the closely related alphavirus, Sindbis. This observation prompted us to examine a wide range of alphaviruses. All 14 of the alphaviruses tested showed reactivity with F13. In addition, nine out of nine flaviviruses tested showed F13 reactivity. F13 binding is not a phenomenon associated with all virally infected cells—no signal was seen in cells infected with influenza (an orthomyxovirus), vesicular stomatitis virus (a rhabdovirus), or Uukuniemi virus (a bunyavirus).

A

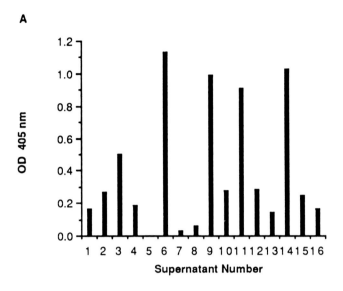

B

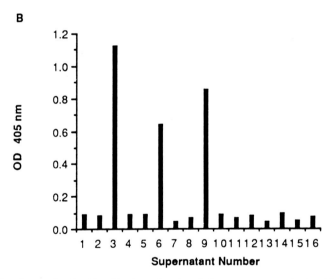

FIG. 13. Screen for the identification of anti-antiidiotype responses in the third round, following immunization with F13. Solid-phase binding assays using either the immunizing F13 IgM (A) or the original peptide antigen (B) as substrate are shown for 16 fusion-well supernatants. Note that some supernatants contain antibodies that recognize the F13 immunogen but fail to recognize the peptide; these probably represent "framework" antiidiotypes.

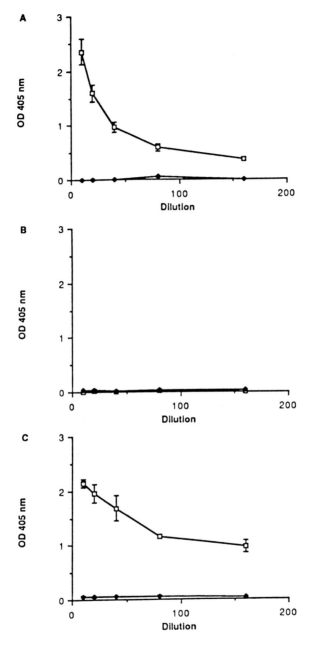

FIG. 14. Binding curves for the polyclonal idiotype (A), monoclonal antiidiotype (B), and monoclonal anti-antiidiotype (C) assayed on the SFV E2 cytoplasmic domain peptide (□) or a control peptide matched for charge and length (●).

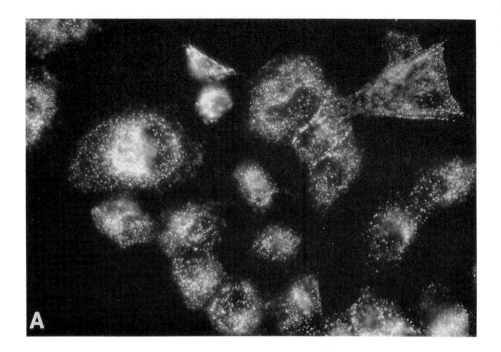

FIG. 15. Indirect immunofluorescent staining with antiidiotype, F13, on fixed permeabilized BHK cells infected with SFV. (A) F13 has been preincubated with a control monoclonal mouse immunoglobulin. (B) F13 has been preincubated with the monoclonal antiantiidiotype, 3G10.

These results closely parallel the observed boundaries of effective pseudo-typing; alphaviruses will only form pseudotypes with other alphaviruses or with flaviviruses (Burge and Pfefferkorn, 1966; Lagwinska et al., 1975). Thus, the range of the F13 epitope correlates well with an independent biological observation on a wide variety of viral types. The implication of the pseudotyping data taken together with the F13 data is that all of the togaviridae use a common receptor–ligand budding mechanism, and that the F13 antibody binds to the active site of the receptor on the nucleocapsid.

Taken together, these results show that there is a specific receptor—ligand-like interaction between the cytoplasmic domain of the E2 glycoprotein and intracellular nucleocapsids, and that this interaction may be accurately mimicked by immunoglobulin molecules.

IV. Conclusion

In this chapter we have described an immunological approach to problems of protein–protein interactions. Two successful uses of this approach to explore protein–protein interactions, which are difficult to characterize by classical methods, have been detailed. The antiidiotype approach is particularly suited to exploring the interactions of the defined sorting and localization signals that have been characterized by studies of protein targeting. Although one must remain aware of the limitations imposed by the structure of the antigen-combining site of an immunoglobulin, the method is a valuable and widely applicable tool for understanding the cell biology of signals in proteins.

References

Amit, A. G., Mariuzza, R. A., Phillips, S. E. V., and Poljak, R. J. (1986). Three dimensional structure of an antigen–antibody complex at 2.8 Angstroms resolution. *Science* **233**, 747–753.

Bentley, G. A., Boulot, G., Riottot, M. M., and Poljak, R. J. (1990). Three-dimensional structure of an idiotype–anti-idiotype complex. *Nature (London)* **348**, 254–257.

Bole, D. G., Hendershot, L. M., and Kearney, J. F. (1986). Posttranslational association of immunoglobulin heavy chain binding protein with nascent heavy chains in non-secreting and secreting hybridomas. *J. Cell Biol.* **102**, 1558–1566.

Booth, C., and Koch, G. L. (1989). Perturbation of cellular calcium induces secretion of luminal ER proteins. *Cell (Cambridge, Mass.)* **59**(4), 729–37.

Bruck, C., Co, M. S., Slaoui, M., Gaulton, G. N., Smith, T., Fields, B. N., Mullins, J. I., and Greene, M. I. (1986). Nucleic acid sequence of an internal image bearing monoclonal anti-idiotype and its comparison to the sequence of the external antigen. *Proc. Natl. Scad. Sci. U.S.A.* **83,** 6578–6582.

Burdette, S., and Schwarz, R. S. (1987). Idiotypes and anti-idiotypic networks. *N. Engl. J. Med.* **317,** 219–224.

Burge, B. W., and Pfefferkorn, E. R. (1966). Phenotypic mixing between group A arboviruses. *Nature (London)* **210,** 1937–1940.

Ceriotti, A., and Colman, A. (1988). Binding to membrane proteins within the endoplasmic reticulum cannot explain the retention of the glucose-regulated protein GRP78 in *Xenopus* oocytes. *EMBO J.* **7**(3), 633–638.

Coutinho, A. (1989). Beyond clonal selection and network. *Immunol. Rev.* **110,** 63–87.

Dean, N., and Pelham, H. R. (1990). Recycling of proteins from the Golgi compartment to the ER in yeast. *J. Cell Bio.* **111**(2), 369–377.

Djabali, K., Portier, M. M., Gros, F., Blobel, G., and Georgatos, S. D. (1991). Network antibodies identify nuclear lamin B as a physiological attachment site for peripherin intermediate filaments. *Cell (Cambridge, Mass.)* **64,** 109–122.

Farid, N. R., and Lo, T. C. Y. (1985). Anti-idiotypic antibodies as probes for receptor structure and function. *Endoc. Rev.* **6,** 1–23.

Freedman, R. B. (1984). Native disuplhide bond formation in protein biosynthesis: Evidence for the role of protein disuiphide isomerase. *Trends Biochem. Sci.* **9,** 438–441.

Fuller, S. D. (1987). The *T* = 4 envelope of Sindbis virus is organized by complementary interactions with a *T* = 3 icosahedral capsid. *Cell (Cambridge, Mass.)* **48,** 923–934.

Fuller, S.D. *et al.,* (1991). In preparation.

Garoff, H., and Simons, K. (1974). Location of the spike glycoproteins of the Semliki Forest virus membrane. *Proc. Natl. Acad. Sci. U.S.A.* **71,** 3988–3992.

Gaulton, G. N., and Greene, M. I. (1986). Idiotypic mimicry of biological receptors. *Annu. Rev. Immunol.* **4,** 253–280.

Haas, I., and Meo, T. (1988). cDNA cloning of the immunoglobulin heavy chain binding protein. *Proc. Natl. Acad. Sci. U.S.A.* **85,** 2250–2254.

Haas, I., and Wabl, M. (1983). Immunoglobulin heavy chain binding protein. *Nature (London)* **306,** 387–389.

Hardwick, K. G., Lewis, M. J., Semenza, J., Dean, N., and Pelham, H. R. (1990). *ERD1,* a yeast gene required for the retention of luminal endoplasmic reticulum proteins, affects glycoprotein processing in the Gogli apparatus. *EMBO J.* **9**(3), 623–630.

Helenius, A., and Kartenbeck, J. (1980). The effects of octylglucoside on the Semliki Forest virus membrane. *Eur. J. Biochem.* **106,** 613–618.

Hurtley, S. M., and Helenius, A. (1989). Protein oligomerization in the endoplasmic reticulum. *Annu. Rev. Cell Biol.* **5,** 277–307.

Jerne, N. K. (1974). Towards a network theory of the immune system. *Ann. Immunol. (Paris)* **125C,** 373.

Jerne, N. K. (1985). The generative grammar of the immune system. *EMBO J.* **4,** 847–852.

Kelly, R. B. (1990). Tracking an elusive receptor. *Nature (London)* **345,** 480–481.

Lagwinska, E., Stewart, C. C., Adles, C., and Schlesinger, S. (1975). Replication of lactic dehydrogenase virus and Sindbis virus in mouse peritoneal macrophages. Induction of interferon and phenotypic mixing. *Virology* **65,** 204–214.

Laver, W. G. Air, G. M., Webster, R. G., and Smith-Gill, S. J. (1990). Epitopes on protein antigens: Misconceptions and realities. *Cell (Cambridge, Mass.)* **61,** 553–556.

Lewis, M. J., and Pelham, H. R. B. (1990). A human homologue of the yeast HDEL receptor. *Nature (London)* **348,** 162–163.

Lewis, M. J., Sweet, D. J., and Pelham, H. R. (1990). The *ERD2* gene determines the specificity of the luminal ER protein retention system. *Cell (Cambridge, Mass.)* **61**(7), 1359–63.

Marriott, S. J., Roeder, D. J., and Consigli, R. A. (1987). Anti-idiotypic antibodies to a polyomavirus monoclonal antibody recognise cell surface components of mouse kidney cells and prevent polyomavirus infection. *J. Virol.* **61**, 2747–2753.

Metsikko, K., and Garoff, H. (1990). Oligomers of the cytoplasmic domain of the p62/E2 membrane protein of Semliki Forest virus bind to the nucleocapsid *in vitro. J. Virol.* **64**, 4678–4683.

Meyer, D. I. (1990). Mimics—or gimmicks. *Nature (London)* **347**, 424–425.

Munro, S., and Pelham, H. R. B. (1986). An hsp70-like protein in the ER: Identity with the 78 kD glucose regulated protein and immunoglobulin heavy chain binding protein. *Cell (Cambridge, Mass.)* **46**, 291–300.

Munro, S., and Pelham, H. R. (1987). A C-terminal signal prevents secretion of luminal ER proteins. *Cell (Cambridge, Mass.)* **48**(5), 899–907.

Murakami, H., Blobel, G., and Pain, D. (1990). Isolation and characterization of the gene for a yeast mitochondrial import receptor. *Nature (London)* **347**, 488–491.

Pain, D., Kanwar, Y. S., and Blobel, G. (1988). Identification of a receptor for protein import into chloroplasts and its localization to envelope contact zones. *Nature (London)* **331**, 232–237

Pain, D., Murakami, H., and Blobel, G. (1990). Identification of a receptor for protein import into mitochondria. *Nature (London)* **347**, 44–449.

Pelham, H. R. B. (1988). Evidence that luminal ER proteins are sorted from secreted proteins in a post-ER compartment. *EMBO J.* **7**, 913–918.

Pelham, H. R. B. (1989). Control of protein exit from the endoplasmic reticulum. *Annu. Rev. Cell Biol.* **5**(1), 1–23.

Pelham, H. R. B. (1990). The retention signal for soluble proteins of the endoplasmic reticulum. *Trends Biochem. Sci.* **15**, 483–486.

Perelson, A. S. (1989). Immune network theory. *Immunol. Rev.* **110**, 5–36.

Reading, C. L. (1982). Theory and methods for immunization in culture and monoclonal antibody production. *J. Immuno. Methods* **53**, 261–291.

Rose, J. K., and Doms, R. W. (1988). Regulation of protein export from the endoplasmic reticulum. *Annu. Rev. Cell Biol.* **4**, 257–288.

Roth, M. A., and Pierce, S. B. (1987). *In vivo* crosslinking of protein disuphide isomerase to immunoglobulins. *Biochemistry* **26**, 4179–4182.

Schlesinger, M., and Schlesinger, S. (1986). The Togaviridae and Flaviviridae. Plenum, New York.

Schrier, M. H., and Lefkovits, I. (1979). Induction of suppression and help during *in vitro* immunisation of mouse spleen cells. *Immunology* **36**, 743–752.

Sege, K., and Peterson, P. A. (1983). Use of anti-idiotypic antibodies as cell-surface receptor probes. *Proc. Natl. Acad. Sci. U.S.A.* **75**, 2443–2448.

Semenza, J. C., Hardwick, K. G., Dean, N., and Pelham, H. R. (1990). ERD2, a yeast gene required for the receptor-mediated retrieval of luminal ER proteins from the secretory pathway. *Cell (Cambridge, Mass.)* **61**(7), 1349–1357.

Simons, K., and Garoff, H. (1980). The budding mechanisms of enveloped animal viruses. *J. Gen. Virol.* **50**, 1–21.

Simons, K., and Warren, G. (1984). Semliki Forest virus: A probe for membrane traffic in the animal cell. *Adv. Protein Chem.* **36**, 79–133.

Vaux, D. (1990a). On the production of anti-idiotype antibodies. *Technique* **2**(1), 18–22.

Vaux, D. (1990b). On the use of *in vitro* immunization for the production of monoclonal antibodies. *Technique* **2**(2), 72–78.

Vaux, D., Helenius, A., and Mellman, I. (1988). Spike-nucleocapsid interaction in Semliki Forest virus reconstructed using network antibodies. *Nature (London)* **336,** 36–42.

Vaux, D., Tooze, J., and Fuller, S. (1990). Identification by anti-idiotype antibodies of an intracellular membrane protein that recognizes a mammalian endoplasmic reticulum retention signal. *Nature (London)* **345,** 495–502.

Weibel, E. R., Staübli, W., Gnägi, H. R., and Hess, F. A. (1969). Correlated morphometric and biochemical studies on the liver cell I. Morphometric model, stereological methods and normal morphometric data for rat liver. *J. Cell Biol.* **42,** 68–91.

Wieland, F. T., Gleason, M. L., Serafini, T. A., and Rothman, J. E. (1987). The rate of bulk flow from the endoplasmic reticulum to the cell surface. *Cell (Cambridge, Mass.)* **50**(2), 289–300.

Williams, W. V., Guy, H. R., Rubin, D. H., Robey, F., Myers, J. F., Kieber-Emmons, T., Weiner, D. B., and Greene, M. I. (1988). Sequences of the cell attachment site of reovirus type 3 and its anti-idiotypic/antireceptor antibody: Modeling of their three dimensional structures. *Proc. Natl. Acad. Sci. U.S.A.* **85,** 6488–6492.

Williams, W. V., Moss, D. A., Kieber-Emmons, T., Cohen, J. A., Myers, J. N., Weiner, D. B., and Greene, M. I. (1989). Development of biologically active peptides based on antibody structure. *Proc. Natl. Acad. Sci. U.S.A.* **86,** 5537–5541.

Chapter 2

In Vivo Protein Translocation into or across the Bacterial Plasma Membrane

ROSS E. DALBEY

Department of Chemistry
The Ohio State University
Columbus, Ohio 43210

I. Introduction

The years 1985 and 1986 were pivotal in the field of bacterial protein export. Several laboratories (Chen *et al.*, 1985; Muller and Blobel, 1984; Geller *et al.*, 1986) reported the *in vitro* translocation of exported proteins into inside-out *Escherichia coli* plasma membrane vesicles. The reaction occurs posttranslational, and requires both ATP and the membrane electrochemical potential. This reconstituted system has played a major role in dissecting the biochemical functions of components of the protein export machinery. Still, the most elegant cell-free systems are those in which pure proteins are solubilized and reconstituted into liposomes. Recently, Driessen and Wickner (1990) have been able to obtain transbilayer translocation

METHODS IN CELL BIOLOGY, VOL. 34

of proOmpA, the precursor form of the major outer membrane protein A, at a very efficient level, which mimics the *in vivo* translocation processes. This proOmpA is incorporated within liposomes containing the SecY, SecE, and SecA proteins, and translocation is energy dependent.

Although the molecular mechanism of prokaryotic protein export has recently been studied using cell-free systems, *in vivo* studies played the major role in studying bacterial export. Pulse-labeling studies demonstrated early on that most proteins exported to the outer membrane or to the periplasmic space are cleaved by leader peptidase at the outer face of the inner membrane. *In vivo* studies, with the use of drugs (Date *et al.*, 1980; Daniels *et al.*, 1981; Enequist *et al.*, 1981) and mutants (Gardel *et al.*, 1987; Oliver and Beckwith, 1982; Ito *et al.*, 1983; Emr *et al.*, 1981; Kumamoto and Beckwith, 1983; Riggs *et al.*, 1988; Dalbey and Wickner, 1985), also revealed that translocation across the membrane requires the membrane potential and the function of *sec* genes.

In this chapter, we describe *in vivo* studies on the export of various proteins into or across the plasma membrane of *E. coli.* This chapter places special emphasis on a number of methods that have been developed to study the fate of a newly synthesized protein in an intact cell. Although the main focus of our lab has been on leader peptidase, we will also discuss work on the export of outer membrane protein A, maltose-binding protein, and the M13 phage coat protein. We will not describe *in vivo* studies that were performed to determine whether proteins are synthesized on free or membrane-bound polysomes.

II. Genetic Manipulation Methods to Study Protein Export

Genetic manipulation methodologies have been crucial in deciphering which regions of a protein are important for its export across or into the plasma membrane. We will illustrate the power of this technique, with leader peptidase, which assembles across the plasma membrane with the topology shown in Fig. 1A. It was believed that proteins such as leader peptidase, which does not have a cleaved amino-terminal leader peptide, have an internal, uncleaved signal sequence that directs insertion and remains in the mature protein, usually as a transmembrane anchor.

We first asked whether leader peptidase contains a short region that is essential for translocation of its large carboxyl-terminal domain. To address this question, we made a series of deletions throughout the leader peptidase molecule that, in sum, cover the entire protein (Fig. 1B). In this site-directed mutagenesis approach (Zoller and Smith, 1983), the gene for leader peptidase is first inserted into the M13 phage vector with the correct orientation. The synthesized mutagenic oligonucleotide is complementary to the single-stranded 5' and 3' regions of the leader peptidase DNA, on the sides where the deletion is going to take place, such that the sequence to be deleted will loopout. Studies of these deletion mutants showed that the second hydrophobic domain was essential for translocation (Dalbey and Wickner, 1987). We next turned to defining the important structural features of this region. Because hydrophobicity is a salient characteristic of leader peptides, and has been shown to be necessary for their function, we next introduced arginine residues at various positions within this region (Fig. 1C). To our surprise, only the second half of H2, which is very hydrophobic, is sensitive to these positively charged residues (Zhu and Dalbey, 1989). In addition, the first half of H2 (residues 62–68) can be entirely deleted from leader peptidase without any measurable effect on translocation, whereas deletion of the second half (residues 70–76) abolishes export (Fig. 1D). That the hydrophobicity is the important factor for signal function is also supported by studies in which apolar domain 2 is replaced with an artificial apolar peptide. When H2 is substituted with 16 consecutive alanyl–leucyl residues (Fig. 1E), leader peptidase can assemble across the membrane with wild-type kinetics (Bilgin *et al.*, 1990). However, this region cannot be too hydrophobic because leader peptidase is more efficiently exported when the apolar core of the uncleaved signal is replaced by consecutive alanyl–leucyl residues, rather than by 16 leucyl residues.

To test if this essential region is part of an uncleaved signal peptide, we asked whether it is interchangeable for a leader peptide (Dalbey *et al.*, 1987). Using recombinant DNA techniques, residues 51–83 of leader peptidase were joined to two exported proteins deprived of their own leader sequences (Fig. 2). This region can translocate the M13 phage coat protein and OmpA across the membrane, indicating that the region containing the second apolar domain functions as a signal peptide. Although this signal is essential for the export of leader peptidase, it is not always sufficient. A deletion removes two-thirds of the carboxy terminus of leader peptidase (Fig. 1B) and also a substitution introduces a single positively charged residue (Fig. 1C) or two negatively charged residues after H2 block insertion. Therefore, an uncleaved signal cannot translocate any sequence carboxylterminal to it.

Although genetic and gene fusion studies have shown that the information for exporting a protein across the membrane is located within the amino-terminal leader peptides of exported proteins (Michaelis and Beckwith, 1982), bacterial leader peptides are usually not sufficient for export.

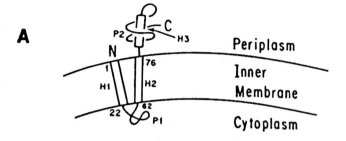

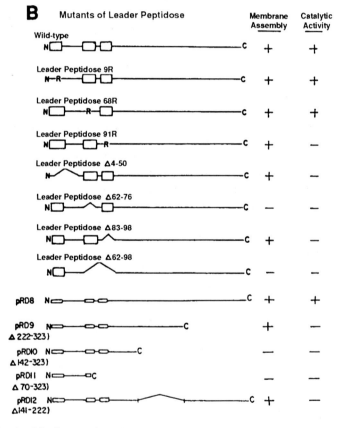

FIG. 1.   Membrane orientation and site-directed mutants of leader peptidase.

C MUTANTS OF LEADER PEPTIDASE AND THEIR MEMBRANE ASSEMBLY PROPERTIES

FIG. 1. (*continued*)

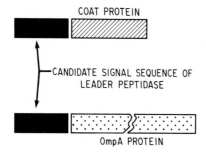

FIG. 2.   DNA that codes for leader peptidase (1–3, 51–83) was joined in-frame to OmpA or coat protein.

┌———leader peptide———┐   ┌———mature region———┐

		-21			+1 +5 +10	processing

1. ProOmpA   RG   MKKTAIAIAVALAGFATVAQA   RGKDNTWYTGAK --- +325   —

2. ProOmpA   RR   MKKTAIAIAVALAGFATVAQA   RRKDNTWYTGAK --- +325   —

3. ProOmpA   GR   MKKTAIAIAVALAGFATVAQA   GRKDNTWYTGAK --- +325   ⁻/+

4. ProOmpA   R5   MKKTAIAIAVALAGFATVAQA   APKDRTWYTGAK --- +325   +

┌———leader peptide———┐   ┌———mature region———┐

-21   +1 +5 +10   processing

1. ProOmpA   GG   MKKTAIAIAVALAGFATVAQA   GGKDNTWYTGAK --- +325   +

2. ProOmpA   EE   MKKTAIAIAVALAGFATVAQA   EEKDNTWYTGAK --- +325   +

3. ProOmpA   EEEE   MKKTAIAIAVALAGFATVAQA   EEEENTWYTGAK --- +325   +

FIG. 3.   Mutants of OmpA with positively or negatively charged residues introduced after the leader peptide.

The early mature region of the exported protein has to have certain features that are compatible with translocation. For example, a positively charged residue carboxylterminal to the leader peptide can have severe effects on export (Li *et al.*, 1988; Yamane and Mizushima, 1988). This is also found with the outer membrane protein A (H. Y. Zhu and R. E. Dalbey, unpublished data). Although four negatively charged residues introduced immediately after the leader peptide do not inhibit export of OmpA, a positively charged residue at position 1 or 2 does block it (Fig. 3). Thus, positively charged residues can prevent transport across the membrane on the immediate C-terminal side of leader peptides.

III.   Exported Proteins Are Synthesized in a Loosely Folded Conformation

Precursor proteins are usually more protease sensitive and hence more loosely folded than are their corresponding mature proteins. (In fact, the majority of periplasmic proteins have evolved such that their structures are protease resistant.) It is widely agreed that an unfolded conformation is a requirement for translocation of proteins across the membrane. The first study that made people take this idea seriously was by Randall and Hardy (1986), who used protease sensitivity as an assay for the folding of the premaltose-binding protein. They showed that there is a correlation between competence for translocation and a lack of a tertiary structure of the mature species. Once the precursor protein obtained a protease-resistant state, it was no longer exported to the periplasm. Similar experiments were done prior to those by Wolfe and Wickner (1984) with the bacterial leader peptidase. This protein accumulated in a more protease-sensitive conformation inside the plasma membrane when uncouplers were added to dissipate the proton motive force. Figure 4 shows that the wild-type and mutant leader peptidases, which accumulate inside the plasma membrane when the membrane potential is blocked, are more protease sensitive than is the mature translocated form of wild-type leader peptidase. The protocol that has proved successful for studying leader peptidase will now be described.

Prior to adding protease in the protease-sensitivity study, cells are pulsed with a radioactive amino acid to label the newly synthesized polypeptide. Often, [^{35}S]methionine or [^{35}S]cysteine is used to label proteins because each is available at high specific activity and is relatively inexpensive. In addition, a drug such as carbonyl cyanide *m*-chlorophenylhydrazone (CCCP) is added just before the labeling to prevent the rapid export of

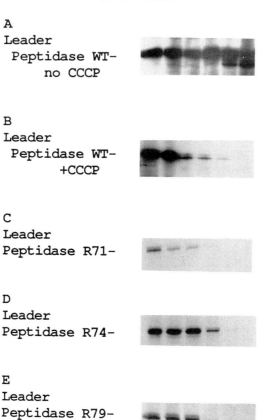

A
Leader
  Peptidase WT–
      no CCCP

B
Leader
  Peptidase WT–
      +CCCP

C
Leader
  Peptidase R71–

D
Leader
  Peptidase R74–

E
Leader
  Peptidase R79–

Trypsin (μg/ml) : 0.01 0.1 1 10 100 1000

FIG. 4.   Protease sensitivity of leader peptidase mutants. Cells with a plasmid coding for wild-type leader peptidase (A and B) and mutant leader peptidase R71 (C), R74 (D), or R79 (E) were pulse-labeled with Trans[35]S label for 1 minute and added to an equal volume of sucrose buffer [40% (w/v) sucrose, 20 mM EDTA, 60 mM Tris-HCl, pH 8.0]. (B) Cells were incubated with CCCP (final concentration 50 μM) for 45 seconds prior to adding the radioactive methionine for 1 minute. After labeling, 100 μl of M9 medium containing methionine (5 mg/ml) and chloramphenicol (10 mg/ml) was added to the cultures on ice. Cells were mixed with an equal volume of ice-cold sucrose buffer, treated with lysozyme (1 mg/ml), lysed by the addition of 10 volumes of ice-cold water, and then digested for 60 minutes with various concentrations of trypsin. Trypsin digestion was terminated by the addition of trypsin inhibitor (1.25 mg/ml) and PMSF (5 mM). The samples were then analyzed by leader peptidase immunoprecipitation and subjected to sodium dodecyl sulfate–gel electrophoresis and fluorography. From Zhu and Dalbey (1989).

radioactive export-competent proteins across the plasma membrane ($t_{1/2}$ of approximately 15 seconds). Immunoprecipitation must be performed for many proteins, including leader peptidase, because there are comigrating bands on a sodium dodecyl sulfate (SDS) polyacrylamide gel. Our procedure contains approximately 0.2% SDS after dilution of the sample into the Triton X-100 immunobuffer.

PROTEASE-SENSITIVITY ASSAY

1. The day before, inoculate a single colony into 1 ml of M9 minimal medium (Miller, 1972) supplemented with 0.5% fructose and the 19 amino acids (without methionine) in a 13 × 100-mm sterile culture tube and grow at 37°C in a shaking waterbath to the stationary phase.
2. Back-dilute culture 1 to 50 in M9 medium the next day and shake at 37°C in a waterbath.
3. Add 10 μl of arabinose (0.2% final concentration) at an apparent optical density of 0.2 at 600 nm to induce synthesis of leader peptidase.
4. After 30 minutes, pulse label cultures (0.5 ml) with 100 μCi/ml of Trans^{35}S label (ICN Radiochemicals), which is 85% [^{35}S]methionine and 15% [^{35}S]cysteine. For cells expressing wild-type leader peptidase, cells were treated with 50 μM CCCP (in ethanol) for 45 seconds prior to labeling.
5. After a 1-minute pulse label, cells (0.5 ml) expressing wild-type leader peptidase or mutant leader peptidase are immediately quenched with an equal volume of ice-cold sucrose buffer (60 mM Tris-HCl, pH 8.0, 40% (w/v) sucrose, and 5 mM EDTA).
6. Treat plasmolyzed cells with lysozyme (1 mg/ml) for 10 minutes to digest the cell wall.
7. Add 10 × volumes of ice-cold water to lyse the 100 μl of spheroplasts.
8. Add various concentrations of protease (0.01–1000 μg/ml) to an aliquot (100 μl) of the cell extract. Incubate on ice for 60 minutes. Trypsin (L-1-tosylamide-2-phenylethyl-chloromethylketone treated) and proteinase K are commonly used to assay protease sensitivity, as they are cheap and readily inhibited by phenylmethysulfonyl fluoride (PMSF).
9. Add trypsin inhibitor (1.25 mg/ml) and PMSF (5 mM) to quench the reaction.
10. To concentrate the proteins for sodium dodecyl sulfate–polyacrylamide gel electrophoresis (SDS–PAGE), aliquots are added to an

equal volume of ice-cold 20% trichloroacetic acid. After a 30-minute incubation at 4°C, collect the radiolabeled proteins by centrifugation in a microfuge (15 minutes at 12,000 g). After aspirating the supernatant away, suspend the pellet by vortexing in 1 ml of chilled acetone at 0°C to remove traces of trichloroacetic acid and collect the pellet by centrifugation at 12,000 g for 5 minutes. Repeat this acetone washing step and solubilize the pellet in SDS–buffer (10 mM Tris-HCl, pH 8.0, and 2% SDS).

11. For immunoprecipitation, dilute SDS-solubilized samples (100 μl) in 1.0 ml with 2.5% Triton X-100 buffer (10 mM Tris-HCl, pH 8.0, 5 mM EDTA, 150 mM NaCl, and 2.5% Triton X-100). Add appropriate amounts of antisera to the diluted samples and incubate for at least 30 minutes at 4°C. The immune complexes are precipitated by adding 25 μl of a 10% suspension of glutaraldehyde-fixed *Staphylococcus aureus* (Staph A) and incubating for 30 minutes to overnight. Pellet Staph A-immune complex by centrifuging at 12,000 g for 15 seconds and carefully aspirate the supernatant. Suspend the Staph A pellet by vortexing in 1 ml of 2.5% Triton X-100 buffer. After several additional washing steps, the pellet is resuspended in sample buffer (160 mM Tris-HCl, pH 6.8, 4% SDS, 20% glycerol, 1.5% 2-mercaptoethanol, and 0.2% bromophenol blue) and boiled in a heating block for 3 minutes. Now the samples are ready for SDS–PAGE and fluorography. It must be noted that a preliminary experiment should be performed to determine the levels of antibody and fixed Staph A that are needed to precipitate all of the desired protein in the cell extract. In addition, if the immunoprecipitations have high background, it may be necessary prior to adding antibody to incubate the diluted SDS-solubilized extract with fixed Staph A (25 μl) to remove proteins that nonspecifically bind to it. After incubating for 15 minutes, the fixed Staph A is pelleted; subsequently the supernatant is treated with antibody, as described above. Moreover, the amount of SDS must not go above 0.2% otherwise the amount of immunoprecipitated proteins decreases dramatically.

The protease-sensitive conformation of exported proteins in the cytoplasm is achieved, in many cases, by the binding of newly synthesized proteins to cytosolic factors. These factors, termed molecular chaperones, promote export by keeping the proteins in translocation-competent conformations. So far, the best characterized cytosolic factor is SecB (Kumamoto and Beckwith, 1983; Weiss *et al.*, 1988). It is required for the export of many proteins, including maltose-binding protein, OmpA, OmpF, and

PhoA. We have asked whether this protein is required for the export of leader peptidase. A SecB null strain, kindly provided by Dr. Carol Kumamoto, was transformed by a plasmid containing the leader peptidase gene. We find that leader peptidase rapidly translocates across the membrane and becomes accessible to added protease (methodology described in detail in Section IV) under conditions wherein OmpA accumulated in a precursor form and was resistant to external protease (S. Cheng and R. E. Dalbey, unpublished data). This result indicates that the protease-sensitive conformation of cytosolic leader peptidase is not mediated by binding to SecB.

IV.   Protein Translocation across the Plasma Membrane Often Requires the *sec* Genes and the Membrane Electrochemical Potential

The application of bacterial genetics by Beckwith, Silhavy, Emr, Bassford, and colleagues has enabled many of the components of the export machinery to be identified, including SecA, SecB, SecY (also called PrlA), SecD, and SecE. Thermosensitive mutations in these proteins affect the export both of proteins to the periplasm and outer membrane and of some proteins to the inner membrane. *In vivo* pulse-chase experiments and cell fractionation have been crucial in characterizing the various export-deficient mutants.

Translocation across the membrane requires the membrane electrochemical potential as well as the *sec* genes (Date *et al.*, 1980; Daniels *et al.*, 1981; Enequist *et al.*, 1981). When CCCP an uncoupler of the proton motive force, is added to cells, newly synthesized proteins are not transferred across the membrane. As of yet, the exact role of the membrane potential is not known.

The effects of CCCP or the thermosensitive mutations in the *sec* genes on the export of preproteins can be evaluated by testing whether the precursor form of an exported protein accumulates. It is easy to detect the precursor form by immunoprecipitation and SDS–PAGE. For proteins such as leader peptidase, which do not contain a cleavable leader peptide, it is necessary to measure directly their distribution across the membrane. This can be achieved by determining the accessibility of an exported protein to a protease added from the outside of the cell. If the protein of interest does not require the *sec* genes or the membrane electrochemical potential, then the exported protein will be digested by protease at the nonpermissive

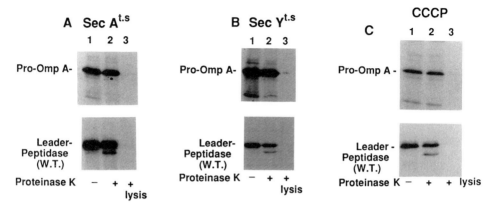

FIG. 5. Leader peptidase requires SecA, SecY, and the electrochemical membrane potential for translocation of its large carboxyl-terminal domain across the plasma membrane. (A and B) Exponentially growing CJ105 (*secA*[ts51]) and CJ107 (*secY*[ts24]) were grown at 30°C under aeration in M9 minimal medium (Miller, 1972) supplemented with 0.5% fructose and all amino acids except methionine. In the early log phase, arabinose (0.2% final concentration) was added to induce synthesis of leader peptidase, then cells were shifted to 42°C. After 4 hours, 200 μCi of Trans[35]S label was added for 2 minutes and then the cells were chilled on ice, converted to spheroplasts, and incubated on ice with or without proteinase K (1 mg/ml). Where indicated, a portion of cells was then treated with 2.0% Triton X-100 prior to the addition of proteinase K to lyse the cell. Subsequently, PMSF (5 mM final concentration) was added to inhibit the protease. Samples were immunoprecipitated with antiserum to leader peptidase or OmpA, loaded on an SDS–polyacrylamide gel, and analyzed by fluorography. (C) For CCCP-treated cells, cultures of exponentially growing MC1061/pRD8 (0.5 ml) expressing leader peptidase were treated with 5 μl of ethanol containing 10 mM CCCP for 45 seconds and then incubated with 50 μCi of Trans[35]S label. After labeling for 2 minutes, the cells were analyzed for protease mapping, as described in the Sec study.

temperature or in the presence of CCCP, respectively. On the other hand, if the exported protein does require these two components, then the labeled protein will be inaccessible to protease.

We find the most reliable protease-accessibility procedure employs spheroplasts (protocol 1) that have been stabilized by MgSO$_4$ (Randall and Hardy, 1986). This method results in very little lysis and usually allows more than 95% accessibility of the outer surface of the inner membrane. Protocol 2, described below, is simpler and is quite useful for certain strains, such as HJM114. When doing these experiments, controls are essential. Outer membrane protein A usually serves as a positive control, except in *sec* and CCCP studies. This protein should be completely digested by protease from the outer surface of the plasma membrane. On the other hand, a cytoplasmic protein (ribulokinase, for example) should remain inaccessible to protease, showing that the plasma membrane remained intact. A control should also be performed in which the membrane bilayer is disrupted with detergent to show that the failure to proteolyze the protein of interest is due to the protection of the membrane. Trypsin and proteinase K are often the proteases of choice. Below is described the labeling protocol that has been successful for the *sec* and CCCP studies of leader peptidase and outer membrane protein A. Figure 5 shows that the translocation of OmpA and the insertion of the large carboxyl-terminal domain of leader peptidase require both SecA and SecY, and the membrane electrochemical potential.

1. sec-LABELING PROTOCOL

1. Grow 1 ml of CJ105 (HJM114, *secA*^{ts}, *leu82::Tn10*) or CJ107 (HJM114, *secY*^{ts}) bearing a pING plasmid coding for wild-type leader peptidase at 30°C under aeration in M9 media, which has been supplemented with all amino acids (50 μg/ml each) except methionine.
2. Dilute overnight cultures 1:50 in 1 ml of fresh medium and, in the early log phase, add arabinose (0.2%) to induce synthesis of pING-encoded leader peptidase. After induction, shift cells to 42°C.
3. After 4 hours, label proteins by adding 50 μl of Trans^{35}S label to the cells.
4. Process cells after 2 minutes of labeling for protease accessibility according to protease-accessibility protocol 1 or 2 (described below).

2. CCCP-LABELING PROTOCOL

1. Back-dilute an overnight culture of MC1061/pRD8 1:50 in 1 ml of M9 minimal media supplemented with 0.5% fructose and the 19 amino acids, except methionine, and grow at 37°C to the mid-log phase.
2. Induce with arabinose (0.2%) and shake for 3 hours at 37°C.
3. Add CCCP (final concentration 50 μM) to 0.5 ml of cells for 45 seconds prior to adding 50 μl of Trans^{35}S label. After 1 minute, add 100 μl of M9 medium containing nonradioactive methionine (5 mg/ml) and chloramphenicol (10 mg/ml) to the culture, and chill it on ice. (Interestingly, different amounts of CCCP need to be added to block the export of different proteins.)
4. Process sample for protease mapping.

3. PROTEASE-ACCESSIBILITY ASSAY

a. Protocol 1

1. Transfer pulse-labeled cells (1 ml) into a prechilled tube on ice.
2. Collect the cells in a microcentrifuge by centrifuging for 45 seconds.
3. Resuspend the pelleted cells in 0.25 ml of chilled 0.1 M Tris-acetate, pH 8.2, 0.5 M sucrose, and 5 mM EDTA.
4. Add 20 μl of lysozyme (2 mg/ml) and immediately add 0.25 ml of ice-cold water, then incubate on ice.
5. After 5 minutes, add $MgSO_4$ to a final concentration of 18 mM (100 μl of 0.2 M $MgSO_4$).
6. Verify with a phase-contrast microscope that >95% of the rod-shaped cells have been converted to round spheroplasts.
7. Collect the spheroplasts by a 40-second centrifugation in a microcentrifuge.
8. Resuspend very carefully the pellet in 0.5 ml of ice-cold 50 mM Tris-acetate, pH 8.2, 0.25 M sucrose, and 10 mM $MgSO_4$. Swirl gently. Do not vortex.
9. Incubate 150-μl portions of spheroplasts with or without 15 μl of trypsin (500 μg/ml) for 60 minutes at 0°C. In another aliquot, add Triton X-100 (final concentration 2%) prior to adding trypsin.
10. Quench trypsin by adding PMSF (5 mM) and trypsin inhibitor (1.25 mg/ml final concentration).
11. Immunoprecipitate sample as described before.

b. Protocol 2

1. Transfer 0.75 ml of radiolabeled cells to an equal volume of chilled sucrose buffer [60 mM Tris-HCl, pH 8.0, 40% (w/v) sucrose, and 20 mM EDTA]. Vortex vigorously for 15 seconds.
2. Treat plasmolyzed cells (200 μl) at 0°C without further addition or with trypsin (500 μg/ml). Add Triton X-100 (2%) to another sample (200 μl) in order to lyse the cells.
3. After quenching with trypsin inhibitor and PMSF as described above, analyze the samples by immunoprecipitation, followed by SDS–PAGE and fluorography.

Another distinct advantage of the *in vivo* protein export methods is that they can be used to measure the kinetics of protein translocation across the inner membrane. Newly synthesized proteins are pulse labeled with an amino acid and then chased with a large excess of the nonradioactive amino acid used in the labeling. Aliquots are removed at various times of chase, then quenched either by adding directly to an equal aliquot of

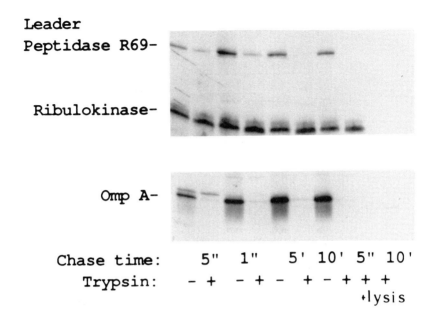

Leader
Peptidase R69-

Ribulokinase-

Omp A-

| Chase time: | 5" | 1" | 5' | 10' | 5" | 10' |
| Trypsin: | − + | − + | − | + − | + + | + |

†lysis

Fɪɢ. 6.   Kinetics of membrane assembly of leader peptidase with an arginine mutation at residue 69. Exponentially growing cells of HJM114 with plasmid expressing leader peptidase R69 were pulse labeled for 30 seconds, chased for 5 seconds and 1, 5, and 10 minutes, and then osmotically shocked by treatment with 40% (w/v) sucrose, 20 m*M* EDTA, and 60 m*M* Tris-HCl, pH 8.0. Portions of these cells were incubated at 0°C with trypsin for 60 minutes. For some samples, the cells were lysed by the addition of 2.0% Triton X-100. Trypsin was inactivated by the addition of trypsin inhibitor (12.5 mg/ml) and PMSF (5 m*M*). Samples were immunoprecipitated with antisera to leader peptidase, ribulokinase, and OmpA and were analyzed by SDS–PAGE and fluorography. Modified from Zhu and Dalbey (1989).

ice-cold 20% trichloroacetic acid or by treating as described in the above protease-accessibility protocols. Figure 6 shows a protease-accessibility study of a mutant wherein the kinetics of translocation of the large carboxyl-terminal domain of leader peptidase are slower than those of the wild-type protein.

V.   The Removal of Leader Peptides by Leader Peptidase

Several leader or signal peptidases, which remove leader peptides from precursor proteins after they have crossed the plasma membrane, have been isolated in pure form in *E. coli*. One, called lipoprotein signal peptidase (Innis *et al.*, 1984), processes only lipoproteins destined to the

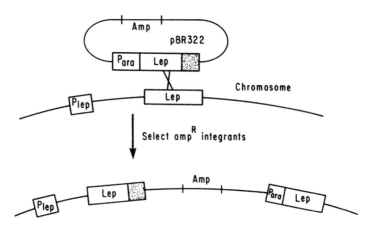

Fig. 7. Integration of the pPR322-derived plasmid pRD9 at the leader peptidase gene locus (*lep*) of the host chromosome. The deleted region of *lep* is represented as shaded areas. From Dalbey and Wickner (1985).

outer membrane. The rest of the preproteins are processed by leader peptidase, whose gene has been cloned (Date and Wickner, 1981), sequenced (Wolfe *et al.*, 1983), and overexpressed (Dalbey and Wickner, 1985). The pure enzyme has very broad specificity and can process many preproteins from a wide variety of sources, including yeast preacid phosphatase, honey bee prepromellitin, and human prehormones (preproinsulin and preinterferon, for example).

To determine the physiological role of this enzyme in protein export, we have developed a method to control the expression of leader peptidase within the cell (Dalbey and Wickner, 1985). It is a variation on the well-known gene disruption technique, in which the normal constitutive leader peptidase promoter is replaced with an inducible arabinose promoter (Fig. 7). Figure 8 shows that precursor to the inner membrane, the periplasm, and the outer membrane accumulates as cells become more deficient in leader peptidase.

Using this method, we can inactivate the chromosomal copy of a gene, and at the same time insert a regulated promoter in front of the introduced gene. This strategy can be used to study whether any gene is essential, but should not be used, of course, if there are any downstream genes in an operon.

1. Construction of a Leader Peptidase-Inducible Strain

Three steps are involved in constructing an arabinose-dependent leader peptidase strain: (1) cloning the leader peptidase gene (*lep*) into a vector

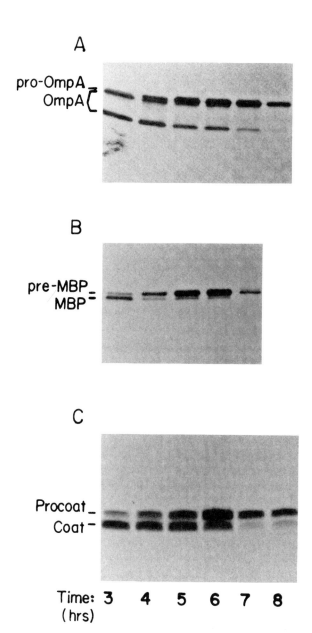

FIG. 8. Precursors of OmpA, MBP, and coat protein accumulate in pRD9/H560 after the shift to arabinose-free medium. Cells were grown in the presence of arabinose and then shifted to arabinose-free medium. Aliquots (0.5 ml) of the cell culture were taken at the indicated times after the arabinose shift, labeled with 50 μCi of [^{35}S] methionine for 1 minute at 37°C, and analyzed by SDS–PAGE and fluorography. (A) Immunoprecipitation of OmpA and its precursor (pro-OmpA). (B) Immunoprecipitation of MBP and its precursor (pre-MBP). A separate culture was infected with M13 at 2 hours at a multiplicity of 50 after the shift in M9 minimal medium with 0.5% glucose. At the indicated times, aliquots of this culture were labeled with [^{35}S] methionine. (C) Immunoprecipitation of procoat and coat. From Dalbey and Wickner (1985).

that expresses leader peptidase under the control of the *araB* promoter, (2) truncation of the gene using the restriction enzyme *Eco*RI to yield an inactive leader peptidase molecule, and (3) recombination of the vector into the chromosome at the *lep* locus via homologous recombination. The last step is achieved by using a *polA⁻*, which lacks a functional, DNA polymerase 1 protein. Polymerase 1 is required for the replication of the vector in transformed cells. Ampicillin is used to select for plasmids that have integrated into the chromosome.

a. Insertion of the Leader Peptidase Gene into a Plasmid with a Promoter That Controls Its Expression

1. Remove the leader peptidase gene by cutting the M13mp8lep (50 μg) vector with the restriction enzymes *Sal*I and *Sma*I. The 1.2-kb leader peptidase fragment is isolated from a 0.7% agarose gel using DEAE paper (Schleicher & Schuell, Inc., Keene, NH).
2. Add the isolated Lep fragment to the purified *Sal*I–*Sma*I cut pING plasmid (a pBR322-derived plasmid carrying the *araB* promoter and the arabinose regulatory elements) and ligate overnight with DNA ligase. The plasmid with the *lepB* gene is called pRD8.
3. Transform the ligated DNA into MC1061 using the calcium chloride procedure of Cohen *et al.*, (1973).
4. Grow transformants in M9 media containing 0.5% fructose and analyze them for arabinose-inducible expression of leader peptidase by pulse-labeling cells with and without arabinose, as described above.

b. Modification of the Leader Peptidase Gene to Inactivate the Protein

1. Cut pRD8 (50 μg) with *Eco*RI, which cleaves at the linker region and within the *lepB* gene.
2. Isolate the truncated plasmid on an agarose gel, then join the ends with DNA ligase to form the vector pRD9.
3. Transform DNA into MC1061, and then analyze for arabinose-dependent expression of a truncated leader peptidase protein.

c. Introduction of the Vector into the Chromosome at the Leader Peptidase Locus via Homologous Recombination

1. Grow 80 ml of H560 (*polA⁻*) to an apparent optical density of 0.5 at 600 nm. (The strain H560 should be tested for methylmethane sulfonate sensitivity on plates (Monk and Kinross, 1972).
2. After chilling the cells, collect them by centrifugation at 6000 rpm for 5 minutes at 4°C.

3. Resuspend the cells in 10 ml of ice-cold 50 mM CaCI$_2$ and incubate for 1 hour at 0°C.
4. Collect cells by centrifugation, as described in step 2, and resuspend in 1 ml of cold 50 mM CaCI$_2$. Hold for 1 hour at 0°C.
5. Incubate cells (0.2 ml) with or without pRD9 (20 μg) for 30 minutes on ice.
6. Heat shock cells by incubating at 37°C for 2-$\frac{1}{2}$ minutes. Add 2 ml of TYE medium and shake cells in a waterbath for 45 minutes.
7. Collect the cells by centrifugation at 5000 rpm. Resuspend cells in 100 μl of TYE and spread all the cells on a TYE ampicillin plate containing 0.2% arabinose.
8. Small colonies appear at about 35 hours when pRD9 is added.
9. Streak out on a TYE plate, confirming that they are arabinose dependent for growth.

With the use of this arabinose-dependent leader peptidase strain, we have found that leader peptidase is not required for preprotein translocation across the membrane. Rather, cleavage is needed to release proteins into the periplasm or to allow them to continue on to the outer membrane. Preproteins are anchored to the membrane by their leader peptide. This was demonstrated for the maltose-binding protein using the osmotic shock procedure, as described by Neu and Heppel (1965), at 0°C. Figure 9 shows

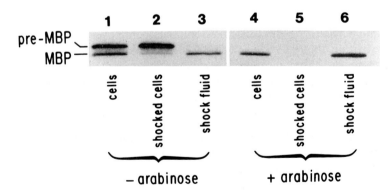

FIG. 9. Localization of pre-MBP by osmotic shock in leader peptidase-deficient cells. A culture of pRD9/H560 was grown in the presence or absence of arabinose. After 9 hours, 10 ml of cells was pulse labeled with 200 μCi of [^{35}S]methionine for 1 minute and isolated by centrifugation (16,000 g, 0°C, 5 minutes). Osmotic shock fluid and shocked cells (cell cytoplasm and membrane) were prepared as described by Neu and Heppel (1965) at 0°C. Aliquots of each sample were immunoprecipitated with antiserum against MBP. Total cells (lanes 1 and 4), shocked cells (lanes 2 and 5), and shock fluid (lanes 3 and 6) were analyzed by SDS–PAGE and fluorography. From Dalbey and Wickner (1985).

that maltose-binding protein, which accumulates in the leader peptidase-deficient cells, remains with the shocked cells and is not released into the shock fluid (periplasmic fraction).

2.   Osmotic Shock Protocol

1. Label cells (10 ml) with 200 μCi of Trans^{35} S label for 1 minute, then transfer to a centrifuge tube containing 10 g of ice.
2. Collect cells by centrifugation, wash with ice-cold Tris buffer (10 mM Tris, pH 8.0), then repeat the centifugation.
3. Collect cells again by centrifuging at 6000 rpm for 10 minutes at 4°C.
4. Resuspend cells in 2 ml of chilled sucrose solution (20% sucrose, 0.03 M Tris, pH 8.0).
5. Add 10 μl of 0.5 M EDTA while stirring and incubate on ice for 15 minutes. Remove an aliquot (200 μl) for total ^{35}S-labeled proteins.
6. Pellet cells and keep the supernatant, which is the periplasmic fraction, and resuspend the shocked cells in 1.8 ml of sucrose solution.
7. Immunoprecipitate part or all the samples.
8. Analyze the cells, shocked cells, and shock fluid by SDS–PAGE and fluorography.

VI.   Conclusion

In this chapter we describe methodologies employed to study protein export within the cell. They include genetic manipulation techniques to determine which regions of an exported protein are essential, or irrelevant, for protein export and membrane assembly, and pulse-chase studies combined with cell fractionation and protease mapping. Taken together, these methods have led to the following view of protein export:

1. Proteins are initially synthesized in a precursor form with a conformation that is more "open" and protease sensitive than the mature form.
2. Translocation across the membrane often requires the assistance of both the Sec proteins and the membrane proton motive force.
3. Translocation is not coupled to leader peptidase processing but is necessary for the release of preproteins from the outer surface of the plasma membrane.
4. The information for translocation exists within leader peptides and uncleaved signals in exported proteins.
5. Sequences carboxyl terminal to signal peptides may prevent translocation across the membrane.

Whereas the *in vivo* studies have been crucial in identifying the components of the export machinery and in showing that the membrane electrochemical potential is important for export, they have fallen short in illuminating both the enzymatic functions of the *sec* genes and the role of the potential. In this regard, the *in vitro* systems have been extremely useful. For example, using a reconstituted system, it has recently been shown that the SecA protein has ATPase activity (Lill *et al.*, 1989) and that the SecB protein has unfolding activity (Collier *et al.*, 1988). Soon biochemical analysis will uncover the function of the SecY and SecE proteins as well. However, the physiological significance of an enzymatic activity must always be determined within the cell. A more complete understanding of the export pathway in *E. coli* will be facilitated by both *in vitro* studies and the *in vivo* studies described here.

ACKNOWLEDGMENTS

I thank Heng-yi Zhu and Jong-In Lee, who contributed to the work on the membrane assembly of leader peptidase. This work was supported by National Science Foundation Grant (DCB-8718578), an American Cancer Society Junior Faculty Award, and a Basil O'Conner starter grant from the March of Dimes.

REFERENCES

Bilgin, B., Lee, J. I., Zhu, H. Y., Dalbey, R. E., and von Heijne, G. (1990). *EMBO J.* **9**, 2717–2722.

Chen, L., Rhoads, D., and Tai, P. C. (1985). *J. Bacteriol.* **161**, 973–980.

Cohen, S. N., Chang, A. C. Y., Boyer, H. W., and Helling, R. B. (1973). *Proc. Natl. Acad. Sci. U.S.A.* **70**, 3240–3244.

Collier, D. N., Bankaitis, V. A., Weiss, J. B., and Bassford, P. J., Jr. (1988). *Cell (Cambridge, Mass.)* **53**, 273–283.

Dalbey, R. E., and Wickner, W. (1985). *J. Biol. Chem.* **260**, 15925–15931.

Dalbey, R. E., and Wickner, W. (1987). *Science* **235**, 783–787.

Dalbey, R. E., Kuhn, A., and Wickner, W. (1987). *J. Biol. Chem.* **262**, 13241–13245.

Daniels, C. J., Bole, D. G., Quay, S. C., and Oxender, D. L. (1981). *Proc. Natl. Acad. Sci. U.S.A.* **78**, 5396–5400.

Date, T., and Wickner, W. (1981). *Proc. Natl. Acad. Sci. U.S.A.* **78**, 6106–6110.

Date, T., Zwizinski, C., Ludmerer, S., and Wickner, W. (1980). *Proc. Natl. Acad. Sci. U.S.A.* **77**, 827–831.

Driessen, A. J., and Wickner, W. (1990) *Proc. Natl. Acad. Sci. U.S.A.* **37**, 3107–3111.

Emr, S. D., Hanley-Way, S., and Silhavy, T. J. (1981). *Cell (Cambridge, Mass.)* **23**, 79–88.

Enequist, H. G., Hirst, T. R., Harayama, S., Hardy, S. J. S., and Randall, L. L. (1981). *Eur. J. Biochem.* **116**, 227–233.

Gardel, C., Benson, S., Hunt, J., Michaelis, S., and Beckwith, J. (1987). *J. Bacteriol.* **169**, 1286–1290.

Geller, B. L., Movva, N. R., and Wickner, W. (1986). *Proc. Natl. Acad. Sci. U.S.A.* **83**, 4219–4222.

Innis, M. A., Tokunaga, M., Williams, M. E., Loranger, J. M., Chang, W. Y., Chang, S., and Wu, H. C. (1984). *Proc. Natl. Acad. Sci. U.S.A.* **81**, 3708–3712.

Ito, K., Wittekind, M., Nomura, M., Shiba, K., Yura, T., Miura, A., and Nashimoto, H. (1983). *Cell (Cambridge, Mass.)* **32**, 789–797.

Kumamoto, C. A., and Beckwith, J. (1983). *J. Bacteriol.* **154**, 253–260.

Li, P., Beckwith, J., and Inouye, H. (1988). *Proc. Natl. Acad. Sci. U.S.A.* **85**, 7685–7689.

Lill, R., Cunningham, K., Brundage, L. A., Ito, K., Oliver, D., and Wickner, W. (1989). *EMBO J.* **8**, 961–966.

Michaelis, S., and Beckwith, J. (1982). *Annu. Rev. Microbiol.* **36**, 435–465.

Miller, J. H. (1972). "Experiments in Molecular Genetics." Cold Spring Harbor Lab., Cold Spring Harbor, New York.

Monk, M., and Kinross, J. (1972). *J. Bacteriol.* **109**, 971–978.

Muller, M., and Blobel, G. (1984). *Proc. Natl. Acad. Sci. U.S.A.* **81**, 7737–7741.

Neu, H. C., and Heppel, L. A. (1965). *J. Biol. Chem.* **240**, 3685–3692.

Oliver, D., and Beckwith, J. (1982). *J. Bacteriol.* **150**, 686–691.

Randall, L. L., and Hardy, S. J. S. (1986). *Cell (Cambridge, Mass.)* **46**, 921–928.

Riggs, P. D., Derman, A. I., and Beckwith, J. (1988). *Genetics* **118**, 571–579.

Weiss, J. B., Ray, P. H., and Bassford, P. J. (1988). *Proc. Natl. Acad. Sci. U.S.A.* **85**, 8978–8982.

Wolfe, P. B., and Wickner, W. (1984). *Cell* (Cambridge, Mass.) **36**, 1067–1072.

Wolfe, P. B., Wickner, W., and Goodman, J. (1983). *J. Biol. Chem.* **258**, 12073–12080.

Yamane, K., and Mizushima, S. (1988). *J. Biol. Chem.* **263**, 19690–19696.

Zhu, H. Y., and Dalbey, R. E. (1989) *J. Biol. Chem.* **264**, 11833–11838.

Zoller, M. J., and Smith, M. (1983). *In* "Methods in Enzymology" (R. Wu, L. Grossman, and K. Moldave, eds.) Vol. 100, pp. 468–500. Academic Press, New York.

Chapter 3

Analysis of Membrane Protein Topology Using Alkaline Phosphatase and β-Galactosidase Gene Fusions

COLIN MANOIL

Department of Genetics
University of Washington
Seattle, Washington 98195

I. Introduction

Understanding the function of a membrane protein in depth is commonly limited by the structural information available for the protein. In this Chapter, I describe a simple genetic method for identifying the disposition of different parts of a polypeptide chain relative to the membrane, the "topology" of the membrane protein. This approach is based on the

61

finding that the specific activities of certain enzymes ("sensor" enzymes), when fused to a membrane protein, reflect the subcellular disposition of the membrane protein fusion site (Manoil *et al.*, 1988). This article describes methods in *Escherichia coli* for using alkaline phosphatase (the PhoA product, normally a periplasmic protein) and β-galactosidase (the LacZ product, normally a cytoplasmic protein) fusions to analyze topologies of cytoplasmic membrane proteins.

The rationale for using gene fusions to study membrane protein topology is shown in Fig. 1. It appears that the subcellular location of alkaline phosphatase or β-galactosidase attached to a membrane protein generally

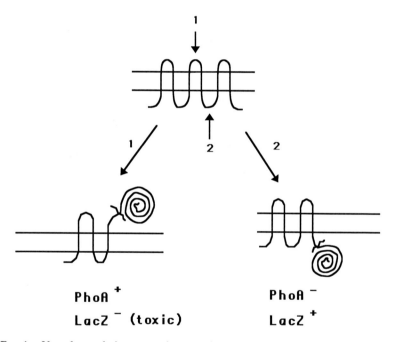

PhoA $^+$

LacZ $^-$ **(toxic)**

PhoA $^-$

LacZ $^+$

FIG. 1. Use of gene fusions to analyze membrane protein topology. Fusion of alkaline phosphatase to a cytoplasmic membrane protein at a periplasmic site (site 1) yields a hybrid protein with its alkaline phosphatase moiety situated in the periplasm, where it is enzymatically active. Fusion of alkaline phosphatase at a cytoplasmic site (site 2) yields an inactive, cytoplasmically disposed enzyme. β-Galactosidase hybrids show reciprocal behavior, with fusions at periplasmic sites yielding proteins with low specific activities when expressed at low levels, which are usually toxic when expressed at high levels. It is not known how much of the β-galactosidase polypeptide is exported to the cytoplasm when it is attached at a periplasmic site. Cytoplasmic sites of β-galactosidase attachment yield high-activity proteins that are relatively nontoxic.

corresponds to the normal location of the junction site in the unfused membrane protein. For example, fusion of alkaline phosphatase at a periplasmic site (site 1) appears to yield a hybrid protein with the alkaline phosphatase moiety in the periplasm, where it is highly active. Fusion at a cytoplasmic site (site 2) yields a cytoplasmic alkaline phosphatase that is inactive. Fusions to β-galactosidase show a reciprocal behavior: periplasmic fusion sites yield hybrid proteins with reduced specific activities that are usually highly toxic if expressed at high levels, whereas cytoplasmic fusion sites yield hybrid proteins with high specific activities. The combined use of alkaline phosphatase and β-galactosidase fusions thus provides high enzyme activity signals for both periplasmic and cytoplasmic sites in cytoplasmic membrane proteins. It is also possible to interconvert alkaline phosphatase and β-galactosidase fusions to compare the activities of the two enzymes fused at a single site (Section II,C).

As a starting point in an analysis of membrane protein topology, the amino acid sequence of the protein is arranged into a model (or models) based on hydrophobicity and the distribution of positively charged amino acid residues (Engelman *et al.*, 1986; von Heijne, 1987). A test of the model using gene fusions then normally requires the following steps: (1) generation of fusions to the plasmid-borne membrane protein gene using *in vivo* and *in vitro* methods; (2) identification of the junction sites of the fusions using restriction mapping and DNA sequencing; (3) assay of the hybrid protein enzymatic activities in permeabilized cells; and (4) determination of the rates at which hybrid proteins are synthesized using pulse-label experiments.

II.   Generation of Gene Fusions

A.   *In Vivo* Methods

Alkaline phosphatase and β-galactosidase fusions can be simply generated using transposon derivatives carrying *phoA* or *lacZ* sequences (Fig. 2) (Manoil and Beckwith, 1985; Gutierrez *et al.*, 1987; Manoil, 1990). The gene fusions generated encode hybrid proteins consisting of different amounts of membrane protein sequences at their N-termini and nearly all of alkaline phosphatase or β-galactosidase at their C-termini, connected by a "linker" sequence of 17–22 amino acid residues encoded mainly by transposon end sequences.

Transposon Tn*phoA* fusions can be generated using either a phage lambda derivative (λTn*phoA*) or a strain carrying Tn*phoA* in an F*lac*

Factor (CC202). Tn*lacZ* fusions are usually generated using a strain carrying a chromosomal insert of Tn*lacZ* (CC170).

Protocols for generating fusions using λTn*phoA* and CC170 are given below. (The protocol for generating Tn*phoA* fusions using CC202 is virtually identical to that for generating Tn*lacZ* fusions from CC170 except for the substitution of CC202 for CC170 and of XP for XG in media and is not presented.) In the protocols that follow, the composition of growth media is that described by Miller (1972) and molecular biological techniques are those described by Maniatis *et al.* (1982).

1. STRAINS, PHAGE, AND PLASMIDS

CC118: Δ(*ara,leu*)7697 Δ*lac*X74 Δ*phoA*20 *galE galK thi rpsE rpoB argE*(am) *recA1*

CC202: F42 *lacI3* zzf-2::Tn*phoA*/CC118

LE392: *supF supE hsdR galK trpR metB lacY tonA*

λTn*phoA*: *b*221 *cI*857 *P*am3 with Tn*phoA* in or near *rex* (λTn*phoA* stocks are grown using LE392 as host)

Target plasmid: The membrane protein whose topology it to be analyzed is normally encoded by a plasmid present at high copy number (e.g., one derived from pBR322). [There are examples in which hybrid protein production is toxic (e.g., see Calamia and Manoil, 1990). In such cases, it is advantageous to generate fusions to membrane proteins whose expression can be regulated. Fusions can then be generated under conditions of nonexpression and detected afterward using a screening method such as replica plating, requiring only minor modifications of the two protocols that follow.]

2. GENERATION OF Tn*phoA* FUSIONS USING λTn*phoA*

1. Start with a strain carrying the target plasmid (encoding the membrane protein of interest) in monomeric form. The strain should be *RecA⁻ PhoA⁻ Kan*ˢ Lambdaˢ, such as CC118.
2. Grow cells at 37°C in L broth (LB) containing 10 mM MgSO$_4$ and antibiotic selective for the plasmid (e.g., ampicillin) to early stationary phase and add λTn*phoA* at a multiplicity of approximately 1.0. Incubate at 30°C for 15 minutes without agitation for phage adsorption, then dilute aliquots 1:10 into LB. Grow a number of separate cultures from each infection to help guarantee the isolation of insertions arising from independent transposition events.
3. Grow cultures 4–15 hours at 30°C with aeration.

4. Plate 0.2-ml aliquots of undiluted cultures onto L agar (LA) containing a plasmid-selective antibiotic (e.g., ampicillin), 300 μg/ml kanamycin and 40 μg/ml XP (5-bromo-4-chloro-3-indolylphosphate p-toluidine salt). Incubate 2–3 days at 30°C.
5. Plates should show growth of colonies (typically 10–100) that range in color from white to dark blue. Pool colonies from individual platings (e.g., by scraping them up together with a toothpick) and prepare plasmid from the pool using an alkaline lysis method. Transform CC118 with this preparation, selecting transformants on LA containing antibiotic selective for the plasmid (e.g. ampicillin), 30 μg/ml kanamycin, and 40 μg/ml XP.
6. After 1–2 days incubation at 37°C, purify cells from faint to dark blue colonies by two rounds of single-colony isolation on LA containing 30 μg/ml kanamycin [Note: low-activity fusions can often be detected after incubation of transformant plates at 4°C for 1–7 days after the 2-day incubation at 37°C (E. Traxler, personal communication).] It can be informative after purification to streak cells onto LA containing the antibiotic selective for the plasmid (e.g., ampicillin) and 40 μg/ml XP but lacking kanamycin. Colonies grown on such plates that show abundant blue/white sectoring usually contain two different plasmids, one with and one without a transposon insertion. An additional transformation using plasmid prepared from such cells is required to isolate cells carrying a single plasmid.
7. Isolate plasmid DNAs from cultures started from the purified transformant colonies. Plasmid DNA is analyzed by restriction mapping and DNA sequencing to identify sites of Tn*phoA* insertion.

3.   GENERATION OF Tn*lacZ* FUSIONS USING CC170

1. Introduce the monomeric target plasmid carrying the membrane protein gene into CC170 by transformation.
2. Resuspend single transformant colonies in LB and plate dilutions onto LA containing antibiotic selective for the plasmid (e.g., ampicillin), 300 μg/ml kanamycin, and 40 μg/ml XG (5-bromo-4-chloro-3-indolylgalactoside). After 1–2 days at 37°C colonies of various sizes and intensities of blue should appear.
3. Make plasmid DNA by alkaline SDS extraction of either (1) pooled blue and white colonies or (2) pooled blue colonies that have been restreaked and grown overnight to provide increased cell mass. Use this preparation to transfrom CC118, selecting transformants on LA containing antibiotic selective for the plasmid (e.g., ampicillin), 30 μg/ml kanamycin, and 40 μg/ml XG.

4. After 1–2 days incubation at 37°C, purify cells from blue colonies by two rounds of single-colony isolation on LA containing 30 μg/ml kanamycin. (Note: transformant colonies showing low β-galactosidase activities often result from out-of-frame *lacZ* fusions.) After purification, streak cells onto LA containing the antibiotic selective for the plasmid (e.g., ampicillin) and 40 μg/ml XG but lacking kanamycin. Colonies grown on such plates that show extensive blue/white sectoring are usually the result of double-transformation events.

5. Isolate plasmid DNA from cultures started from the purified transformant colonies. Plasmid DNA is analyzed by restriction mapping and DNA sequencing to identify sites of Tn*lacZ* insertion.

B. *In Vitro* Methods

Alkaline phosphatase and β-galactosidase fusions can be constructed *in vitro* using plasmid vectors with appropriately placed restriction sites (Casadaban *et al.*, 1983; Hoffman and Wright, 1985; Simons *et al.*, 1987; Gutierrez and Devedjian, 1989). The fusions can be constructed directly at restriction sites or after partial exonucleolytic digestion of the membrane protein genes (Henikoff, 1987).

A method for generating fusions with predefined end points using chemically synthesized oligonucleotides was described by Boyd *et al* (1987). In this method, a plasmid carrying Tn*phoA* (or Tn*lacZ*) sequences 3' to the membrane protein gene is deleted for sequences extending from a point in the membrane protein gene to the beginning of Tn*phoA* (or Tn*lacZ*) sequences. The exact structure of the final fusion is dictated by the sequence of an oligonucleotide able to hybridize to the target gene sequence and the beginning of Tn*phoA* (or Tn*lacZ*). Fusions constructed by this *in vitro* method are "isogenic" to those generated by transposon insertion, because they have the same junction sequence, including the "linker" sequence.

C. Fusion Switching

Fusion switching is a process by which a *phoA* fusion can be converted into *lacZ* fusion, or vice versa. The ability to generate both types of fusions at exactly the same site makes it possible to compare their enzymatic activities in assessing the subcellular location of the site. Methodology for fusion switching using homologous recombination has been described recently (Manoil, 1990). A method for fusion switching *in vitro* at a restriction site has also been developed (C. Manoil, unpublished results).

D. Choice of Methods for Generating Gene Fusions

The goal of a fusion analysis is to create an informative set of fusions as efficiently as possible. The generation of fusions *in vivo* by Tn*phoA* and Tn*lacZ* insertion is simple and inexpensive, but has the disadvantage that membrane protein genes may have insertion "hot spots" that limit the number of different fusions obtained. For example, in an analysis of the 417-amino acid *Lac* permease using Tn*phoA* fusions, 21 of 65 insertions analyzed were at two sites (Calamia and Manoil, 1990). The use of other transposon derivatives in addition to Tn*phoA* and Tn*lacZ* may help alleviate this problem (Berg *et al.,* 1989; R. Kolter, personal communication). In spite of the existence of insertion hot spots, a fusion set sufficient to test models for proteins predicted to have simple topologies (e.g., one or two spanning segments) can usually be obtained by transposition alone. However, for proteins with a large number of spanning segments, a complete analysis generally requires the construction of fusions *in vitro*. For example, an analysis of *Lac* permease that implies a 12-span structure for the protein required the construction of five *phoA* fusions in addition to fusions at 27 different sites generated by Tn*phoA* insertion. Note that it is ideal to have each cytoplasmic and periplasmic segment of a membrane protein tagged by more than one fusion, because there are occasional cases in which individual fusions show unusually high or low specific activities (Section V).

E. Identifying Fusion Junction Positions

The positions of sequence junctions are identified by a combination of restriction mapping and DNA sequencing. Useful restriction sites for positioning Tn*phoA* insertions in plasmids are *Dra*I and *Eco*RI, and useful sites for positioning Tn*lacZ* inserts are *Bam*HI, *Eco*RV, and *Sac*I (Fig. 2). These sites are situated near the left ends of the fusion transposons and therefore can generate small left-end junction fragments after additional cutting at restriction sites near the N-terminal coding regions of the membrane protein genes being analyzed. Such small fragments simplify accurate mapping of the transposon inserts.

The exact positions of fusion junctions are determined by DNA sequencing using a double-stranded DNA template (Tabor and Richardson, 1987). Single-strand oligonucleotide primers used for such sequencing hybridize to the 5' end of the *phoA* or *lacZ* gene and allow polymerization of DNA through the transposon left ends into target gene sequences. Sequences of primers that can be used for this purpose are presented in the legend to Fig. 2.

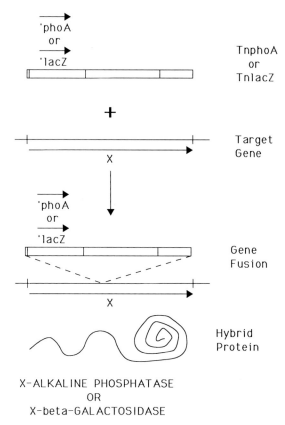

X-ALKALINE PHOSPHATASE
OR
X-beta-GALACTOSIDASE

FIG. 2.   Generation of gene fusions by Tn*phoA* and Tn*lacZ* transposition. Insertion of Tn*phoA* or Tn*lacZ* into a target (gene "X") in the appropriate orientation and translational reading frame generates a gene fusion in which C-terminal sequences of the target gene product are replaced by alkaline phosphatase or β-galactosidase sequences. At the junction of each hybrid protein (i.e., between X protein sequences and alkaline phosphatase or β-galactosidase sequences) is a "linker" sequence of 17 (Tn*phoA*) or 22 (Tn*lacZ*) amino acid residues encoded mainly by transposon end sequences. Useful restriction sites for positioning Tn*phoA* inserts in plasmids are *Dra*I (with a single site 0.255 kb from the left end) and *Eco*RI (sites at 0.77 and 1.10 kb), and useful sites for positioning Tn*lacZ* inserts are *Bam*HI (sites at 0.06, 3.13, and 5.7 kb), *Eco*RV (1.16 kb), and *Sac*I (2.0 kb). For DNA sequence analysis, oligonucleotide primers are used that hybridize to the 5' ends of the *phoA* or *lacZ* sequences in the transposons and direct polymerization through the transposon left ends into target gene sequences. The primer used for Tn*phoA* insert sequencing is 5'-AATATCGCCCTGAGCA-3' and the primer used for Tn*lacZ* insert sequencing is either 5'-CGCCAGGGTTTTCC-CAGTCACGAC-3' (available commercially from New England Biolabs) or 5'-CGGG-ATCCCCCTGGATGG-3'. The alkaline phosphatase fused by Tn*phoA* insertion lacks its signal sequence and five additional amino acids, and the β-galactosidase fused by Tn*lacZ* insertion lacks nine N-terminal amino acids.

III. Enzyme Assays

The alkaline phosphatase and β-galactosidase activities of hybrid proteins are measured in permeabilized whole cells. An alkaline phosphatase assay protocol is presented below, and the β-galactosidase assay is described in Miller (1972).

1. ASSAY OF ALKALINE PHOSPHATASE ACTIVITY IN PERMEABILIZED CELLS

1. Grow 5-ml duplicate overnight cultures of cells to be assayed at room temperature with aeration in LB supplemented with an antibiotic selective for the gene fusion plasmid.
2. Dilute overnight cultures 1/100 into fresh media and incubate with aeration at 37°C until cultures reach midexponential growth (usually 1.5–2 hours).
3. Centrifuge 1 ml of each culture in a microcentrifuge 3–5 minutes (16,000 g) at 4°C. Wash cells once in cold 10 mM Tris-HCl, pH 8.0, 10 mM MgSO$_4$, and resuspend final pellet in 1 ml of cold 1 M Tris-HCl, pH 8.0.
4. Dilute 0.1 ml of cells into 0.4 ml of cold 1 M Tris-HCl, pH 8.0, for an OD$_{600}$ reading. [Correct for this 1/5 dilution in the calculation of alkaline phosphatase activity (step 9).]
5. Add 0.1 ml of washed cells (either undiluted or from the diluted cells in step 4, depending on the level of alkaline phosphatase activity expected) into 0.9 ml 1 M Tris-HCl, pH 8.0, 0.1 mM ZnCl$_2$ in 13-mm × 100-mm glass tubes. Include a blank without cells. Add 50 μl 0.1% SDS and 50 μl chloroform, vortex, and incubate at 37°C for 5 minutes to permeabilize cells. Place tubes on ice 5 minutes to cool.
6. Add 0.1 ml of 0.4% p-nitrophenyl phosphate (in 1 M Tris-HCl, pH 8.0) to each individual tube, agitate, and place in 37°C water bath. Note time.
7. Incubate each tube until pale yellow, then add 120 μl 1:5 0.5 M EDTA, pH 8.0, 1 M KH$_2$PO$_4$, and place the tube in an ice-water bath to stop the reaction. Note time.
8. Measure OD$_{550}$ and OD$_{420}$ for each sample, using an assay mixture without cells as a blank.
9. Units activity $= \dfrac{[OD_{420} - (1.75 \times OD_{550})]1000}{\text{time (min)} \times OD_{600} \times \text{vol. cells (ml)}}$

2. COMMENTS

1. For fusions requiring induction for expression, growth conditions prior to assay must be adjusted. For example, for assay of *lacY–phoA* fusions, cells are diluted 400-fold in step 2, grown for 2 hours and exposed to isopropyl thiogalactoside (a *lac* operon inducer) for 1 hour prior to assay (Calamia and Manoil, 1990).

2. Cytoplasmic forms of alkaline phosphatase are sometimes activated in the course of the assay procedure. All cases of this activation observed thus far have involved relatively short hybrid proteins. The activation can be eliminated by the addition of 1 mM iodoacetamide to assay buffers (A. Derman and J. Beckwith, personal communication).

3. Cells not permeabilized by chloroform–SDS treatment as part of the alkaline phosphatase assay show lower, less reproducible activities than those that have been permeabilized.

4. Lewis *et al.* (1990) have recently found that the rate of growth of cells on polyphosphate as phosphate source is a measure of alkaline phosphatase activity.

IV. Rates of Hybrid Protein Synthesis

In analyzing the topology of a protein using gene fusions, it is important to measure the relative rates of synthesis of representative hybrid proteins. This measurement is essential to show that the differences in enzymatic activities observed for different hybrids are not due to differences in expression levels. Examples of differences in expression for fusions to the same protein have been observed for both alkaline phosphatase and β-galactosidase fusions (Froshauer *et al.*, 1988; San-Millan *et al.*, 1989; Manoil, 1990). For putative periplasmic domain alkaline phosphatase fusions, it is also necessary to ascertain that the specific activity of the hybrid protein is comparable to that of alkaline phosphatase or to an alkaline phosphatase hybrid known to be exported (e.g., a β-lactamase–alkaline phosphatase hybrid). Determining the steady-state level of hybrid protein (such as by Western blotting) is insufficient for these purposes, because there are frequently differences in the rates of degradation of hybrid proteins with different fusion junctions (for further discussion, see San-Millan *et al.*, 1989).

DETECTION OF NEWLY SYNTHESIZED HYBRID PROTEINS BY PROTEIN
LABELING FOLLOWED BY IMMUNOPRECIPITATION

1. This procedure is modified from Ito *et al.*, 1981. Grow cells over-night with aeration at room temperature or 37°C in the minimal medium to be used for radioactive labeling of the protein, e.g., M63 supplemented with growth requirements (no methionine!) and an antibiotic selective for the plasmid carrying the gene fusion.

2. Dilute the overnight culture in the same medium to a cell density of about 5×10^7 cells/ml and grow at 37°C with aeration to a density of $(2-5) \times 10^8$ cells/ml.

3. Add 0.5 ml of the cell culture to a prewarmed tube containing 10–15 μCi [^{35}S]methionine (~1000 Ci/mmol) and incubate 1 mi-nute at 37°C. (It is not necessary to agitate cells for aeration during this short labeling period.)

4. Add 0.5 ml of cold 10% trichloroacetic acid, vortex mix, and place on ice for at least 15 minutes.

5. Centrifuge 10 minutes 16,000 g in a microcentrifuge. The pellet is small and may be spread out on the side of the tube.

6. Wash the pellet twice with 1 ml of cold acetone, then dry it under vacuum.

7. Add 50 μl of SDS buffer (10 mM Tris-HCl, pH 8.0, 1% SDS, 1 mM EDTA, and 5% β-mercaptoethanol) to dissolve the pellet. Heat in a boiling waterbath for 2–3 minutes followed by vortex mixing. Con-tinue heating and mixing until visible pellet has dissolved.

8. Allow the mixture to cool, then add 450 μl of cold KI buffer (50 mM Tris-HCl, pH 8.0, 150 mM NaCl, 2% Triton X-100 (w/v), and 1 mM EDTA). Vortex, then centrifuge 10 minutes at 16,000 g at 4°C.

9. Take 200 μl from the top of the supernatant and add to 300 μl of cold KI buffer. Add antibody. (Note: the amount of antibody required to be in antibody excess over antigen should be determined in a sepa-rate experiment in which the amount of antibody is varied. Anti-body directed against alkaline phosphatase is commercially available from 5′→3′, Inc., West Chester, PA, and antibody directed against β-galactosidase is commercially available from Promega, Madison, WI).

10. Incubate at 4°C 2.5–18 hours.

11. Add 50–100 μl of a suspension of formalin-treated, heat-killed *Staphlococcus aureus* cells (e.g., "IgGsorb," The Enzyme Center, Inc., Malden, Ma, reconstituted in KI buffer according to the

supplier's instructions). Incubate at 4°C with mixing every 5 minutes for 20 minutes.

12. Centrifuge 1 minute at 16,000 g to pellet. Remove supernatant by aspiration and discard.

13. Wash pellet twice with 0.5 ml of high-salt buffer [50 mM Tris-HCl, pH 8.0, 1 M NaCl, 1% Triton X-100 (w/v)], and once with 0.5 ml of 10 mM Tris-HCl, pH 8.0.

14. Resuspend final pellet in 50 μl of gel sample buffer (Laemmli, 1972). Heat in boiling waterbath 5 minutes.

15. Centrifuge 5 minutes at 16,000 g. Use supernatant as a sample in SDS–polyacrylamide gel electrophoresis (Laemmli, 1972).

16. After electrophoresis, gels are prepared for autoradiography by soaking them in 7.5% acetic acid (v/v) for 15 minutes at 37°C, followed by a rinse with water and soaking in 1M sodium salicylate for 1 hour at 37°C. The gel is then dried and used to expose X-ray film.

17. Protein bands positioned using the autoradiogram are cut out of the dried gel, rehydrated in 100 μl water, and shaken gently for 2 days at 42°C in a solution of 7.5% Protosol in Econofluor (Dupont) to elute radioactive material, which is quantitated by liquid scintillation analysis.

V.   Interpretation of Alkaline Phosphatase Fusion Results

Rules that appear to govern the activities of alkaline phosphatase fusions to membrane proteins are illustrated by the properties of representative *lac* permease–alkaline phosphatase hybrid proteins (Fig. 3) (Calamia and Manoil, 1990).

1. Periplasmic segment alkaline phosphatase fusion typically show 10- to 100-fold greater alkaline phosphatase activity than do cytoplasmic segment fusions. The activities of periplasmic segment fusions tend to decrease somewhat with length, and long fusions tend to be more toxic than short fusions.

2. For fusions with junctions within putative membrane-spanning rather than in cytoplasmic or periplasmic segments, the alkaline phosphatase activity depends on how many apolar residues are present and how the spanning segment is oriented. For segments oriented with their N-termini in the periplasm ("incoming" spanning segments), low alkaline phosphatase activity can require not only all of the apolar residues, but also polar

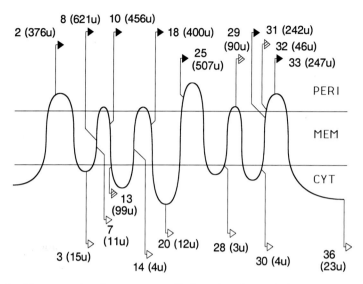

FIG. 3. Representative *Lac* permease–alkaline phosphatase fusions. Sites of alkaline phosphatase attachment to *Lac* permease are drawn relative to a 12-span model for the membrane protein (Calamia and Manoil, 1990). The fusion number and units of alkaline phosphatase activity for each are shown (e.g., cells carrying fusion 2 express 376 units of alkaline phosphatase activity). Solid arrowheads, high activity; open arrowheads, low activity; striped arrowheads, intermediate activity.

residues of the cytoplasmic segment C-terminal to the spanning segment (Boyd *et al.*, 1987) (e.g., compare fusions 13 and 14). For segments with their N-termini in the cytoplasm ("outgoing" spanning segments), about half (9–11 apolar residues) of the segment suffices for high alkaline phosphatase activity (compare fusions 7 with 8 amino acids of helix 3 and fusion 8 with 12 residues of helix 3). These rules may differ for spanning segments that do not consist entirely of apolar amino acid residues.

3. Two of the periplasmic segment *LacY–PhoA* fusions showed intermediate rather than high alkaline phosphatase activity. The junction of one of them (fusion 29) was positioned in a periplasmic segment following a particularly hydrophilic putative membrane-spanning segment, which may account for a lower efficiency of export of the alkaline phosphatase moiety than normal. In fact, when the spanning segment was made less hydrophilic by changing an arginine residue to an alanine residue by mutagenesis, the alkaline phosphatase activity of fusion 29 increased to that of a typical periplasmic fusion (J. Calamia and C. Manoil, unpublished data). This membrane-spanning segment may normally require interaction with C-terminal sequences of *Lac* permease (missing in fusion 29) for

efficient membrane insertion. The intermediate alkaline phosphatase activity of a second periplasmic segment fusion (fusion 32) is enigmatic, particularly because it is bordered closely on each side by typical periplasmic segment high-activity fusions (fusions 31 and 33). The existence of fusions such as fusion 32 illustrates the importance of analyzing multiple fusions to periplasmic and cytoplasmic segments for which individual fusions give intermediate activities or activities incompatible with the overall fusion set.

VI.   Conclusions and Cautions

The analysis of membrane protein topology using gene fusions has the advantages that it is applicable to even the shortest of cytoplasmic or periplasmic segments, does not rely on the presence and reactivity of any particular amino acid, and is technically simple. The validity of the technique has been established in studies of proteins for which topological information for the unfused protein was available, e.g., from protease accessibility, antibody binding, or side-chain modification studies (Manoil and Beckwith, 1986; Chun and Parkinson, 1988; San Millan *et al.*, 1989; Manoil, 1990; Calamia and Manoil, 1990), and is suggested for other proteins analyzed by its having given a self-consistent topological picture (Boyd *et al.*, 1987; Akiyama and Ito, 1987).

The successful use of gene fusions to analyze membrane protein topology requires that replacing C-terminal sequences of a membrane protein with alkaline phosphatase or β-galactosidase from a site of fusion not alter the normal subcellular location of the site in the unfused membrane protein. It has seemed unlikely that such behavior in fusions would be universal, and indeed, a few exceptions have been oberved for proteins with complex topologies (Section V). It may be possible largely to eliminate such exceptions through the construction of "sandwich" fusions in which alkaline phosphatase is inserted into the entire membrane protein rather than attached to membrane protein N-terminal sequences (Ehrmann *et al.*, 1990). Nevertheless, the existence of exceptions underscores the importance of regarding the use of gene fusions to study membrane protein topology as a supplement to, rather than as a replacement for, more traditional biochemical and immunological methods (Jennings, 1989) for elucidating and establishing membrane protein topology.

ACKNOWLEDGMENTS

I wish to acknowledge helpful methodological discussions with J. Beckwith, D. Boyd, J. Calamia, M. Carson, A. Derman, C. Lee, and E. Traxler.

REFERENCES

Akiyama, Y., and Ito, K. (1987). *EMBO J.* **6,** 3465–3470.

Berg, C., Berg, D., and Groisman, E. (1989). *In* "Mobile DNA" (D. Berg and M. Howe, eds.) pp. 879–925. ASM Press, Washington, D.C.

Boyd, D., Manoil, C., and Beckwith, J. (1987). *Proc. Natl. Acad. Sci. U.S.A.* **84,** 8525–8529.

Calamia, J., and Manoil, C. (1990). *Proc. Natl. Acad. Sci. U.S.A.* **87,** 4937–4941.

Casadaban, M., Martinez-Arias, A., Shapira, S., and Chou, J. (1983). *In* "Methods in Enzymology" (R. Wu, L. Grossman, and K. Maldave, eds.) Vol. 100, pp. 293–308. Academic Press, New York.

Chun, S., and Parkinson, J. S. (1988). *Science* **239,** 276–278.

Ehrmann, M., Boyd, D., and Beckwith, J. (1990). *Proc. Natl. Acad. Sci. U.S.A.* **87,** 7574–7578.

Engelman, D., Steitz, T., and Goldman, A. (1986). *Ann. Rev. Biophys. Biophys. Chem.* **15,** 321–353.

Froshauer, S., Green, N., Boyd, D., McGovern, K., and Beckwith, J. (1988). *J. Mol. Biol.* **200,** 501–511.

Gutierrez, C., and Devedjian, J. (1989). *Nucleic Acids Res.* **17,** 3999.

Gutierrez, C., Barondess, J., Manoil, C., and Beckwith, J. (1987). *J. Mol. Biol.* **195,** 289–297.

Henikoff, S. (1987). *In* "Methods in Enzymology" (R. Wu, ed.), Vol. 155, pp. 156–165. Academic Press, Orlando, Florida.

Hoffman, C., and Wright, A. (1985). *Proc. Natl. Acad. Sci. U.S.A.* **82,** 5107–5111.

Ito, K., Bassford, P., and Beckwith, J. (1981). *Cell* **24,** 707–717.

Jenning, M. L. (1989). *Annu. Rev. Biochem.* **58,** 999–1027.

Laemmli, U.K. (1972). *Nature* (London) **277,** 680–685.

Lewis, M., Chang, J., and Simoni, R. D. (1990). *Biol. Chem.* **265,** 10541–10550.

Maniatis, T., Fritsch, E., and Sambrook, J. (1982). "Molecular Cloning: A Laboratory Manual". Cold Spring Harbor Lab., Cold Spring Harbor, New York.

Manoil, C. (1990). *J. Bacteriol.* **172,** 1035–1042.

Manoil, C., and Beckwith, J. (1985) *Proc. Natl. Acad. Sci. U.S.A.* **82,** 8129–8133.

Manoil, C., and Beckwith, J. (1986). *Science* **233,** 1403–1408.

Manoil, C., Boyd, D., and Beckwith, J. (1988). *Trends Genet.* **4,** 223–226.

Miller, J. H. (1972). "Experiments in Molecular Genetics." Cold Spring Harbor Lab. Cold Spring Harbor, New York.

San-Millan, J., Boyd, D., Dalbey, R., Wickner, W., and Beckwith, J. (1989). *J. Bacteriol.* **171,** 5536–5541.

Simons, R., Houman, F., and Kleckner, N. (1987). *Gene* **53,** 85–96.

Tabor, C., and Richardson, C. C. (1987). *Proc. Natl. Acad. Sci. U.S.A.* **84,** 4767–4771.

von Heijne, G. (1987). "Sequence Analysis in Molecular Biology," pp. 97–112. Academic Press, Orlando, Florida.

Chapter 4

Expression of Foreign Polypeptides at the Escherichia coli Cell Surface

MAURICE HOFNUNG

Unité de Programmation Moléculaire et Toxicologie Génétique
CNRS UA 1444, Institut Pasteur
75015 Paris, France

Garde toi, tant que tu vivras,
De juger les gens sur la mine
[La Fontaine]

METHODS IN CELL BIOLOGY, VOL. 34

I. Reasons and Obstacles

Beyond the esthetic satisfaction of changing the outside appearance of a cell in a heritable manner, there are at least three reasons why it is interesting to express foreign polypeptides at the surface of a bacterium such as *Escherichia coli*. First, expressing a foreign polypeptide at the *E. coli* cell surface is part of the general problem of targeting the location of proteins within the cell; its solution should help us to understand the codes and mechanisms for targeting (Lee *et al.*, 1989; for recent reviews, see Randall *et al.*, 1987; Verner and Schatz, 1988; Pugsley, 1989; Nikaido and Reid, 1990; see also other Chapters, this volume). Second, the bacterial cell surface plays an essential role in the interactions between the bacterium and its environment, including its potential hosts; by expressing chosen polypeptide sequences at the surface, one may be able to control some of these interactions (Mims, 1987). Third, the construction of bacteria expressing selected polypeptides at their surface opens the way to a number of applications (Hofnung, 1988a,b; Hofnung *et al.*, 1988; see also Section V).

The envelope of a Gram-negative bacterium such as *E. coli* includes two membranes: the inner membrane and the outer membrane (Fig. 1). These two membranes enclose an aqueous layer, the periplasm. The inner membrane, the outer membrane, and the periplasm constitute different compartments to which proteins are specifically directed.

We will use an operational definition and say that a polypeptide is exposed at the cell surface when it can be detected from the outside on intact cells by a procedure that does not detect polypeptides known to be located in the periplasm or in the inner membrane. This means that we consider that polypeptides carried by cell surface appendages such as fimbriae, pili, or flagellae are cell surface exposed. However, we will distinguish between polypeptides that are released in the medium by the bacteria and that we do not consider as surface exposed (secreted polypeptides) and polypeptides that remain tightly bound to the cell: this distinction is not always easy because some protein species are released in part in the medium. As long as a large fraction (over 50%) remains bound to the cell we will consider that it is cell surface exposed.

Fig. 1. A schematic drawing of the *E. coli* K12 envelope. The lipopolysaccharide of *E. coli* K12 presents only short chains of O-polysaccharides, so that outer membrane proteins are relatively accessible from the outside. Other strains of Gram-negative bacteria usually present longer chains and may be covered with a capsule (Hancock and Poxton, 1988, and references therein). Drawing by Jean-Marie Clément.

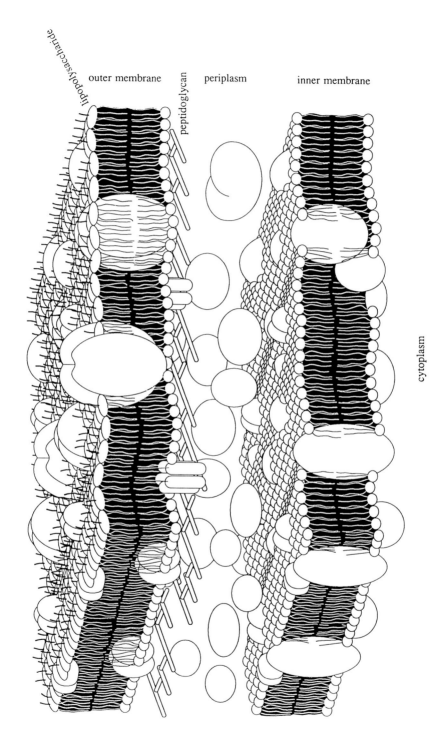

Modification of the bacterial cell surface can be obtained by expressing complete cell surface elements from other bacterial species, including proteins, LPS, or capsules (Jones *et al.*, 1972; Engleberg *et al.*, 1984; Yamamoto *et al.*, 1985; Formal *et al.*, 1981; Agterberg *et al.*, 1988; O'Callaghan *et al.*, 1988; Attridge *et al.*, 1990).

In this chapter we are especially interested in procedures that allow expression of foreign peptides from distant organisms (viruses, mammals, etc.) at the bacterial surface. We will indicate a number of strategies to achieve this goal. These strategies are complemented by detailed protocols developed in the use of the LamB protein as a vector protein (see Sections II and IV). The reader who is interested in protocols for other known vector proteins (listed in Section IV) or who wants to develop new vectors is invited to modify these protocols according to the biological properties of the vector protein used.

There are a number of serious obstacles in achieving cell surface expression of a chosen polypeptide. In addition to those due to the gaps in our knowledge on protein export, there are difficulties inherent to the complexities of the cell envelope and of the cell surface.

Although the inner membrane is a typical phospholipid bilayer, the outer membrane is made of two leaflets of a different nature. The inner leaflet is composed of phospholipids and the outer leaflet is composed essentially of lipopolysaccharide (Fig. 1). The lipopolysaccharide is an elaborate structure that comprises cell surface-exposed polysaccharide chains of variable length, part of which constitutes the O-antigen (Hitchcock *et al.*, 1986). Depending of the length of the polysaccharide chain and the detection method, a polypeptide expressed at the surface of the lipidic part of the outer membrane may or may not be detected. It is known that O-polysaccharide can mask phage receptors in the outer membrane of the bacteria (Schwartz and Le Minor, 1975; van der Ley *et al.*, 1986), can influence antibody binding to proteins at the bacterial surface (Bentley and Klebba, 1988; Agterberg *et al.*, 1988; A. Charbit, unpublished), and can exclude high-molecular-weight maltosaccharides from binding to maltoporin (Ferenci and Lee, 1986).

There are also problems due to the available methods for determining the exact localization of a molecule in the cell envelope. Methods based on cell disruption and fractionation are generally not appropriate to determine cell surface exposure because they most often destroy the surface and, in addition, may lead to difficulties in dentifying the origin of some components in a given fraction (Tommassen, 1986; Voorhoot *et al.*, 1988). When one operates with whole cells, its is essential to control thoroughly cell integrity, otherwise proteins from other compartments may become detectable from the outside of the cells (Bayer *et al.*, 1990).

II. "Permissive" Sites

One may expect that, in the future, completely artificial proteins will be devised that will be targeted to the cell surface so as to expose chosen sequences. This will require a more advanced knowledge of export pathways and of the folding rules for proteins. We already know at least that the mechanisms for targeting proteins (and other cell surface components) are rather general and "portable."

At present, the general approach to cell surface expression consists in starting from a protein already located at the cell surface—we will call it the vector protein—and grafting genetically a heterologous polypeptide—we will call it the passenger polypeptide, or the foreign sequence, or the insert—in such a way that the hybrid protein constructed will expose the foreign sequences at the cell surface. This means that the insert should satisfy two conditions: (1) it should not perturb extensively the export and the folding of the hybrid protein; we designate the corresponding insertion sites as "permissive" sites (Hofnung *et al.,* 1988); (2) it should be positioned at or near the cell surface in the hybrid.

A priori the behavior of the hybrid protein should depend on the site of insertion and on the nature of the insert. Results obtained so far indicate that the most critical factor is the site of insertion. In most cases, if a foreign sequence is inserted without precautions within an exported protein, the resulting hybrid will be unstable or toxic to the bacteria (Bouges-Boquet *et al.,* 1984). Most sites leading to cell surface expression accept a variety of inserts as long as their size does not exceed a certain length and that their composition is not extreme in terms of hydrophobicity or charged residues (Charbit *et al.,* 1988; Agterberg *et al.,* 1990).

In order to find sites appropriate for insertion and cell surface exposure, one may proceed with an educated guess or with an experimental approach. The more one knows about the protein, the more one may expect the educated guess approach to work. If nothing is known about the protein but that the gene has been cloned, one may use directly an experimental approach to determine permissive sites.

A. Educated Guess

The educated guess approach consists in trying to predict permissive sites, i.e. regions of the protein that are likely to be flexible enough to accommodate inserts without damage to the protein biogenesis, final localization, and folding—and among them to select those that are likely to be cell surface exposed and test for accessibility of the insert from the outside.

If the sequence of the vector protein is known, hydrophilic sequences or regions predicted as turns are good candidates for permissive sites. These regions can be determined with a number of prediction methods (Doolittle, 1986). Then predicting the cell surface-exposed sites depends on modeling the folding and membrane insertion of the protein.

If the sequence of the vector protein is known in different species, one may look at regions in which the sequence varies the most between species. It is known that sequences located at the surfaces of proteins (and even more so at the cell surface) are more prone to mutational changes. Because these regions can vary without affecting the protein localization and folding, they are good candidates for inserting a passenger and expressing it at the cell surface (see examples in Section IV).

If one knows the structure of the vehicle protein, it appears reasonable to try loops that are predicted or known to be cell surface exposed. Indeed, loops are generally quite tolerant to the insertion of foreign sequences. However, deciding on the sole basis of the final structure does not guarantee that the loop region is not playing a critical role in the export or folding and that the insert will not thus be deleterious. If the COOH-terminal ends of the protein are known to be exposed, it could be a good choice, because the folding constraints are probably reduced. In the case of the NH_2-terminal end, the situation is more complex, because most exported proteins are made with a signal peptide: the insertion has to be made after the region of the signal peptide, but it may perturb cleavage of the signal peptide.

B. Experimental Approach

The principle consists in determining by a genetic approach permissive sites as regions of the proteins that can accept a foreign insertion without deleterious consequences for the protein or for the bacterial cell. Then one examines whether the foreign sequence is detectable at the cell surface.

In practice, the approach includes two steps. As already mentioned, detailed protocols will take the outer membrane LamB *protein* from *E. coli* as example (Boulain *et al.*, 1986; Charbit *et al.*, 1986, 1991).

Step 1.   Random linker insertion.   In the first step, an oligonucleotide—called olig 1, 6 to 12 bp long—encoding a restriction site and a few amino acid residues is inserted at random into the structural gene of the vector protein, and screening is performed for clones that express a stable, nontoxic mutant protein. The idea is that if the corresponding sites can accept a few residues, they will, at least in a number of cases, correspond to flexible regions of the protein and will be able to accept large inserts. Detection of the mutant vector proteins can be made by gel

SDS–PAGE if it is abundant enough, by immunodetection if an antibody is available, or by an activity test if there is one. The procedure for insertion of a short oligonucleotide is quite efficient, reaching up to 100% of the clones recovered (see protocol 1 below). For unknown reasons, possibly contaminating exonucleolytic activity, this method generates concomitant deletions in a number of cases (Boulain *et al.*, 1986; Duplay *et al.*, 1987).

The structural gene for the vector protein is carried by a plasmid and the restriction site encoded by the oligonucleotide is chosen so as to correspond to a new restriction site on the plasmid. This site can thus be used to map the insert within the gene and to proceed to the second step, namely, to clone a larger insert.

1. PROTOCOL 1. RANDOM LINKER MUTAGENESIS IN GENE *lamB*

This protocol is from Boulain *et al.* (1986). The 12-bp linker comprising the *Bam*HI site (sequence 5′-CGCGGATCCGCG-3′) was purchased from New England Biolabs. Plasmid pACl carrying the wild-type *lamB* gene was opened at random by partial digestion with DNase I in the presence of Mn^{2+}. Linearized plasmid band was purified from 5% TEB acrylamide gel and resuspended in TEB at 1 μg/2 μl (TEB, Tris-Cl, 10 mM, pH 8; EDTA, 1 mM, pH 8). The extremities were then filled in using the Klenow fragment of *E. coli* DNA polymerase.

a. Cloning. The plasmid DNA was religated in the presence of an excess of unphosphorylated *Bam*HI linker (Lathe *et al.*, 1984); 2 μg of plasmid vector (4 μl) was mixed with 2 μg of *Bam*HI linker (2 μl) and ligation was carried out in 20 μl (final concentration) at 4°C overnight (4 μl vector plus 2 μl double-stranded oligonucleotide plus 2 μl T4 ligase special buffer plus 2 μl T4 DNA ligase plus 10 μl H_2O).

b. Removal of Unligated Oligonucleotides. To the 20-μl ligation mixture add 380 μl of a solution of dimethyl sulfoxide, 10%, in TE, and heat 2 minutes at 70°C. Add 10 μl of spermine 4 HCL, 100 mM (2.5 mM final concentration), and add 1 μl of dextran at 10 mg/ml. Freeze immediately in liquid nitrogen and then leave on ice for at least 30 minutes. Centrifuge at 13000 rpm for 30 minutes. Wash the pellet once in ethanol (75%), TEB (25%), and twice in ethanol (70%). The dry pellet is finally resuspended in 250 μl STE (Tris-Cl, 10 mM, pH 8; EDTA, 1 mM, pH 8; NaCl, 0.1%) and stored at −20°C. For transformation, 50 μl of the ligation mixture is normally sufficient (about 400 ng and >100 transformants). The LamB-negative strain pop6510 (Boulain *et al.*, (1986) was used as a recipient for transformation. Selection of the transformants was performed on complete solid medium containing ampicillin (100 μg/ml final concentration).

They were then screened for sensitivity toward phage λhh* (a two-step host-range mutant of phage lambda) by cross-streaking on ML ampicillin plates. Out of 550 transformants, 260 (approximately 50%) were sensitive to λhh*. Growth of λhh* is only prevented by mutations that affect drastically the LamB protein, such as nonsense mutations or deletions (Braun-Breton, 1984). Consequently, most of the clones resistant to λhh* were expected to carry important modifications in the *lamB* gene and thus to have lost all function. This is why we retained essentially λhh*-sensitive clones. These clones were further tested by restriction analysis. Double digestions of plasmid DNAs with *Bam*HI and enzymes cutting only once in the plasmid, such as *Nco*I, *Stu*I, and *Cla*I, examined by 1.5% agarose gel, permitted us to localize the *Bam*HI insertions. Over 80% of the transformant clones had inserted a *Bam*HI linker. Twelve transformants (approximately 5%) had inserted the *Bam*HI linker within *lamB*. Further studies revealed that nine of them were different and that seven of them were permissive for insertion of the C3 epitope from poliovirus (Charbit *et al.*, 1991).

    c.   *Other Procedure: Insertion at Natural Restriction Sites.*   Instead of using DNase I for introducing random double-stranded cuts, one may use restriction endonucleases such as *Hpa*II in the presence of ethidium bromide to linearize the plasmid DNA and to introduce an hexanucleotide linker at the corresponding sites (Barany, 1985). This method has also been used to introduce linker mutations (insertions of two residues in the LamB protein) (Heine *et al.*, 1988).

    Step 2.   *Cloning of the foreign peptide.*   For the second step, it is essential to insert a sequence for which one has a sensitive detection test. A convenient solution is to use a sequence corresponding to a continuous epitope against which a monoclonal antibody with sufficient affinity is available. An oligonucleotide—called olig 2—encoding the sequence of interest is inserted into the new restriction site and the monoclonal antibody can be used to screen for clones that express the epitope at the cell surface (for assays, see Section III). The procedure will be described in the case of the insertion of the C3 epitope from poliovirus in the LamB protein from *E. coli* (Charbit *et al.*, 1986, 1991).

    Because insertion of olig 1 was made at random, the exact position of the restriction cut with respect to the reading frame is not known. In order to insert olig 2 in the proper reading frame, there are at least two solutions. One solution consists in determining the sequence at the insertion site and synthesizing olig 2 with the correct frame. Because it is likely that olig 1 inserts will be found in the three possible frames, three different versions of olig 2 will be needed for the various sites (Fig. 2). Another solution consists in synthesizing the three different frames of olig 2, using the mix-

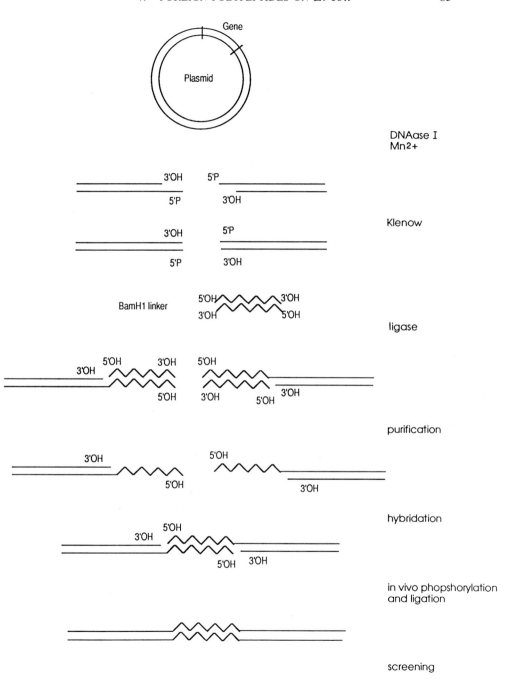

FIG. 2. Random linker insertion procedure. The successive phases of the procedure are represented (see text for details).

ture for insertion, and screening directly for detection of the epitope. We found that this procedure was useful because it does not require determination of the DNA sequence at step 1. DNA sequencing can then be performed only at the final stage on the clones retained for further work (Charbit *et al.*, 1991).

2. PROTOCOL 2. C3 EPITOPE INSERTION

The double-stranded oligonucleotides corresponding to the three different reading frames of the C3 epitope, numbered phase 1, 2, and 3, (Fig. 3), were purified on a 20% acrylamide–urea denaturing gel (Charbit *et al.*, 1986).

We knew from previous studies that an intact COOH-terminal end was required for LamB activities (Boulain *et al.*, 1986). All the *Bam*HI inserts studied here conserved at least one of the activities of the protein. This indicated that the reading frame was not shifted by the presence of the linker. Thus, three double-stranded oligonucleotides with an exact number of codons (13 codons in each case) were used.

Plasmids DNAS were linearized by digestion with the restriction enzyme *Bam*HI and purified on 5% acrylamide gels. Because the oligonucleotides were unphosphorylated, each transformant obtained after subcloning of the C3 oligonucleotides corresponded either to the parental construction (no insertion) or to the insertion of a single double-stranded oligonucleotide. Statistically, if all the transformants corresponded to the insertion of a double-stranded oligonucleotide, one clone out of six should have inserted the correct phase of the C3 epitope in the correct orientation (correct frame, one in three; and correct orientation, one in two). In all

Phase 1

```
                  93                                          103
       asp pro asp asn pro ala ser thr thr asn lys asp lys
5 ' GAT CCG GAT AAC CCG GCG TCG ACC ACT AAC AAG GAT AAG 3 '
       GC CTA TTG GGC CGC AGC TGG TGA TTG TTC CTA TTC CTA G
```

Phase 2   Phase 3

```
          93        103                          93        103
      ser asp ... lys                        ile asp ... lys
5 ' GA TCC GAT ... AAA G      3 '   5 ' G ATC GAT ... AAA GG      3 '
      G CTA ... TTT CCT AG                    CTA ... TTT CCT AG
```

FIG. 3. The C3 epitope in the three different reading frames.

cases, we obtained between 30 and 70% of transformants with an insertion. Twenty clones from each transformation, tested by immunoblotting after heat denaturation with anti-C3 monoclonal antibody, were usually sufficient to select the good construction. In agreement with what was expected, in most cases approximately 10–20% of the clones assayed expressed the C3 epitope. This indicated that the subcloning procedure using a mixture of three phases of the oligonucleotides encoding the C3 epitope was efficient.

An equimolecular mixture of double-stranded oligonucleotides corresponding to the three different phases of the C3 sequence was used. After transformation into strain pop6510, DNA restriction analysis and immunoblotting of total extracts with anti-LamB monomer serum and anti-C3 monoclonal antibody were performed in parallel on the clones generated. All the LamB–C3 hybrid proteins constructed could be detected by both an anti-LamB polyclonal serum and an anti-C3 monoclonal antibody.

a. Purification and Annealing of the Oligonucleotides. Each unphosphorylated oligonucleotide (single stranded) was purified from 8 to 20% acrylamide–urea denaturing gel and resuspended in TE buffer at a concentration of 1 μg/μl. Then, 5 μl of each complementary strand was mixed so that the concentration of double-stranded oligonucleotide was 1 μg/μl. The mixture was heated at 70°C for 2 minutes and then cooled in water at 60°C on ice.

b. Purification of the Vector and Subcloning Techniques. These are as described in protocol 1.

c. DNA Sequence Determination. Oligonucleotides corresponding to nine segments of the *lamB* gene were used as primers for DNA sequence analysis. In order to allow convenient reading of the gels, these segments were separated by intervals of 150 bases. The first oligonucleotide started at the first base corresponding to the mature LamB protein. The sequences of the nine primers (16mers) were as follows:

5'-GTTGATTTCCACGGCT-3'; 5'-AAGTGTGGAAAGAGGG-3';
5'-CTCCACCATCTGGGCA-3';
5'-TCCTCTGAAGCTGGTG-3'; 5'-GTCGTGCCAACTTGCG-3'; 5'-
GGGTAAAGGGCTGTCG-3';
5'-ATGTACCAGGATATCA-3'; 5'-AGAACAATCAGTACAA-3';
5'-CGCGAACTTCGGCAAA-3'

The sequence was determined on the construct with the C3 epitope insertion and the sequence of the linker insertion was deduced. The sequencing procedure was derived from the dideoxynucleotide method by using rapid preparations of the plasmid DNA (Boulain *et al.*, 1986) from the different mutants as templates. In each case, the approximate location of the insertion point was first determined by restriction enzyme analysis, and then the appropriate primer was chosen for elongation. Biolabs AMV reverse transcriptase was used.

3. WHY NOT A ONE-STEP PROCEDURE?

Instead of proceeding in two steps, why not proceed in one step, that is, insert directly at random an oligonucleotide encoding the foreign epitope and screen for detection of the epitope on the cell surface? We found that there were difficulties with a one-step approach. First, random insertion of a longer oligonucleotide (30–50 bp) was less efficient: we usually obtained on the order of 1% inserts among the clones recovered. Second, the proportion of inserts with the correct orientation and the correct reading frame is only 1/18 (1/3 for the reading frame of the epitope, 1/3 for the reading frame distal to the epitope, and 1/2 for the orientation). Thus the proportion of correct inserts is only about 1/1800 of the clones recovered, which we found was too low for convenient detection of permissive sites (A. Charbit, R. Ekwa, E. Dassa, J. Ronco, unpublished results).

III.   Detection Methods

We will examine essentially methods involving the use of intact cells. As already indicated, it is critical to include proper controls for cellular integrity, otherwise nonexposed polypeptides may become accessible from the outside.

Procedures involving cell disruption followed by fractionation generally lead at best to the conclusion that a polypeptide is associated with the outer membrane. They usually do not allow us to conclude the exact position of the peptide with respect to the membrane. In addition, as already mentioned, they are subject to artifacts (Tommassen, 1986). However, methods involving double immunolabeling of membrane vesicles may allow us to reach solid conclusions: two antibodies are used, one targeted at the peptide of interest, the other one at a control protein with known cellular localization (den Blaauwen and Nanninga, 1990).

A necessary condition to have the foreign sequence located at the cell surface is that the vector protein (with its passenger) be normally positioned. In most cases, at least some of the biological activities of the vector protein require correct localization. Those properties that are not abolished by the presence of the passenger can be used for verification. For example (see Section IV), in the case of the LamB protein, phage sensitivity and ability to grow on maltodextrins were simple, available tests (Charbit *et al.*, 1986) (protocol 2 above), and bacterial motility could be used in the case of flagellin (Kuwajima *et al.*, 1988; Newton *et al.*, 1989).

Depending on its properties, different assays may be used to ascertain the presence of a polypeptide at the bacterial cell surface.

A.   Chemical Labeling

If the peptide can be specifically labeled by a procedure that does not operate through the membrane, its cell surface expression can be readily demonstrated. This can be done by iodination (^{125}I) of tyrosyl residues in presence of lactoperoxidase (Marchalonis *et al.*, 1971; King and Swanson, 1978). This technique requires that the passenger contains a tyrosine and that it can be identified after labeling, which is straightforward if the vector protein does not expose tyrosyl residues.

B.   Protease Accessibility

One examines if the polypeptide is accessible to proteases added from the outside to intact cells. The action of the protease can be monitored by looking at the cleavage of the polypeptide by SDS–PAGE, or by examining if other properties of the polypeptide are affected (enzyme activity, antigenicity, etc.). Even proteins that are very resistant to proteases, such as the outer-membrane proteins LamB or OmpA, become generally sensitive upon insertion of a foreign sequence. To conclude, on cell surface exposure, it is important to show (by protein sequencing, for example) that cleavage occurs within the insert and not elsewhere in the vector protein (Ronco *et al.*, 19990; Freudl *et al.*, 1986). The procedure is illustrated herein in the case of an insert of the C3 epitope of poliovirus after residue 153 of mature LamB protein (hybrid protein LamB-153–C3).

PROTOCOL 3

The hybrid protein was expressed from plasmid pAJC264 in strain pop6510 (Chabit *et al.*, 1986). Cells were grown at 37°C in liquid minimal medium 63B1 with 0.2% glucose as a carbon source, 100 μl/ml ampicillin, and supplemented with threonine and leucine at 0.01% and methionine at 0.001%. IPTG was added ($10^{-3}M$ final concentration) to exponential cultures at $OD_{600} = 0.5$. Ten minutes after induction with IPTG, cells were labeled with [^{35}S]methionine (10 μCi/ml of cell culture) for 40 minutes. Bacteria were harvested by centrifugation, washed twice in minimal medium plus chloramphenicol (100 μg/ml) at 4°C, and resuspended in Tris

(50 mM pH 8). Whole cells and were incubated for 1 hour at 37°C. Two concentrations of trypsin were used, either 200 μg/ml (final concentration) for digestions of whole cells, or 100 μg/ml (final concentration) for digestions of bacterial lysates. Control samples (without enzyme) were incubated in the same conditions. Reactions were stopped by addition of TCA (5% final concentration), and stored at 4°C for 15 minutes.

After treatment, the TCA precipitates were washed once with acetone and dried. The pellets were dissolved in lysis buffer (Tris, 25 mM PH8; EDTA, 10 mM; glucose, 50 mM; lysozyme, 4 mg/ml; SDS, 2%), followed by a 10-minute incubation at 100°C. The samples were then diluted 10-fold in Tris (50 mM pH8; Triton X-100, 0.9%; EDTA, 5 mM; ovalbumin, 0.1%). Incubations with the immune serum were as described in Charbit *et al.* (1986).

Peptide Purification and Sequencing. LamB153–C3 was prepared essentially according to the procedure described previously for LamB (Gabay and Yasunaka, 1980): 1 liter of bacterial culture of strain AJC 264-C3 (expressing P153–C3) in minimal medium M63 B1 was induced with IPTG (10^{-3} M) at OD$_{600}$ = 0.4 and growth continued until OD$_{600}$ = 1.0. Cells were then collected by centrifugation, and the pellet was incubated for 30 minutes at 60°C in extraction buffer (2% SDS, 10% glycerol, 2 mM MgCl2, 10 mM Tris-HCl, pH 7.4). The suspension was centrifuged at 100,000 g for 1 hour. The pellet containing the LamB–C3 hybrid protein trapped in the peptidoglycan complex was then solubilized by incubation for 30 minutes at 37°C in Tris-HCl (10 mM, pH 7.4) containing 2% Triton X-100. After centrifugation, the solubilized protein recovered in the supernatant was concentrated to 2 ml (by ethanol precipitation).

The 300 μl of "purified" LamB153–C3 was treated with 100 μg/ml trypsin for 30 minutes at 37°C. The reaction was stopped by addition of sample buffer and boiled 5 minutes at 100°C. The sample was applied to SDS–20% polyacrylamide discontinuous gel. After electrophoresis, the proteins contained in the gel were transferred electrophoretically to PVDF membrane (Immobilon Transfer-Millipore) for 40 minutes at 0.25 A in transfer buffer (10 mM 3-{cyclohexylamino}-1-propanelsufonic acid, 10% methanol, pH 11), as described in Matsudaira (1987). The PVDF membrane was then stained with 0.1% Coomassie blue R-250 in 50% methanol for 5 minutes. The size of the well used to apply the sample was such that it allowed cuting each stained band into two halves. Each half was utilized for the immunodetection either with anti-LamB serum or with anti-C3 monoclonal antibody. The N-terminal sequence of each peptide was determined by using a gas-phase sequencer (Applied Biosystem, model A470) with on-line HPLC identification of amino acid residues.

This approach has two limitations. First, a negative result (no cleavage) may be due to a lack of sensitivity of the polypeptide to the protease used. However, a number of proteases are available, so there is some chance of finding one that is active on the sequence studied. Second, a positive result requires that proper controls are performed to be sure that the protease activity did not penetrate into the periplasm. Indeed, drastic conditions (large amounts of protease, long incubation periods, etc.) may result in the cleavage of nonsurface-exposed sequences either because the cells are damaged or because some permeability of the outermembrane to the protease exists or has been generated by the treatment.

C.   Enzymatic Activity

If the polypeptide displays enzymatic activity, one may use it to demonstrate cell surface expression. This can be conveniently done if a substrate unable to cross the outer membrane is available: nitrocefin is such a substrate for β-lactamase (O'Callaghan *et al.*, 1972; Kornacker and Pugsley, 1990). It is important to ensure that the outer membrane is indeed impermeable to the substrate when the hybrid protein is expressed.

D.   Antigenicity

In many cases antibodies against the foreign sequence are available, and the most widely used types of assays are immunological assays. They have at least two limitations. First, the hybrid protein may be constrained in conformations where the passenger polypeptide it is not detected by the antibody used (Charbit *et al.*, 1986; MacIntyre *et al.*, 1988). Second, if the antibody is targeted to a short peptide within the passenger (for example, an epitope included within 10 residues), the results will only give information on this epitope; thus a positive result may indicate that only this short peptide is exposed, whereas a negative result may indicate that part of the epitope is not accessible, which does not mean that some other part of the passenger is not exposed.

Binding of the antibodies to the bacteria can be examined with a number of different techniques (reviewed in Hancock and Poxton, 1988). The sensitivities of these assays differ widely and depend on the properties of the serum used and on the conformation and accessibility of the foreign epitope. In presence of a positive signal, specificity controls are absolutely needed: they consist in performing the same experiment using a preimmune serum, or better, on an identical strain with an immunoligically unrelated passenger polypeptide.

1.   Bacterial agglutination.   In its simplest form, agglutination consists in mixing bacteria with dilutions of serum and looking at clumping of cells under the microscope. A related technique, called coagglutination, consists in coating specific antibodies on a particulate support and mixing it with the cells.

2. *Immunofluorescence.* Antibody is labeled with a fluorescent dye and the cells are observed with an ultraviolet microscope.

3. *ELISA with intact cells.* The cells are immobilized on a solid support, labeled with a specific antibody, and then revealed with an enzyme-conjugated second antibody.

1. PROTOCOL 4

The assays were performed in microtitration plates (Nunc). Cells were grown in minimal medium and harvested in the late exponential phase ($OD_{600} = 1$), washed, and resuspended in PBS buffer. Plates were coated with 5×10^6 bacteria (100 μl per well of a bacterial dilution at $OD_{600} = 0.1$ in PBS buffer). After overnight incubation at 37°C, the excess of antigen was discarded and well were saturated with 250 μl of PBS containing 0.5% gelatin, for 1 hour at 37°C. Then 100-μl of dilutions of antibodies were added (at a final dilution of 1/5000 in the case of anti-LamB mAb E-302, E-72, E-177, and E-347) (Molla *et al.*, 1989; Charbit *et al.*, 1991). After extensive washes, antigen–antibody complexes were developed by peroxidase-labeled antimouse antibodies (1 hour at 37°C) and ABTS (2,2'-azino-di-(3-ethylbenzthiazolin sulfonic acid) 20 minutes at room temperature. Values were recorded at 420 nm. In the case of LamB, assays were performed simultaneously on LamB-positive (strain ACl) and -negative (pop6510) controls (Charbit *et al.*, 1991). Three determinations were done for each assay.

4. *RIA with intact cells.* Cells are labeled with a specific antibody and revealed with ^{125}I-labeled protein A.

2. PROTOCOL 5

The assays were performed in 1.5-ml Eppendorf tubes on liquid cultures prepared as described above. 50 μl of each culture at $OD_{600} = 1$ was mixed with 50 μl of PBS with 0.5% BSA (abbreviated as PBS–BSA), washed by centrifugation, and resuspended in 100 μl of PBS–BSA so that the final OD_{600} per assay was 0.5.10 μl of a dilution of antibody was added (1/100 dilution of C3mAb in PBS–BSA in the case of the C3 epitope inserted in the LamB protein) (Charbit *et al.*, 1991), and the mixture incubated for 1 hour at 37°C under agitation. After two washes with 250 μl of PBS–BSA, the pellets were resuspended in 250 μl of a dilution of ^{125}I-labeled protein

A (from Amersham) in the same buffer, containing approximately 100,000 cpm. The resuspended pellets were incubated 1 hour at room temperature, and then washed twice in PBS–BSA. The pellets were then counted in a gamma counter. It is important to check that the Fc fragment of antibody used was well recognized by the protein A.

5. *Immunoelectron microscopy.* Binding of the antibody is revealed by coupling of electron-dense material to a specific reagent (protein A, secondary antibody). Immunoelectron microscopy can be performed on intact cells or thin sections. The main problems have been associated with the fixation procedure, which is responsible for antigen denaturation, and the choice of an electron-dense material, leading to good resolution under the electron microscope (review in Ryter, 1987).

6. *Targeted action of complement.* *Escherichia coli* K12 strains expressing the LamB protein in its outer membrane are specificaly killed by anti-LamB serum in the presence of complement. This is also true when anti-LamB MAb specific for epitopes exposed at the cell surface and corresponding to isotypes that fix complement are used (Desaymard *et al.,* 1986, and references). This procedure should at least theoretically allow to test for the presence of a foreign epitope onn *E. coli* K12 if an adequate specific antibody is available.

IV.   Passengers and Vectors

Three classes of proteins that have been used as vectors will be described below: outer-membrane proteins, subunits from cells appendages, and proteins that are normally secreted into the culture medium. The upper size limit for an insert within the vector varies from a few residues to over 60, depending on the site and the nature of the insert. COOH-terminal fusions are expected to accept larger insertions; this indeed appears to be the case.

A.   Passengers

As passenger, a reasonable choice consists in using a sequence issued from a protein that can be exported by the same system as the vector: the idea is that the sequence is then compatible with this export pathway. The passengers that were successfully exposed so far (see examples in the rest of this section) were generally well-defined antigenic determinants included within a short peptide (continuous epitopes) and corresponding to viral envelope (exported) proteins or domains of bacterial-exported proteins (β-lactamase or cholera toxin). To our knowledge, alkaline phosphatase, a periplasmic enzyme, and β-galactosidase, a cytoplasmic enzyme, which are both used extensively as reporter enzymes, have not been expressed at the cell surface despite several attempts. Because of the increasing variety of export mechanisms that are being discovered and our present inability to predict the fate of most fusions, we cannot exclude that this will be possible in the future.

B.   Outer Membrane Proteins

These abundant proteins span the outer membrane a number of times. LamB, PhoE, and OmpA have two unusual properties for membrane proteins: they are rather hydrophilic and include mostly β-extended regions.

1.   LamB

The LamB protein from *E. coli* was the first vector developed for cell surface expression of foreign peptides. LamB is a trimeric outer membrane protein (the monomer comprises 421 residues) that is involved in the diffusion of maltose and maltodextrins through the outer membrane and is the cell surface receptor for phage lambda. LamB synthesis is inducible by maltose. The initial experimental search for permissive sites revealed two sites (after residues 153 and 374) leading to cell surface exposure of the well-defined C3 epitope from poliovirus (Charbit *et al.*, 1986). This epitope is included within residues 93–103 of the VP1 protein of the virus. Cell surface expression was shown by immunoelectron microscopy and by immunofluorescence on intact cells treated with the anti-C3 monoclonal antibody. A simple radioimmunoassay with ^{125}I-labeled protein A was also developed (Charbit *et al.*, 1991) (see Section II,D,4).

Site LamB153 can accept inserts of a wide variety of sequences up to a size of at least 55–60 residues, without loss of all LamB functions (Charbit *et al.*, 1988). More recently, another cell surface-exposed permissive site (after residue 253) has been detected (Charbit *et al.*, 1991). It is coherent that all three sites correspond to regions predicted as cell surface-exposed loops on the folding model of this protein (Charbit *et al.*, 1988) (Fig. 4).

2.   PhoE

The phosphate-limitation-inducible outer membrane protein PhoE from *E. coli* was used to express antigenic determinants from the foot-and-mouth disease virus (FMDV) at the bacterial cell surface. PhoE is active as

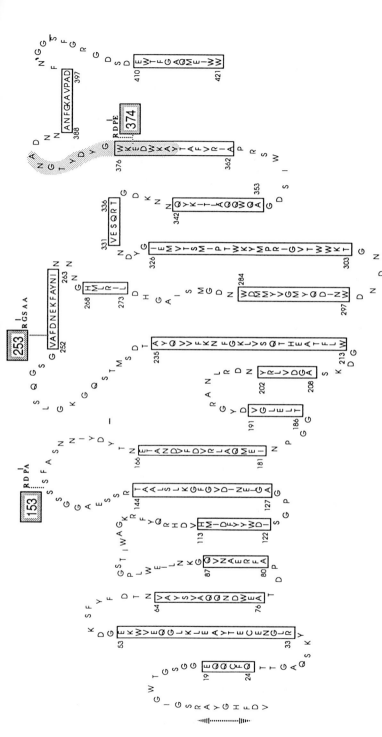

FIG. 4. Cell surface-exposed permissive sites in the LamB protein and on cell surface. The sequence of the mature LamB protein is represented in the one-letter code and according to the present folding model (Charbit *et al.*, 1988). When the C3 epitope from poliovirus was inserted after residues 154, 253, and 374, it was detected both on the isolated native protein and on the cell surface with the C3 monoclonal antibody (Charbit *et al.*, 1986; 1991). Additional residues due to the linker insertion at step one (see text) are indicated in bold letters near the insertion site. At site 374, a deletion occurred upon insertion of the linker (indicated by grey area).

□ C3 detected on native hybrid protein and on cell surface

a trimer (the monomer includes 330 residues). Insertions were made, following an educated guess, into four regions predicted to be exposed to the outside: the second, fourth, fifth, and eighth exposed loops of PhoE. Cell surface exposure was shown by means of an ELISA performed with intact cells. The size limits for the inserts were approximately 50 and 30 residues for the second (after residue 73) and eighth loops (after residue 314), respectively. Simultaneous expression of a determinant in the second and eighth region was performed (Agterberg et al., 1987, 1990).

3.   OmpA

The outer membrane OmpA protein increases the efficiency of F' mediated conjugation and is a phage receptor. It seems to be active as a monomer (325 residues). Insertion of 15 residues between residues 153 and 154, predicted to be cell surface exposed, rendered the protein sensitive to proteinase K added from outside the cell (Freudl et al., 1986). It was proposed that this site could serve to express foreign epitopes at the surface.

4.   TraT

TraT is an outer membrane oligomeric lipoprotein specified by the traT gene of plasmids of the IncF group (223 residues). This protein reduces the ability of host cells to act as recipients in conjugation with bacteria harboring closely related plasmids (surface exclusion) and mediates resistance to the bactericidal activity of the serum. The C3 epitope of poliovirus was inserted at a number of natural restriction sites. At three of them (after residues 61, 200, and 216), the epitope could be detected at the E. coli surface by an ELISA (Taylor et al., 1990; Harrison et al., 1990).

C.   Subunits from Cell Appendages

Cell appendages such as flagellae or fimbriae (or pili) generally carry major antigenic determinants that vary among strains and correspond to regions that are good candidates for the insertion of a foreign sequence.

1.   FLAGELLIN

The motility of flagellated bacteria is due to the rotation of the external flagellar filament, which is a polymer of a single protein, flagellin. The flagellin protein is generally about 500 residues long; whereas the NH_2- and

COOH-terminal ends are conserved among strains, the central part is variable (especially 350 bp) and can be replaced by foreign sequences. A nice test for expression of the foreign sequence is immobilization of the bacteria by an antibody directed toward the foreign sequence. The flagellin from *E. coli* K12 was used to express an 11-residue sequence from hen egg-white lyzozyme (Kuwajima *et al.*, 1988) and the flagellin from *Salmonella typhimurium* was used to express a number of various epitopes up to at least 48 residues long (Newton *et al.*, 1989; Wu *et al.*, 1989).

2.   FIMBRIAE

Fimbriae are long and flexible filamentous structures that mediate adherence of bacteria to a number of surfaces. There are 100–200 such structures on the surface of a single *E. coli* cell. They include numerous copies (about 1000) of a major constituent (fimbrillin, 128 residues in the case of *E. coli*) as well as minor components. Their biogenesis depends on several genes (usually eight). Comparison of sequences from related fimbriae reveal highly variable peptides within fimbrillin, which usually correspond to the major antigenic determinants. Four different fimbriae have been used.

 a.   Type 1 Fimbriae.   In the case of the type 1 fimbriae from *E. coli*, epitopes corresponding to up to 18 residues could be expressed in two sites (after residues 28 and 57), so that they were detectable on the surface of the fimbriae (Hedegaard and Klemm, 1989; Klemm and Hedegaard, 1990).

 b.   P fimbriae.   Exposure of epitopes of at most of 14 residues was performed by insertion in two out of the five variable regions (HR1 and HR4). As shown by ELISA, exposure was more favorable in HR1 (van Die *et al.*, 1990).

 c.   K88 Fimbriae.   Heterologous epitopes of 11 residues were inserted by replacement into regions 163–173: the epitopes were detectable by ELISA and immunoelectron microscopy on the bacterial fimbriae, but these were expressed at a greatly reduced level (Bakker *et al.*, 1990).

 d.   Type 4 Fimbriae.   The fimbrial subunit of the Gram-negative *Bacteroides nodosus* was used to express a foreign epitope from FMDV (residues 144–159) at the surface of *Pseudomonas aeruginosa* by fusion at the COOH-terminal end of fimbrillin; internal additions did not lead to the formation of fimbriae (Jennings *et al.*, 1989).

D.   Secreted Proteins

Extracellular proteins are released from the cell at the end of one of several pathways, some of which involve a number of secretion-specific genes. Genetic fusions to such proteins may result in long-lasting or transient expression of the passenger at the cell surface.

1.   IgA Protease

The IgA protease is an extracellular protein secreted by pathogenic *Neisseria* species. This protease is able to cleave human IgA_1 antibody. In contrast to most other exported proteins of Gram-negative bacteria, the *iga* gene is sufficient by itself to direct secretion of the protease from *Neisseria*, but also from *E. coli* and *Salmonella* species. The mature protease (106 kDa) is processed in several steps from a precursor (169 kDa). It was shown that one subregion of the *Neisseria* IgA protease precursor, the β-domain (IGaβ), could direct a covalently attached periplasmic protein to the surface of *Salmonella typhimurium*. The passenger chosen was the β-subunit of cholera toxin. Demonstration of cell surface exposure was accomplished by protease accessibility experiments (trypsin and IgA protease) and immunolabeling (indirect immunofluorescence microscopy and immunogold electron microscopy). It is important to note that the strain of *S. typhimurium* was a GalE mutant so as to avoid shielding of the passenger by the O antigen of *Salmonella* (Klauser *et al.*, 1990, and references therein).

2.   Pullulanase

Pullulanase is a maltose-inducible starch-debranching lipoprotein that is released into the medium by the Gram-negative strain *Klebsellia pneumoniae*. This protein exists as a cell surface-bound intermediate before it is released into the medium from early stationary growth onward. Cell surface anchoring is due to the N-terminal fatty acid modification. In contrast to IgA protease, pullulanase secretion depends on the products of at least eight specific secretion genes. When these genes are transferred into *E. coli*, these bacteria secrete pullulanase exactly like *Klebsiella pneumoniae*. One hybrid protein where β-lactamase (Bla) was fused after nucleotide 1500 of *pulA* resulted in exposure of β-lactamase at the cell surface. This was demonstrated by showing that the β-lactamase activity could be assayed on intact cells with nitrocefin, by immunofluorescence with antibodies against PulA and Bla, and by immunoelectron microscopy. Like the wild-type PulA protein, the hybrid was released from the cell during the stationary phase, so that surface expression was transient. Fusions at other sites or fusion of another passenger (alkaline phosphatase) did not lead to cell surface exposure (Kornacker and Pugsley, 1990).

V.   Applications

Bacteria presenting a foreign antigen at their surface carry a new label. This label could be used as a means to detect the strain within a mixture of other strains and to help monitor the dissemination of this strain in the environment. The fact that the antigen is on the bacterial surface should facilitate screening or enrichments procedures (see Section V,C). But probably the most interesting feature for practical purposes is that these strains provide a source of self-replicating "immobilized" (poly)peptide. Production in large amounts is neither difficult nor expensive and may lead to various applications. We will discuss briefly three of them.

A.   Vaccines

It is generally believed that presentation of an antigen at the surface of a bacterium is favorable for inducing an immune response. This view relies in part on the fact that cell surface-exposed structures are often powerful immunogens, for example, repetitive sequences in parasites (review in Anders, 1988), bacterial outer membrane proteins (Udhayakumar and Muthukkaruppan, 1987), constituents of cell appendages (van Die *et al.*, 1990, and references therein), or O antigen in *Salmonella* (Hitchcock *et al.*, 1986, and references therein). Also, the fact that a number of bacteria (or parasites) escape the immune defenses of their host by antigenic variation of some surface component is in agreement with a critical role of surfaces in the interactions with the immune system (review in Seifert and So, 1988).

A number of recent studies have shown that a good antibody response could indeed be elicited by peptides expressed at the surface of live recombinant bacteria (Charbit *et al.*, 1987). Most importantly, antibodies were raised that were able to neutralize the pathogen from which the peptide was derived (see, for example, Charbit *et al.*, 1990; O'Callaghan *et al.*, 1990, and other articles in the same volume; Newton *et al.*, 1989).

On another hand, it was also shown that presentation at the cell surface is not a prerequisite to obtain an antibody response against a foreign epitope (Leclerc *et al.*, 1990). Further studies are now needed to compare quantitatively and qualitatively (classes and isotypes) the antibody response, but also other components of the immune response (secretory antibodies and cellular responses) against an epitope, depending on its location in the bacterial cell, the nature of the bacterial cell, and the infection route.

In the perspective of vaccination, attenuated mutants of virulent strains such as *Salmonella*, which are not pathogenic but have kept some potential for invasion, are explored for delivery of foreign epitopes (review in Curtiss, 1990). One goal is to develop strains expressing different foreign epitopes in the hope of obtaining multivalent vaccines. There is no obvious reason why it should not be possible to use different vector proteins in the same strain and express either the same antigenic determinant or different ones at several different sites in the same bacterial cell.

B. Diagnosis

The idea is to use a bacterial cell expressing a foreign peptide at its surface as a reagent to monitor the presence of antibodies against this peptide. This presence can be detected by simple tests such as agglutination or immunofluorescence. One potential problem is background due to antibodies reacting with other components of the bacterial surface. We found that, with bacteria expressing chosen peptides from the HIV virus within an exposed site of the LamB protein, monitoring of the antibody response against these peptides in human sera could be done by immunoblot from whole bacteria, with no background problems (Charbit *et al.*, 1990).

C. Search for Active Peptides

Peptides with a chosen biological activity can be selected from a large population of randomly generated sequences. It was recently shown that filamentous phages could be used to display a library of randomly generated sequences and to screen for peptides presenting affinity toward a given monoclonal antibody (Scott and Smith, 1990; Cwirla *et al.*, 1990). This approach should be useful in the analysis of the specificity of antibodies, in the search for ligands binding to receptors, and possibly in the quest for mimetic drugs. Bacteria expressing foreign sequences at their surface in an exposed loop of the LamB protein can similarly be used to display a library of random peptides (A. Siccardi, personal communication). Such libraries may allow exposure of larger peptides and use of screening methods adapted to bacteria (A. Charbit, P. Métézeau, M. Goldberg, A. Siccardi and M. Hofnung, unpublished).

D. Immunopurification

It is also possible to use intact bacteria expressing a foreign sequence at their surface as "affinity beads" to affinity purify monospecific antibodies against this sequence even without purification of the corresponding protein (Engleberg *et al.*, 1984).

VI. Perspectives

Most of the work on bacterial surface expression of foreign peptides has been performed so far in *E. coli*. The vector proteins used for expression at the *E. coli* cell surface are expected to be generally functional in the other species of Enterobacteriacae and in the other Gram-negative bacteria. There are, however, two important precautions that we would like to emphasize. First, the genes needed to target the vector protein have to be present in the new host bacteria (naturally or artificially). Second, the lipopolysaccharide or another cell surface structure (capsule) should not prevent not prevent accessibility to the foreign antigen. Another solution consists in searching directly for permissive sites in cell surface proteins from these other bacterial species.

The objective of expressing foreign peptides at the surface of Gram-positive bacteria could presumably be fulfilled by the same kind of approach, provided cell surface-exposed proteins are available from these bacteria.

In conclusion, our control on gene expression increases rapidly not only so as to allow us to obtain the product of a gene at the moment and in the amount desired, but also at the exact cellular (or extracellular) location planned. Expression of foreign sequences at the bacterial surface is one example of this emerging ability to target gene products. We have started to use the messages for cellular localization, although we do not yet understand many of them. These early constructions should, however, help us to understand the codes and mechanisms for cellular targeting and to use them for studying interactions among cells and their environment, including their potential hosts, and for a number of applications, some of which have been indicated here.

ACKNOWLEDGMENTS

I thank Alain Charbit for useful comments on the manuscript and editing of the protocols. Work in the author's laboratory is supported by grants from the Association pour le Développement de la Recherche sur le Cancer, the Ligue Nationale contre le Cancer, the Fondation pour la Recherche Médicale, and the World Health Organization (Transdisease Vaccinology Programme).

REFERENCES

Agterberg, M., Adriaanse, H., and Tommassen, J. (1987). Use of outer-membrane protein PhoE as a carrier for the transport of a foreign antigenic determinant to the cell surface of *Escherichia coli* K-12. *Gene* **59**, 145–150.

Agterberg, M., Fransen, R., and Tommassen, J. (1988). Expression of *Escherichia coli* PhoE protein in avirulent *Salmonella typhimurium aroA* and *galE* strains. *FEMS Microbiol. Lett.* **50,** 295–299.

Agterberg, M., Adriaanse, H., Lankhof, H., Meloen, R., and Tommassen, J. (1990). Outer-membrane PhoE protein of *Escherichia coli* as a carrier for foreign antigenic determinants : Immunogenicity of epitopes of foot-and-mouth disease virus. *Vaccine* **8,** 85–91.

Anders, R. F. (1988). Antigens of *Plasmodium falciparum* and their potential as components of a malaria vaccine. *In* "The Biology of Parasitism". (P. T. Englund and, A. Sher, eds.), MBL Lect. Biol., Vol. 9, p. 201. Alan R. Liss, New York.

Attridge, S. R., Daniels, D., Morona, J. K., and Morona, R. (1990). Surface coexpression of *Vibrio cholerae* and *Salmonella typhi* O-antigens on Ty21a clone EX210. *Microb Pathog.* **8,** 177–188.

Bakker, D., van Zijderveld, F. G., van der Veen, S., Oudega, B., and de Graaf, F. K. (1990). K88 fimbriae as carriers of heterologous antigenic determinants. *Microb Pathog.* **8,** 343–352.

Barany, F. (1985). Two-codon insertion mutagenesis of plasmid genes by using single-stranded hexameric oligonucleotides. *Proc. Natl. Acad. Sci. U. S .A.* **82,** 4202–4206.

Bayer, M. H., Keck, W., and Bayer, M. (1990). Localization of penicillin binding protein 1b in *Escherichia coli*: Immunoelectron microscopy and immunotransfer studies. *J. Bacteriol.* **172,** 125–135.

Bentley, A., and Klebba, P. E. (1988). Effect of lipopolysaccharide structure on reactivity of antiporin monoclonal antibodies with the bacterial cell surface. *J. Bacteriol.* **170,** 1063–1068.

Bouges-Bocquet, B., Villaroya, H., and Hofnung, M. (1984). Linker mutagenesis in the gene of an outer membrane protein of *E. coli*. LamB. *J. Cell. Biochem. 24,* 217–228.

Boulain, J.-C., Charbit, A., and Hofnung, M. (1986). Mutagenesis by random linker insertion into the *lamB* gene of *E. coli* K12. *Mol. Gene. Genet.* **205,** 339–348.

Braun-Breton, C. (1984). Screening for *lamB* missense mutations altering all the activities of the lambda receptor in *Escherichia coli* K12. *Ann Microbiol. (Paris)* **135A,** 181–190.

Charbit, A., Boulain, J.-C., Ryter, A., and Hofnung, M. (1986). Probing the topology of a bacterial membrane protein by genetic insertion of a foreign epitope. Expression at the cell surface. *EMBO J.* **5,** 3029–3037.

Charbit, A., Sobczak, E., Michel, M. L., Molla, A., Tiollais, P., and Hofnung, M. (1987). Presentation of two epitopes of the preS2 region of hepatitis B on live recombinant bacteria. *J. Immunol.***139,** 1658–1664.

Charbit, A., Molla, A., Saurin, W., and Hofnung, M. (1988). Versatility of a vector for expressing foreign polypeptides at the surface of Gram–bacteria. *Gene* **70,** 181–189.

Charbit, A., Molla, A., Ronco, J., Clément, J.-M., Favier, V., Bahraoui, E. M., Montagnier, L., Le Guern, A., and Hofnung, M. (1990). Immunogenicity and antigenicity of conserved peptides from the envelope of HIV1 expressed at the surface of recombinant bacteria. *AIDS* **4,** 545–551.

Charbit, A., Ronco, J., Michel, V., Werts, C., and Hofnung, M. (1991). Permissive sites and the topology of an outer-membrane protein with a reporter epitope, *J. Bacteriol.* **173,** 262–275.

Curtiss, R., III (1990). Antigen delivery systems for analysing host immune responses and for vaccine development. *Trends Biotechnol.* **8,** 237–240.

Cwirla, S. E., Peters, E. A., Barrett, R. W., and Dower, W. J. (1990). Peptides on phage: A vast library of peptides for identifying ligands. *Proc. Natl. Acad. Sci. U. S. A.* **87,** 6378–6382.

denBlaauwen, T., and Nanninga, N. (1990). Topology of penicillin-binding protein 1b of

Escherichia coli and topography of four antigenic determinants studied by immunolabeling electron microscopy. *J. Bacteriol.* **172**, 71–79.

Desaymard, C., Débarbouillé, M., Jolit, M., and Schwartz, M. (1986). Mutations affecting antigenic determinants of an outer membrane protein of *Escherichia coli*. *EMBO J.* **5**, 1383–1388.

Doolittle, R. F. (1986). "Of Urfs and Orfs. A Primer on How to Analyze Derived Amino Acid Sequences." University Science Books, U. S. A.

Duplay, P., Szmelcman, S., Bedouelle, H., and Hofnung, M. (1987). Silent and functional changes in the periplasmic maltose binding protein of *Escherichia coli* K12. I : Transport of maltose. *J. Mol. Biol.* **194**, 663–673.

Engleberg, N. C., Pearlman, E., and Eisenstein, B. I. (1984). *Legionella pneumophila surface antigens cloned and expressed in Escherichia coli* are translocated to the host cell surface and interact with specific anti-*Legionella* antibodies. *J. Bacteriol.* **160**, 199–203.

Ferenci, T., and Lee, K. S. (1986). Exclusion of high-molecular weight maltosaccharides by lipopolysaccharide O-antigen of *Escherichia coli* and *Salmonella typhimurium*. *J. Bacteriol.* **167**, 1081–1082.

Formal, S. B., Baron, L. S., Kopecko, D. J., Washington, O., Powell, C., and Life, C. A. (1981). Construciton of a potential bivalent vaccine strain: Introduction of *Shigella sonnei* form I antigen genes into the *galE Salmonella typhi* Ty21 a typhoid vaccine strain. *Infect. Immun.* **34**, 746–750.

Freudl, R., MacIntyre, S., Degen, M., and Henning, U. (1986). Cell surface exposure of the outer membrane protein OmpA of *Escherichia coli* K-12. *J. Mol. Biol.* **188**, 491–494.

Gabay, J., and Yasunaka, K. (1980). Interaction of the lamB protein with the peptidoglycan layer in *Escherichia coli* K12. *Eur. J. Biochem.* **104**, 13–18.

Hancock, I. C., and Poxton, I. R. (1988). "Bacterial Cell Surface Techniques." Wiley, New York.

Harrison, J. L., Taylor, I. M., and O'Connor, C. D. (1990). Presentation of foreign antigenic determinants at the bacterial cell surface using the TraT lipoprotein. *Res. Microbiol.* **141**, 1009–1012.

Hedegaard, L., and Klemm, P. (1989). Type 1 fimbriae of *Escherichia coli* as carriers of heterologous antigenic sequences. *Gene* **85**, 115–124.

Heine, H.-G., Francis, G., Lee, K.-S., and Ferenci, T. (1988). Genetic analysis of sequences in maltoporin that contribute to binding domains and pore structure. *J. Bacteriol.* **170**, 1730–1738.

Hitchcock, P. J., Leive, L., Mäkelä, P. H., Rietschel, E. T., Strittmatter, W., and Morrison, D. C. (1986). Lipopolysaccharide nomenclature—Past, present, and future. *J. Bacteriol.* **166**, 699.

Hofnung, M. (1988a). Expression of foreign antigenic determinants on bacterial envelope proteins. *Antonie van Leeuwenhoek* **54**, 441–445.

Hofnung, M. (1988b). Engineered protein fusions and live recombinant bacterial vaccines. *In* "Proceedings of the 8th International Biotechnology Symposium" (G. Durand, L. Bobichon, and J. Florent, eds.), Vol. II, pp. 713–724. Société Française de Microbiologie, Paris.

Hofnung, M., Bedouelle, H., Boulain, J.-C., Clément, J.-M., Charbit, A., Duplay, P., Gehring, K., Martineau, P., Saurin, W., and Szmelcman, S. (1988). Genetic approaches to the study and the use of proteins: Random point mutations and random linker insertions. *Bull. Inst. Pasteur (Paris)* **86**, 95–101.

Jennings, P. A., Bills, M. M., Irving, D. O., and Mattick, J. S. (1989). Fimbriae of *Bacteroides nodosus*: Protein engineering of the structural subunit for the production of an exogenous peptide. *Protein Eng.* **2**, 365–369.

Jones, R. T., Koeltzow, D. E., and Stocker, B. A. D. (1972). Genetic transfer of *Salmonella typhimurium* and *Escherichia coli* lipopolysaccharide antigens to *Escherichia coli* K-12. *J. Bacteriol.* **11**, 758–770.

King, G., and Swanson, J. (1978). Studies on gonococcus infection. XV. Identification of surface proteins of *Neisseria gonorrhoeae* correlated with leukocyte association. *Infect. Immun.* **21**, 575–584.

Klauser, T., Pohlner, J., and Meyer, T. F. (1990). Extracellular transport of cholera toxin B subunit using *Neisseria* IgA protease β-domain: Conformation-dependent outer membrane translocation. *EMBO J.* **9**, 1991–1999.

Klemm, P., and Hedegaard, L. (1990). Fimbriae of *Escherichia coli* as carriers of heterologous antigenic sequences. *Res. Microbiol.* **141**, 1013–1017.

Kornacker, M. G., and Pugsley, A. P. (1990). The normally periplasmic enzyme beta-lactamase is specifically and efficiently translocated through the *Escherichia coli* outer membrane when it is fused to the cel surface enzyme pullulanase. *Mol. Microbiol.* **4**(7), 1101–1109.

Kuwajima, G., Asaka, J.-I., Fujiwara, T., Fujiwara, T., Nakano, K., and Kondoh, E. (1988). Presentation of an antigenic determinant from hen egg-white lysozyme on the flagellar filament of *Escherichia coli*. *Bio/Technology* **6**, 1080–1083.

Lathe, R., Kieny, M. P., Skory, S., and Lecocq, J. P. (1984). Laboratory methods linker tailing: Unphosphorylated linker oligonucleotides for joining DNA termini. *DNA* **3**, 173–182.

Leclerc, C., Martineau, P., Van Der Werf, S., Deriaud, E., Duplay, P., and Hofnung, M. (1990). Induction of virus neutralizing antibodies by bacteria expressing the C3 poliovirus epitope in the periplasm. *J. Immunol.* **144**, 3174–3182.

Lee, C., Li, P., Inouye, H., Brickman, E. R., and Beckwith, J. (1989). Genetic studies on the inability of β-galactosidase to be translocated across the *Escherichia coli* cytoplasmic membrane. *J. Bacteriol.* **171**, 4609–4616.

MacIntyre, S., Freudl, R., Eschbach, M.-L., and Henning, U. (1988). An artificial hydrophobic sequence functions as either an anchor or a signal sequence at only one of two positions within the *Escherichia coli* outer membrane protein OmpA. *J. Biol. Chem.* **263**, 19053–19059.

Marchalonis, J. J., Cone, R. E., and Santer, V. (1971). Enzymic iodination. A probe for accessible surface proteins of normal and neoplastic lymphocytes. *Biochem. J.* **124**, 921–927.

Matsudaira, P. (1987). Sequence from picomole quantities of proteins electroblotted onto polyvinylidene difluoride membranes. *J. Biol. Chem.* **262**, 10035–10038.

Mims, C. A. (1987). "The Pathogenesis of Infectious Disease." Academic Press, Orlando, Florida.

Molla, A., Charbit, A., Le Guern, A., Ryter, A., and Hofnung, M. (1989). Antibodies against synthetic peptides and the topology of LamB, an outer-membrane protein from *Escherichia coli* K12. *Biochemistry* **28**, 8234–8241.

Newton, S. M. C., Jacob, C. O., and Stocker, B. A. D. (1989). Immune response to cholera toxin epitope inserted in *Salmonella* flagellin. *Science* 244, 70–72.

Nikaido, H., and Reid, J. (1990). Biogenesis of prokaryotic pores. *Experientia* **46**, 174–180.

O'Callaghan, C. H., Morris, A., Kirby, S. M., and Shingler, A. H. (1972). New method for the detection of beta-lactamases by using a chromogenic cephalosporin substrate. *Antimicrob. Agents Chemother.* **1**, 283–288.

O'Callaghan, D., Maskell, D., Beesley, J. E., Lifely, R., Roberts, I., Boulnois, G., and Dougan, G. (1988). Characterization and *in vivo* behaviour of a *Salmonella typhimurium aroA* strain expression *Escherichia coli* K1 polysaccharide. *FEMS Microbiol. Lett.* **52**, 269–274.

O'Callaghan, D., Charbit, A., Martineau, P., Leclerc, C., Van Der Werf, S., Nauciel, C., and Hofnung, M. (1990). Immunogenicity of foreign peptide epitopes expressed in bacterial envelope proteins. *Res. Microbiol.* **141**, 963–969.

Pugsley, A. P. (1989). "Protein Targeting." Academic Press, San Diego, California.

Randall, L. L., Hardy, S. J., and Thom, J. R. (1987). Export of protein: A biochemical view. *Annu. Rev. Microbiol.* **41**, 507–541.

Ronco, J., Charbit, A., and Hofnung, M. (1990). Creation of targets for proteolytic cleavage in the LamB protein of *E. coli* K12 by genetic insertion of foreign sequences: Implications for topological studies. *Biochimie* **72**, 183–189.

Ryter, A. (1987). Contribution of immunogold labelling to study of the outer membrane of Gram-negative bacteria. *Microbiol. Sci.* **4**, 270–273.

Schwartz, M., and Le Minor, L. (1975). Occurrence of the bacteriophage Lambda receptor in some enterobacteriaceae. *J. Virol.* **15**(4), 679–685.

Scott, J. K., and Smith, G. P. (1990). Searching for peptide ligands with an epitope library. *Science* **249**, 386–390.

Seifert, H. S., and So, M. (1988). Genetic mechanism of antigenic variation. *Microbiol. Rev.* **52**, 327.

Taylor, I. M., Harrison, J. L., Timmis, K. N., and O'Connor, C. D. (1990). The TraT lipoprotein as a vehicle for the transport of foreign antigenic determinants to the cell surface of *Escherichia coli* K-12: Structure function relationships in the TraT protein. *Mol. Microbiol.* **4**(8), 1259–1268.

Tommassen, J. (1986). Fallacies of *E. coli* cell fractionations and consequences thereof for protein export models. *Microb. Pathog.* **1**, 225–228.

Udhayakumar, V., and Muthukkaruppan, V. R. (1987). Protective immunity induced by outer membrane proteins of *Salmonella typhimurium* in Mice. *Infect. Immun.* **55**, 816–821.

van der Ley, P., de Graaff, P., and Tommassen, J. (1986). Shielding of *Escherichia coli* outer membrane proteins as receptors for bacteriophages and colicins by O-antigenic chains of lipopolysaccharide. *J. Bacteriol.* **168**, 449–451.

van Die, I., van Oosterhout, J., van Megen, I., Bergmans, H., Hoekstra, W., Enger-Valk, B., Barteling, S., and Mooi, F. (1990). Expression of foreign epitopes in P-fimbriae of *Escherichia coli*. *Mol. Gen. Genet.* **222**, 297–303.

Verner, K., and Schatz, G. (1988). Protein translocationacross membranes. *Science* **241**, 1307–1313.

Voorhout, W., de Kroon, T., Leunissen-Bijvelt, J., Verkleij, A., and Tommassen, J. (1988). Accumulation of LamB–LacZ hybrid proteins in intracytoplasmic membrane-like structures in *Escherichia coli* K12. *J. Gen. Microbiol.* **134**, 599–604.

Wu, J. Y., Newton, S., Judd, A., Stocker, B., and Robinson, W. S. (1989). Expression of immunogenic epitopes of hepatitis B surface antigen with hybrid flagellin proteins by a vaccine strain of *Salmonella*. *Proc. Natl. Acad. Sci. U.S.A.* **86**, 4726–4730.

Yamamoto, T., Tamura, U., and Yokota, T. (1985). Enteroadhesion fimbriae and enterotoxin of *Escherichia coli*: Genetic transfer to a streptomycin-resistant mutant of the galE oral-route live-vaccine *Salmonella typhi* Ty21a. *Infect. Immun.* **50**, 925–928.

Chapter 5

In Vitro Biochemical Studies on Translocation of Presecretory Proteins across the Cytoplasmic Membrane of Escherichia Coli

SHOJI MIZUSHIMA, HAJIME TOKUDA, AND SHIN-ICHI MATSUYAMA

Institute of Applied Microbiology
The University of Tokyo, Yayoi
Tokyo 113, Japan

I. Introduction
II. Preparation of Presecretory Proteins for *in Vitro* Translocation Studies
   A. Presecretory Proteins
   B. Preparation of Presecretory Proteins as Substrates for *in Vitro* Translocation
III. Preparation of Everted Membrane Vesicles Exhibiting Efficient Translocation Activity
   Method of Preparation
IV. *In Vitro* Translocation Assay
   Assay Procedures
V. Methods for Identification and Characterization of Translocation Intermediates
VI. Overproductions and Purifications of Protein Components Involved in the Translocation Reaction
   A. General Methods for Overproduction
   B. Overproductions of Sec Proteins
   C. SecE-Dependent Overproduction of SecY
   D. Purifications of SecA, SecE, and SecY
VII. Methods for Analysis of SecA Functions
   A. SecA-Dependent *in Vitro* Translocation Assay
   B. Assay of Translocation ATPase Activity of SecA
   C. Cross-Linking Studies
   D. Denaturation and Renaturation of SecA
   E. Reconstitution of SecA Analogs from Truncated Fragments

METHODS IN CELL BIOLOGY, VOL. 34

I. Introduction

The translocation of secretory proteins across the cytoplasmic membrane in prokaryotes shares common features with secretory protein translocation across the endoplasmic reticulum membrane in eukaryotes. Bacterial cells, especially *Escherichia coli* cells, are advantageous for studying such translocation mechanisms in that genetic and gene engineering methods can easily be applied to these organisms. In fact, a number of secretion mutants of *E. coli* have been isolated and characterized genetically (Beckwith and Ferro-Novick, 1986).

Biochemical studies on prokaryotes were once behind those on eukaryotes, largely due to lack of a suitable *in vitro* system for precise biochemical studies. But during the last few years, great progress has been made in the biochemical studies on the mechanisms underlying translocation of presecretory proteins across the cytoplasmic membrane of *E. coli*. This has mainly been due to the development of an *in vitro* system exhibiting efficient protein translocation. Furthermore, recently developed procedures for overproduction of larger amounts of protein components involved in protein translocation and reconstitution of a protein translocation system have made possible precise biochemical studies on the components involved in the protein translocation reaction.

This chapter describes methods for *in vitro* biochemical studies on protein translocation across the cytoplasmic membrane of *E. coli*. Most of the methods described are those developed in our laboratory. For details of other biochemical methods, the reader should refer to other chapters also dealing with *in vitro* biochemical methods for study of the prokaryotic translocation system.

II.   Preparation of Presecretory Proteins for *in Vitro* Translocation Studies

A.   Presecretory Proteins

A system for *in vitro* protein translocation in prokaryotes requires everted (inside-out) cytoplasmic membrane vesicles, which are equivalent to microsome vesicles prepared from eukaryotic cells, and presecretory proteins to be used as substrates. So far, only a few species of presecretory proteins have been used as substrates for *in vitro* translocation: proOmpA, proMalE, proLamB, proBla, proOmpF–Lpp, and mutant proteins derived from these presecretory proteins. ProOmpF–Lpp is a model presecretory protein, in which the signal peptide and the cleavage site are derived from proOmpF and the following mature region from the major lipoprotein (Yamane *et al.*, 1987, 1988; Yamane and Mizushima, 1988). On the other hand, very little, if any, translocation of proOmpC, proOmpF, or proPhoA, presecretory proteins with respect to the cytoplasmic membrane, was detected *in vitro*. The reason for the inertness of these proteins is unclear.

Some properties of the presecretory proteins that are effectively translocated *in vitro* are summarized in Table I, the amino acid sequences of their signal peptide regions are shown in Table II, and more detailed information on some of these proteins is given below.

1.   ProOmpA

OmpA is a major outer membrane protein with a molecular mass of 35 kDa. ProOmpA can be efficiently translocated into everted membrane vesicles when it is properly unfolded by treatment with urea or in the presence of the trigger factor (Crooke *et al.*, 1988a). Although the translocation is appreciably enhanced in the presence of a proton motive force ($\Delta\bar{\mu}H^+$), the requirement is not essential (Yamada *et al.*, 1989a). Several analogs of proOmpA with deletions (from the carboxyl terminus) have been constructed and all were found to be efficiently translocated into everted membrane vesicles (M. Kato and S. Mizushima, unpublished).

2.   ProOmpF–Lpp

ProOmpF–Lpp is a model presecretory protein comprising the signal peptide and the amino-terminal two-amino acid residues of outer membrane protein OmpF and the carboxyl-terminal 57-amino acid residues of

TABLE I

Presecretory Proteins Used for *in Vitro* Translocation Studies

Presecretory protein and derivatives	Molecular mass of mature domain (kDa)	Signal peptide cleavage site[a]	Other remarks	
ProOmpA	35	Ala-AlaPro	Precursor of an outer membrane protein	Rate of *in vitro* translocation: D26 > D51 > intact proOmpA
ProOmpA D51	17	Ala-AlaPro	—	
ProOmpA D26	7.7	Ala-AlaPro	Highly efficient substrate	
ProMalE	41	Ala-LysIle	Precursor of maltose-binding protein in periplasm	Translocation is stimulated by SecB
ProLamB	47	Ala-ValAsp	Precursor of lambda phage receptor in outer membrane	Translocation is stimulated by SecB
ProBla	29	Ala-HisPro	Precursor of periplasmic β-lactamase	Translocation is stimulated by GroEL
ProOmpF-Lpp	6.5	Ala-AlaGlu	—	Requirement of $\Delta\bar{\mu}H^+$ is less strict when Pro is near the signal cleavage site
ProOmpF-Lpp-L1	6.4	Ala-AlaPro	—	
ProOmpF-Lpp-L4	6.5	Ala-ProGlu	Signal peptide is uncleavable	
ProOmpF-Lpp-[Un]	6.4	Phe-ProGly	Translocation efficiency is very high; signal peptide is uncleavable	
ProOmpF-Lpp[K,R]$_m$[Un]	6.4	Phe-ProGly	The amino terminus of the signal peptide contains different numbers ($n = 0$~4) of Lys/Arg	
ProOmpF-Lpp[L]$_n$[Un]	6.4	Phe-ProGly	The hydrophobic domain of the signal peptide is composed of a polyleucine stretch ($n = 0$~20); can be used for *in vitro* assay only after sonication in 6–8 M urea	

[a] Cleavage sites are denoted by hyphens.

TABLE II

AMINO ACID SEQUENCES OF SIGNAL PEPTIDES[a]

Presecretory protein	Signal sequence
ProOmpA	M K K T A I A I A V A L A G F A T V A Q A*A P
ProMalE	M K I K T G A R I L A L S A L T T M M F S A S A L A K I
ProLamB	M M I T L R K L P L A V A V A A G V M S A Q A M A V D
ProBla	M S I Q H F R V A L T P F F A A F C L P V F A H P
ProOmpF–Lpp	M M K R N ┆I L A V I V P A L L V A┆G T A N A A E
ProOmpF–Lpp[Un]	M M K R N ┆I L A V I V P A L L V A┆G T A N F P G
ProOmpF–Lpp[K]$_m$[Un]	MM[K]$_m$[N]$_{4-m}$┆I L A V I V P A L L V A┆G T A N F P G
ProOmpF–Lpp[R]$_m$[Un]	MM[R]$_m$[N]$_{4-m}$┆I L A V I V P A L L V A┆G T A N F P G
ProOmpF–Lpp[L]$_n$[Un]	M M K R N ┆[L]$_n$(n = 0–20) ┆G T A N F P G

[a] The asterisk denotes the signal peptide cleavage site; [Un] indicates that the signal peptide is uncleavable in these proteins. The region between the two broken lines is the central hydrophobic domain of signal peptides.

the lipoprotein. A large number of mutant proOmpF–Lpps possessing different amino acid residues around the signal cleavage site were constructed and were found to differ in their efficiency of translocation and in their requirement of $\Delta\bar{\mu}H^+$ (Lu *et al.*, 1991).

1. ProOmpF–Lpp mutants possessing Ala-Pro at the signal peptide cleavage site are not attacked by signal peptidase upon translocation (and hence are called uncleavable proOmpF–Lpps), whereas all the mutants possessing Ala-X (except Pro) at the site are attacked by the peptidase.

2. A mutant protein possessing Phe-Pro at the site, which is also signal peptidase resistant, exhibits appreciably high translocation activity.

3. Mutant proteins possessing Pro near the signal peptide cleavage site can be translocated rather efficiently, even in the absence of $\Delta\bar{\mu}H^+$, whereas the translocations of other mutant proteins are highly $\Delta\bar{\mu}H^+$ dependent.

These findings suggest that the structure around the signal peptide cleavage site has something to do with not only recognition by signal peptidase but also translocation efficiency.

3.   PRESECRETORY PROTEINS POSSESSING ARTIFICIAL SIGNAL PEPTIDES

Signal peptides usually possess one or more positively charged amino acid residues at the amino terminus, followed by a stretch of hydrophobic residues (von Heijne, 1985). All prokaryotic signal peptides so far reported have a positive charge, suggesting that this is essential for protein secretion. The importance of the amino-terminal positive charge in protein secretion has also been suggested *in vivo* by means of gene manipulation (Inouye *et al.*, 1982; Vlasuk *et al.*, 1983; Iino *et al.*, 1987; Bosch *et al.*, 1989).

For studying the effect of a positive charge on the rate of *in vitro* translocation, the number of positive charges in the amino terminus of proOmpF–Lpp was altered by substituting different numbers of Lys, Arg, or His (Table I). Translocation of mutant proOmpF–Lpps was found to depend on the positive charge, the rate of translocation being roughly proportional to the number of positively charged groups, irrespective of the amino acid species that donated the charge (Sasaki *et al.*, 1990). The importance of the charged region in interaction of presecretory proteins with SecA, a component of the protein secretory machinery, has been demonstrated (Akita *et al.*, 1990). The interaction was enhanced when the number of positively charged amino acid residues was increased.

The replacement of the hydrophobic domain of the signal peptide by different lengths of polyleucine residues or polymers with alternate leucine and alanine residues was also performed to determine the role of this domain in protein translocation. The rate of translocation was greatly affected by the number of these residues with a maximum at [Leu]$_8$ or [Ala + Leu]$_{9-10}$ (C. Hikita and S. Mizushima, unpublished). This suggests that the hydrophobic domain is recognized specifically by the secretory machinery rather than nonspecifically by the hydrophobic region of the phospholipid bilayer in the membrane and that the total hydrophobicity of the hydrophobic region of the signal peptide is an important determinant of the substrate specificity.

B.   Preparation of Presecretory Proteins as Substrates for *in Vitro* Translocation

1.   *In Vitro* Radiolabeling of Presecretory Proteins

a.   Transcription Reaction.   Although some of the *E. coli* presecretory protein genes cloned on a plasmid can be used for *in vitro* transcription, some cannot be used due to lack of necessary transactivator proteins or to inefficiency of their own promoters. For efficient and reproducible transcription, it is advisable to use an efficient and well-characterized promoter. The SP6 promoter together with SP6 RNA polymerase are effective for this purpose. Most of the genes encoding presecretory proteins listed in Table I have been cloned downstream of the SP6 promoter and were found to be efficiently transcribed *in vitro*. A typical example of the *in vitro* transcription reaction in 100 μl at 40°C for 2 hours as follows (Melton *et al.*, 1984):

40 mM *Tris-HCl (pH 7.5)*
6 mM MgCl$_2$
2 mM spermidine
10 mM NaCl
10 mM dithiothreitol
100 U   RNasin (RNase inhibitor)
0.5 mM each of ATP, GTP, CTP, and UTP
5 μg   linearized DNA template
50 U   SP6 RNA polymerase

b.   Translation Reaction.   The mRNAs thus transcribed can be translated *in vitro* by conventional procedures. [^{35}S]Methionine is preferable for labeling presecretory proteins, if they possess Met in the mature domain. A typical example of the *in vitro* translation reaction in 100 μl at 37°C for 40 minutes is as follows (Yamane *et al.*, 1987). This reaction mixture has been used for the transcription-coupled translation, and hence some of the reagents, such as CTP and UTP, may be omitted, although we have not performed the reaction in their absence.

10 μl   V buffer
10 μl   amino acid mixture
10 μl   nucleotide mixture
20 μl   S100
0.25 μg   pyruvate kinase
10 μg   creatine kinase
20–40 μCi Tran^{35}S-label (Met; 1000 Ci/mmol)
5–10 μl   transcription mixture
     Fill up to 100 μl with H_2O

V BUFFER
  100 mM MgOAc
  75 mM CaOAc
  580 mM KOAc
  550 mM Tris-OAc (pH 8.2)
  330 mM NH_4OAc
  8 mM spermidine

AMINO ACID MIXTURE (pH 8)
3.5 mM each of all 19 amino acids except Met

NUCLEOTIDE MIXTURE (pH 8)
  210 mM phosphoenolpyruvate
  0.5 mM folinic acid
   1 mg/ml tRNA from *E. coli*
  0.3 mM NADPH
  0.3 mM FAD
  22 mM ATP
  5.6 mM GTP
  5.6 mM CTP
  5.6 mM UTP
  10 mM NADH
  300 mM creatine phosphate

S100 FRACTION
  For the translation reaction, S100 is prepared as follows: *E. coli* K002 (Yamane *et al.*, 1987) is grown in 4 μl of a medium containing (per liter) K_2HPO_4, 26 g; KH_2PO_4, 5 g; yeast extract, 9 g; and glucose, 9 g at 37°C until the cell density reaches 5×10^8 cells/ml. Cells (about 20 g) are washed with 50 mM Tris-acetate (pH 7.8) and 10% sucrose, suspended in 19 ml of buffer A containing 10 mM Tris-acetate (pH 7.8), 14 mM magnesium acetate, 60 mM potassium acetate, and

0.1 mM dithiothreitol, frozen at $-80°C$ and then thawed at room temperature. The suspension is mixed with 15 mg of DNase I and 30 g of glass beads (0.1 mm in diameter), treated with Vibrogen cell mill Vi-4 10 times for 30 seconds with 30-second intervals, and centrifuged at 10,000 g for 10 minutes at 4°C. The resultant supernatant is further centrifuged at 80,000 g for 1 hour at 4°C and the supernatant (about 20 ml) is dialyzed against buffer A after mixing with 250 μg of pyruvate kinase. The preparation is then centrifuged at 80,000 g for 60 minutes at 4°C, mixed with 14 OD$_{260}$ units of sucrose-washed 70 S ribosome, which is prepared as described in Traub *et al.* (1971), and stored at $-80°C$. The concentration of protein in the final preparation is about 30 mg/ml.

2. PURIFICATION OF RADIOLABELED PRESECRETORY PROTEINS

In vitro translocation of presecretory proteins into the everted membrane vesicles can be observed using the translation mixture without further treatment. In many experiments, however, one or more of the following treatments may be required.

   a.   Removal of Small Molecules.   Translocation of presecretory proteins into everted membrane vesicles requires two types of energy, ATP and $\Delta\bar{\mu}H^+$ (Geller *et al.*, 1986; Yamane *et al.*, 1987). The latter can be generated through respiratory oxidation of NADH or succinate by the membrane. For study of the roles of these or other small molecules, which may be involved in the translocation reaction, their removal from the translation mixture is required, and this can easily be carried out by gel filtration as initially shown by Chen and Tai (Chen and Tai, 1985). The reaction mixture (350 μl) after translation is directly applied (Yamane *et al.*, 1988) to a Sephadex G-75 column (3.5-ml bed volume) equilibrated with 50 mM potassium phosphate (pH 7.5). The presecretory protein, which is eluted in the void volume fractions, is collected and used for translocation experiments. Translocation of the presecretory protein thus fractionated absolutely requires ATP and is appreciably enhanced upon the addition of NADH or succinate (Yamane *et al.*, 1988). Small molecules can also be removed by TCA precipitation of proteins (Crooke *et al.*, 1988b; Tani *et al.*, 1990).

   b.   Purification of Presecretory Proteins by Means of Immunoaffinity Column Chromatography.   Immunoaffinity column chromatography may be the best method for purification of small amounts of radiolabeled presecretory proteins. We have purified proOmpF–Lpp through an immunoaffinity column as follows [the original method was reported by Matsuyama and Mizushima (1989)].

Rabbit antilipoprotein IgG is covalently linked to protein A–Sepharose CL-4B (Pharmacia) with dimethylsuberimidate by the method recommended by the manufacturer. The [^{35}S]methionine-labeled proOmpF–Lpp is synthesized in 350 μl of reaction mixture with 100 μCi of Tran^{35}S-label (ICN). The translation mixture is then applied to an immunoaffinity column (0.5 ml) preequilibrated with 20 mM sodium phosphate (pH 7.2) and 0.145 M NaCl. The column is washed with 10 ml of 20 mM Tris-HCl (pH 7.5) and 0.5 M NaCl, and then with 5 ml of 10 mM Tris-HCl (pH 7.5) and 10 mM NaCl. The precursor, which is bound to the column, is eluted with 0.2 M acetic acid and 0.1% BSA (pH 2.4). The presence of BSA is important: in its absence, the recovery of the presecretory protein is very low. The radioactive peak fraction is dialyzed against 50 mM potassium phosphate (pH 7.5). About 10^7 dpm/800 μl of the presecretory protein is obtained.

   c.   Preparation of Highly Purified Presecretory Proteins.   Although the purification of a presecretory protein on an immunoaffinity column usually results in the removal of as much as 99% of the proteins in the reaction mixture, the purified preparation may still be contaminated by several proteins in more than equimolar ratios. In particular, proteins specifically interacting with the presecretary protein may be coimmunoprecipitated. To remove these possible contaminants, we have purified pro-OmpF–Lpp by a combination of immunoaffinity column chromatography and SDS–polyacrylamide gel electrophoresis as follows (Matsuyama and Mizushima, 1989):

   Four volumes of acetone are added to the [^{35}S]proOmpF–Lpp (4 × 10^7 dpm) from the immunoaffinity column in the presence of BSA. The precipitate is collected by centrifugation, dissolved in 50 μl of 100 mM sodium phosphate (pH 7.2), 0.5% SDS, 1% β-mercaptoethanol, and 0.002% bromophenol blue and is boiled for 5 minutes. This solution is then applied to an SDS–polyacrylamide gel for electrophoresis as described (Yamane *et al.,* 1987). Autoradiography is carried out and the band of the presecretory protein is cut out from the gel, sliced, and incubated in 1 ml of 100 mM sodium phosphate (pH 7.2), 0.5% SDS and 0.1% BSA, at room temperature for 10 hours. The gel is removed and acetone is added to the eluate. The resulting precipitate is collected by centrifugation, dissolved in 50 μl of 20 mM sodium phosphate (pH 7.2), 0.145 M NaCl, and 0.5% SDS and is boiled for 5 minutes and then diluted to 500 μl with 20 mM sodium phosphate (pH 7.2), 0.145 M NaCl, and 1% Triton X-100. This solution is applied to a fresh immunoaffinity column. The presecretory protein is eluted with 0.2 M acetic acid and 0.1% BSA (pH 2.4), and then dialyzed against 50 mM potassium phosphate (pH 7.5). The yield of presecretory protein is about 2 × 10^7 dpm/400 μl.

3. LARGE-SCALE PREPARATION OF PRESECRETORY PROTEINS

For biochemical studies, substantial amounts of presecretory proteins, not small amounts of radiolabeled protein, may often be required, e.g., for study of molecular interactions. Presecretory proteins can be accumulated *in vivo* when they are overproduced. Two examples of such overproduction and purification of these proteins are given below.

a. Preparation of ProOmpA. pTac–OmpA is a plasmid in which the *ompA* gene is placed under the control of the *tac* promoter–operator. Strain JM103Lpp⁻ harboring pTac–OmpA is grown at 37°C. When OD_{660} reaches 1.0, IPTG is added to a final concentration of 2 mM and the cultivation is continued for a further 2 hours (Tani *et al.*, 1990).

ProOmpA can be prepared from the cells by the following procedure (Crooke *et al.*, 1988b): A cell paste is suspended in an equal weight of 50 mM Tris-HCl (pH 7.5) and 10% (w/v) sucrose and is then frozen as small nuggets by rapid pipetting into liquid nitrogen and stored frozen at −80°C. The frozen cell suspension (128 g) is thawed, mixed with lysozyme (0.9 mg/ml), and incubated for 5 minutes at 23°C. After addition of $MgCl_2$ (5 mM) and DNase I (4 μg/ml), the suspension is incubated for an additional 5 minutes at 23°C. Extraction buffer [360 ml, 1.5% (w/v) Sarkosyl, 50 mM citrate, titrated to pH 6.0 with solid Na_2HPO_4] is then added and the mixture is subjected to Dounce homogenization. The homogenate is allowed to stand for 30 minutes at 23°C and then is centrifuged (27,000 g, 15 minutes at 23°C). This extraction is repeated twice. The final pellet is suspended in 22 ml of buffer C, containing 50 mM Tris-HCl (pH 8.0), 2 mM dithiothreitol, and 8 M urea, by continuous vortexing for 5 minutes at 23°C, followed by centrifugation for 1 minute at 27,000 g to recover purified proOmpA in the supernatant.

b. Preparation of ProOmpF–Lpp. (Kimura *et al.*, 1991). pK025 is a plasmid carrying the *tac* promoter- and operator-controlled *ompF–lpp* gene (Yamane and Mizushima, 1988). *Escherichia coli* JM103Lpp⁻, harboring pK025, is grown in L broth in the presence of 1 mM IPTG and disrupted in a French pressure cell. The total membrane fraction of the cells is collected by centrifugation, and proOmpF–Lpp is prepared from it by the same method as that used for preparation of native lipoprotein (Hirashima *et al.*, 1973). Briefly, the membrane fraction is dissolved in 10 mM sodium phosphate buffer (pH 7.2) containing 3% SDS and mixed with two volumes of 15% TCA. To the supernatant obtained on centrifugation is added solid urea to a final concentration of 8 M, and then the solution is dialyzed against 10 mM sodium phosphate buffer (pH 7.2) to remove SDS, TCA, and urea. The dialyzed extract is treated with 80% cold acetone to precipitate proteins. Finally, the precipitate is dissolved in 50 mM potassium phosphate buffer (pH 7.5) containing 0.5 mM dithiothreitol and 8 M urea.

4. METHODS FOR KEEPING PRESECRETORY PROTEINS TRANSLOCATION COMPETENT

Presecretory proteins translated *in vitro* often tend to become translocation incompetent. Two possible reasons for this may be considered. The mature domain of secretory proteins may initiate folding into a translocation-incompetent conformation immediately after or even during translation, although the signal peptide attached to the amino terminus is reported to retard the rate of such folding (Park *et al.*, 1988). Another possible reason is that signal peptide regions, which are highly hydrophobic, may interact with each other to form aggregates, which are translocation incompetent.

Several soluble factors have been found to stimulate the *in vitro* translocation reaction: SecB for MalE and LamB (Weis *et al.*, 1988; Watanabe and Blobel, 1989; Hartl *et al.*, 1991), GroEL for Bla (Bochkareva *et al.*, 1988), and the trigger factor for OmpA (Crooke *et al.*, 1988a; Lill *et al.*, 1988). These factors have been suggested to maintain presecretory proteins in a translocation-competent state. These factors are not discussed here; for details, the reader should refer to other chapters of this volume.

Urea treatment has been used to make presecretory proteins translocation competent or to maintain their competence. For example, highly purified proOmpA stored in 8 M urea remains translocation competent, but on dilution or dialysis of this solution it gradually becomes translocation incompetent (Crooke *et al.*, 1988a). Urea treatment is also effective for making TCA-precipitated proOmpA translocation competent (Tani *et al.*, 1990). [^{35}S]ProOmpA synthesized *in vitro* is precipitated with 10% TCA, washed successively with acetone and ethyl ether, and dried. The denatured and dried proOmpA thus obtained exhibits efficient *in vitro* translocation when it is dissolved in 8 M urea and 50 mM potassium phosphate (pH 7.5) and is then diluted more than 10-fold. With this preparation, the rate of translocation is comparable to that with newly synthesized proOmpA, when sufficient amounts of ATP and NADH/succinate are supplied.

It should be noted, however, that treatment with 8 M urea is not always effective in keeping presecretory proteins translocation competent. Mutant proOmpF–Lpps with an artificial signal peptide comprising more than seven Leu residues as the hydrophobic stretch very rapidly form aggregates that are completely translocation incompetent (C. Hikita and S. Mizushima, unpublished); treatment with 8 M urea did not restore the competence. The competence could, however, be restored by sonicating the presecretory proteins in the presence of 6 to 8 M urea (C. Hikita and S. Mizushima, unpublished). The sonicated preparation exhibited efficient translocation upon dilution to make the final concentration of urea as low as 0.6 M, and their translocation-competent state was maintained for at least 3 hours after dilution. The *in vitro* translocation reaction is not inhibited by this concentration of urea.

III. Preparation of Everted Membrane Vesicles Exhibiting Efficient Translocation Activity

A system for *in vitro* protein translocation in prokaryotes requires everted cytoplasmic membrane vesicles. Methods for preparing everted membrane vesicles of *E. coli* were first developed by Hertzberg and Hinkle (1974) and Futai (1974). Such vesicles have been used successfully in *in vitro* biochemical studies of the translocation process. With these *in vitro* systems, however, the efficiency of protein translocation into membrane vesicles is usually not high enough. A large fraction of mature protein that has been processed for signal peptides is often found not to be translocated into membrane vesicles. This was assumed to be due to premature processing, before translocation of precursor proteins, by a signal peptidase on the right-side-out membrane vesicles or on membrane fragments.

We have developed a method for preparing membrane vesicles exhibiting efficient and quantitative protein translocation (Yamada *et al.*, 1989a). The procedure involves the disruption of spheroplasts in a French pressure cell and subsequent fractionation by sucrose gradient centrifugation as described in detail in the next section. The method is a slight modification of the original one.

Method of Preparation

Escherichia coli K002 and K003 (Δunc) are used as sources of everted membrane vesicles. It is often useful to use a Δunc strain, which lacks F_0F_1-ATPase, because in the absence of this enzyme the roles of ATP and $\Delta \bar{\mu}H^+$, both of which are required for the translocation reaction, can be studied independently.

Cells are grown in medium containing, per liter, Bacto-yeast extract, 10 g; Bacto-tryptone, 10 g; KH_2PO_4, 5.6 g; K_2HPO_4, 28.9 g; and glucose, 10 g; cells are harvested when the cell density reaches 5×10^8–7×10^8 cells/ml. About 5 g of cells is obtained from 1 liter of culture. The cells are then converted to spheroplasts as described by Mizushima and Yamada (1975). Tris-acetate buffer (pH 7.8, 4°C) is used in place of Tris-HCl buffer. The mixture is then cooled in an ice-water bath for 30 minutes. After addition of DNase (50 μg/ml of reaction mixture), the spheroplasts are passed once through an Aminco French pressure cell (FA-073) at 250 kg/cm^2 and then homogenized in a Teflon homogenizer to shear the released DNA. The cell lysate is centrifuged at 8000 g for 15 minutes to remove intact cells. It is then recentrifuged at 150,000 g for 2 hours and the pelleted membrane vesicles are suspended in a small volume of 50 mM potassium phosphate (pH 7.5).

The membrane vesicles are then layered over 4.4 ml of a linear gradient of 30–44% (w/w) sucrose in 50 mM potassium phosphate (pH 7.5) and dithiothreitol–glycerol buffer, centrifuged at 60,000 g for 15 hours at 4°C, and then fractionated into fractions 1–5 from the bottom to top of the gradient and the pellet fraction (see Fig. 1). Most of the outer membranes are collected in the pellet under these conditions.

Each fraction is diluted with five volumes of 50 mM potassium phosphate (pH 7.5) and 10% glycerol, and then the membrane vesicles are collected by centrifugation at 150,000 g for 2 hours, dispersed in a minimal volume of the buffer, and stored at -80°C. Membrane vesicles can be frozen and thawed several times without any loss of protein-translocating activity.

With the membrane vesicles of fraction 3 or 4, an efficient and quantitative translocation of proOmpF–Lpp into the vesicles can be observed (Fig. 1A). On the other hand, fractions 1 and 5 and the crude membrane before the sucrose gradient centrifugation do not exhibit such high translocation activities, although significant conversion of the precursor to the mature form takes place.

The gel electrophoretic analyses shown in Fig. 1B reveal that the protein profiles of membrane fractions 3 and 4 correspond to that of the cytoplasmic membrane, whereas the lower fractions are heavily contaminated by the outer membrane. The membrane fractions exhibiting efficient protein translocation contain membrane vesicles 0.3–0.5 μm in diameter, whereas the heavier fractions contain much smaller vesicles.

It is interesting that everted membrane vesicles prepared from spheroplasts exhibit higher translocation efficiency than ones directly prepared from intact cells, despite the fact that the cytoplasmic membrane of spheroplasts is reported to be less efficiently everted than that of intact cells with the French pressure cell treatment (Yamato *et al.*, 1978). On the other hand, the diameter of the former vesicles (0.3–0.5 μm) is considerably larger than that of the latter ones. This may somehow be related to their higher efficiency of translocation.

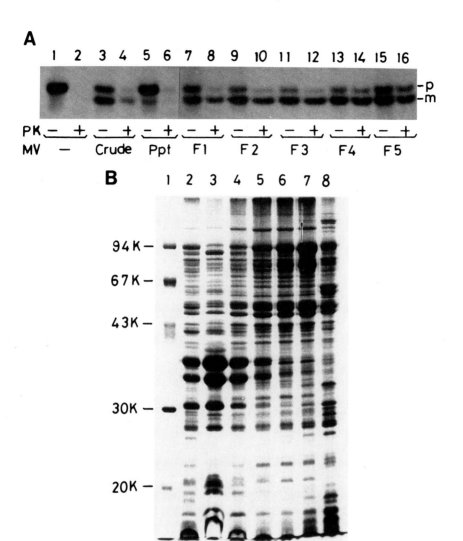

FIG. 1. Translocation activities and protein profiles of crude and fractionated membrane vesicles. Membrane vesicles prepared from K002 (*unc*[+]) were fractionated by sucrose gradient centrifugation. (A) *In vitro* translocation of proOmpF–Lpp was carried out using the following membrane vesicles (MV) (4 μg of protein): no membrane vesicles (lanes 1 and 2), crude membrane vesicles (lanes 3 and 4), the pellet fraction (Ppt) (lanes 5 and 6), fraction 1 (lanes 7 and 8), fraction 2 (lanes 9 and 10) fraction 3 (lanes 11 and 12), fraction 4 (lanes 13 and 14), and fraction 5 (lanes 15 and 16). The samples for lanes 2, 4, 6, 8, 10, 12, 14, and 16 were then treated with proteinase K (PK). All samples were analyzed by polyacrylamide gel electrophoresis and subsequent fluorography. The positions of the precursor (p) and mature (m) forms of proOmpF–Lpp are indicated. (B) Protein profiles of the membrane preparations (40 μg of protein) were analyzed on Laemmli's 12.5% polyacrylamide gel by electrophoresis. The gel was stained with Coomassie brilliant blue. The samples analyzed were molecular-weight markers (lane 1), the crude membrane preparation (lane 2), the pellet fraction (lane 3), fraction 1 (lane 4), fraction 2 (lane 5), fraction 3 (lane 6), fraction 4 (lane 7), and fraction 5 (lane 8). From Yamada *et al.* (1989a).

IV. *In Vitro* Translocation Assay

With a combination of the fractionated membrane vesicle preparation
and an appropriate secretory protein, precise kinetic studies can be made
on the process of protein translocation across the cytoplasmic membrane.
The basic idea of the *in vitro* system is that protein that has been translocated into the everted membrane vesicles is resistant to externally added
protease and can be digested only in the presence of a detergent such as
Triton X-100. Proteinase K is usually used as the protease. The assay
procedure currently used in our laboratory is described below. This assay
procedure has been successfully used for the wide variety of presecretory
proteins listed in Table I.

Assay Procedures

1.   DIRECT USE OF THE TRANSLATION REACTION MIXTURE

To 10 μl of the translation reaction mixture is added 1 μl of 4% cold
methionine to terminate translation. The translocation reaction is then
initiated by the successive additions of 2.0 μl of 62.5 mM NADH and
3–4 μl of membrane vesicles (5 μg protein). Further addition of ATP
(1–5 mM) is often required. The addition of an ATP-generating system
consisting of creatine phosphate and creatine kinase may also appreciably
enhance the rate of translocation, especially when the concentration of
ATP is low (Shiozuka *et al.*, 1990). The total volume is adjusted to 25 μl
with water. The translocation reaction is carried out at 37°C and terminated by cooling the mixture on ice. A portion (14 μl) of the mixture is
used for analyses.

2.   USE OF THE GEL-FILTERED TRANSLATION REACTION MIXTURE

Translocation is assayed at 37°C in 25 μl of 50 mM potassium phosphate
(pH 7.5) containing 5 μg of protein from membrane vesicles, 2 mM
MgSO$_4$, 1–5 mM ATP, 5 mM NADH or succinate, and [^{35}S]methionine-
labeled presecretory protein (about 2×10^5 dpm) that has been gel-
filtered.

In both cases, translated and translocated proteins are analyzed on
polyacrylamide gel with or without treatment with proteinase K at a concentration of 1 mg/ml in the presence or absence of Triton X-100 for
20 minutes at 4°C. The incubation is stopped by adding a final concen-

tration of 6–10% TCA. The precipitate is then successively washed with acetone and ethyl ether and dissolved in 10 μl of a solubilizer comprising 10 mM sodium phosphate (pH 7.2), 1% SDS, 10% glycerol, 1% β-mercaptoethanol, and 0.002% bromophenol blue. Polyacrylamide gel electrophoresis in SDS and subsequent fluorography for determination of translocated proteins can be carried out by a conventional method.

An example of the results of kinetic studies showing the involvements of ATP and $\Delta\bar{\mu}H^+$ in protein translocation is shown in Fig. 2 (Mizushima and Tokuda, 1990).

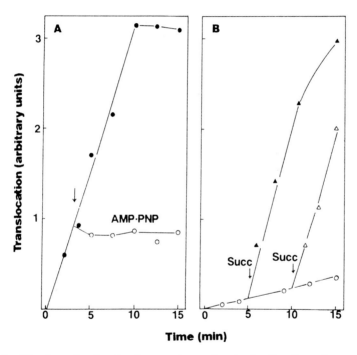

FIG. 2. Kinetic studies showing that protein translocation by *E. coli* membrane vesicles requires both ATP and $\Delta\bar{\mu}H^+$. Translocation of [^{35}S]Met-labeled proOmpF–Lpp was examined at 37°C in 25 μl of 50 mM potassium phosphate, pH 7.5, containing 2 mM MgSO$_4$, 1 mM ATP, and 5 μl of protein from everted membrane vesicles prepared from *E. coli* K003 ($\Delta uncB$-C). (A) The reaction mixture also contained 5 mM NADH (●) to generate $\Delta\bar{\mu}H^+$. After 3 minutes, AMP–PNP was added at a final concentration of 10 mM to inhibit ATP hydrolysis (○). The results indicate that hydrolysis of ATP is essential for protein translocation even in the presence of $\Delta\bar{\mu}H^+$. (B) Translocation was examined in the absence of respiratory substrates (○). As indicated by arrows, succinate (5 mM) was added after 5 (▲) or 10 (△) minutes to generate $\Delta\bar{\mu}H^+$. The results indicate that the rate of translocation was remarkably stimulated by the generation of $\Delta\bar{\mu}H^+$. From Mizushima and Tokuda (1990).

V. Methods for Identification and Characterization of Translocation Intermediates

After establishing of an efficient *in vitro* system for protein translocation, we were able to analyze the translocation process in detail and to identify and characterize translocation intermediates.

Although $\Delta\bar{\mu}H^+$ is not obligatory for the *in vitro* translocation of many presecretory proteins, including proOmpA (see Table I), the rate of translocation is significantly reduced in its absence. In its absence, we observed the transient accumulation, for a few minutes, of a translocation intermediate (Tani *et al.*, 1989). Our method is as follows.

The translation reaction mixture containing [^{35}S]proOmpA is gel filtered through a Sephadex G-75 column to remove small molecules, including NADH and ATP (see Section II,B,2,a). On use of the resulting filtrate as the substrate for the *in vitro* translocation reaction in the absence of $\Delta\bar{\mu}H^+$ (without NADH), a possible translocation intermediate accumulates for a few minutes, its level being maximal about 3 minutes after the initiation of the reaction. This intermediate is detectable on a polyacrylamide gel as a proteinase K-resistant band that accounted for 72% of the mature domain from the amino terminus.

This band does not appear in the absence of ATP or in the presence of AMP–PNP. Upon the addition of NADH, which energizes the membrane, the intermediate is converted to the translocated form of OmpA, even in the presence of AMP–PNP. These results indicate that ATP is absolutely required for the early stage, i.e., up to the stage of the intermediate accumulation, whereas $\Delta\bar{\mu}H^+$ is more critically required for the late stage of the translocation. It is also noticeable that the conversion of the intermediate, which no longer possesses the signal peptide, to the completely translocated OmpA most likely does not require ATP; namely, the process is not coupled with ATP hydrolysis. Geller and Green (1989) recently reached a similar conclusion.

Another form of translocation intermediate can be detected on a polyacrylamide gel in the absence of NADH when the disulfide bridge-possessing proOmpA is used (Tani *et al.*, 1990). OmpA possesses cysteine residues at positions of 290 and 302 from the amino terminus and these residues quantitatively form an intramolecular disulfide bridge when proOmpA is treated with ferricyanide. This oxidized proOmpA can be translocated completely in the presence of $\Delta\bar{\mu}H^+$, indicating that the disulfide bridge loop comprising 13 amino acid residues does not prevent proOmpA from the complete translocation. In the absence of $\Delta\bar{\mu}H^+$, on the other hand, the translocation is blocked at or near the disulfide bridge domain. The accu-

mulation of the disulfide bridge-containing intermediate is ATP dependent, whereas its conversion to the translocated mature form is not blocked
by the presence of AMP–PNP. Thus, the results with the disulfide bridge-
containing proOmpA also suggest that the early and late stages of the
translocation reaction differ in their requirement for ATP and $\Delta\bar{\mu}H^+$.

VI.   Overproductions and Purifications of Protein Components Involved in the Translocation Reaction

For extensive biochemical studies, large quantities of purified protein
components are usually essential. With recent progress in recombinant
DNA technology, cells overproducing certain proteins can now be constructed. But not all proteins can be overproduced: efforts to overproduce
certain proteins, especially those in membranes, have been unsuccessful,
most likely because their overproductions, even though at only severalfold
the normal level, are lethal.

A.   General Methods for Overproduction

Extensive genetic studies on the cellular components of *E. coli* required
for translocation of secretory proteins across the cytoplasmic membrane
have revealed the existence of six genes, *secA* (Oliver and Beckwith, 1981,
1982a), *secB* (Kumamoto and Beckwith, 1983), *secD* (Gardel *et al.*, 1987,
1990), *secE* (Riggs *et al.*, 1988), *secF* (Gardel *et al.*, 1990), and secY (Shiba
et al., 1984). Further extensive studies failed to increase the number of the
genes involved in the translocation process, suggesting that the proteins
encoded by these genes comprise the putative secretory machinery in *E. coli*
(Riggs *et al.*, 1988). SecB is a cytosolic protein that most likely participates
in the translocation of particular presecretory proteins and so is not considered to be a common component of the secretory machinery. The reader
must refer to other chapters for information on SecB.

A general way to construct an overproducing strain is (1) to clone the
relevant gene, (2) to place it under the control of a high-expression promoter on a plasmid, and (3) to transform *E. coli* cells with the plasmid.
It is advisable to use a controllable promoter to avoid unnecessary cell lysis
due to constant overproduction. The *tac* promoter–operator, which is commercially available, is certainly one of the best for this purpose. Use of a
runaway plasmid, whose replication is depressed at a high temperature,
is also advisable (Yasuda and Takagi, 1983).

We have so far succeeded in overproducing SecA (Kawasaki *et al.*, 1989), SecE (Matsuyama *et al.*, 1990b), SecD and SecF (S. Matsuyama, Y. Fujita, and S. Mizushima, unpublished), and SecY (Matsuyama *et al.*, 1990b) with the *tac* promoter–operator on the plasmid, as described in the next two sections.

B.   Overproductions of Sec Proteins

1.   OVERPRODUCTION OF SECA

A lambda phage clone, 15B10 (Kohara *et al.*, 1987), can be used as a source of the *secA* gene, encoding SecA. As outlined in Fig. 3, the *Hpa*I–*Bam*HI fragment carrying the entire *secA* gene except the promoter and the 5' terminus of the *secA* coding region is isolated and ligated with a synthetic oligonulcleotide linker. The linker consists of the upstream of the initiation codon on expression vector pUSI2 (Shibui *et al.*, 1988), containing a typical SD sequence (GGAGG), and the 5' terminus of the *secA* coding region. The ligated fragment is inserted downstream of the *tac* promoter–operator in pUSI2, which also possesses the *lacI* gene coding for the *lac* repressor, to construct the *tac–secA* gene.

Overproduction of SecA is achieved upon induction with IPTG of *E. coli* RR1 carrying pMAN400 constructed in this way (Kawasaki *et al.*, 1989). SecA constitutes more than 20% of the total cellular proteins synthesized under the conditions. Although SecA is a peripheral cytoplasmic membrane protein that is loosely attached to the membrane, the overproduced SecA is mainly localized in the cytosol.

2.   OVERPRODUCTION OF SECE

The *secE* gene has unique *Xma*III and *Kpn*I sites in its upstream and downstream regions, respectively (Downing *et al.*, 1990). A 0.79-kb DNA fragment containing the *Xma*III–*Kpn*I fragment is isolated from chromosomal DNA of MC4100 by agarose gel electrophoresis and then cloned into pMAN802, which was constructed from pBR322 through the conversion of the *Pvu*II site to a *Kpn*I site. The clones are then used to tranform a *secE* cold-sensitive mutant, PR520 (*secEcsE501*), and transformants that can grow at 20°C are obtained. Thus pMAN803 carrying the entire *secE* gene is obtained.

pMAN809 carrying the *tac–secE* gene and the *lacI* gene is then constructed (Matsuyama *et al.*, 1990b). The principle of the procedure is the same as that for the *tac–secA* gene shown in Fig. 3. *Escherichia coli* JM83 is then transformed with pMAN809 and SecE overproduction is observed

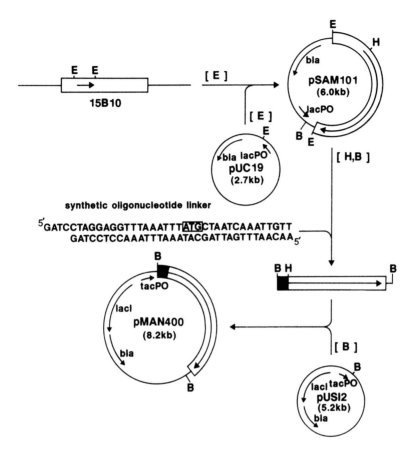

FIG. 3. Construction of pMAN400 carrying a highly expressible *secA* gene. The open box denotes chromosomal DNA. The arrow in the open box shows the coding region and the direction of transcription of the *secA* gene. The closed box denotes the synthetic oligonucleotide linker. The initiation codon ATG of the *secA* gene is boxed. The restriction endonucleases used are shown in parentheses with the following abbreviations: E, *Eco*RI; H, *Hpa*I; B, *Bam*HI. Cleavage sites are also shown. From Kawasaki *et al.* (1989).

upon the addition of IPTG. The overproduced SecE is localized in the cytoplasmic membrane as in cells that do not overproduce SecE. In overproducing cells, as much as 15% of the total cytoplasmic membrane protein is usually SecE, representing about a 500-fold increase in SecE production (Matsuyama *et al.*, 1990b).

3. OVERPRODUCTION OF SECD AND SECF

The *secD* gene and the *secF* gene constitute an operon with the gene order *secD–secF* (Gardel *et al.*, 1990). The *secD* and *secF* genes are cloned from the *E. coli* chromosome and placed under the control of the *tac* promoter–operator on pUSI2, which also possesses the *lacI* gene. Thus pMAN832 carrying the *tac–secD* gene and pMAN828 carrying the *tac–secF* gene are constructed (S. Matsuyama, Y. Fujita, and S. Mizushima, unpublished). The principle of the construction is the same as that for the *tac–secA* gene shown in Fig. 3. Each plasmid is then transferred into *E. coli* W3110 M25 (OmpT). Both SecD and SecF are overproduced in the membrane fraction in the presence of IPTG. The overproduced SecD and SecF usually amount to about 5% of the total membrane protein.

C. SecE-Dependent Overproduction of SecY

Genetic studies have strongly indicated the importance of SecY in translocation of secretory proteins across the cytoplasmic membrane (Ito *et al.*, 1983; Shiba *et al.*, 1984). Because of its potential importance, great efforts have been made to isolate this membrane protein in large quantity. None of such efforts has been successful, however. This has largely been due to difficulty in its overproduction, even with extensive use of gene engineering techniques. Rapid breakdown of pulse-labeled SecY has been observed (Akiyama and Ito, 1990), possibly explaining why it was unsuccessful. We found recently, however, that the overproduction of SecY can be achieved with simultaneous overproduction of SecE (Matsuyama *et al.*, 1990b). The method is summarized as follows.

The *secY* gene cloned in pNO1576 (Ito *et al.*, 1983) is engineered as in the case of SecA so that the gene is placed under the control of the *tac* promoter–operator. Thus, pMAN510 carrying the *tac–secY* gene is constructed. When this plasmid is transferred into *E. coli* JM83 together with pUSI2 carrying the *lacI* gene, no overproduction of SecY is observed even in the presence of IPTG. When pMAN510 is transferred into the same host strain together with pMAN809 carrying the *tac–secE* gene and the *lacI* gene, on the other hand, more than 40-fold overproduction of SecY in the cytoplasmic membrane is observed with concomitant overproduction (about 300-fold) of SecE. These findings strongly suggest that there is a firm interaction between these two membrane components involved in protein secretion and that the interaction stabilizes SecY in the membrane. The existence of such interaction has recently also been suggested genetically (Bieker and Silhavy, 1990) and biochemically (Brundage *et al.*, 1990). In these conditions, overproduced SecY accounts for about 6.4% of the cytoplasmic membrane protein.

SecY thus overproduced is rapidly degraded *in vitro*, especially after solubilization of the membrane with a detergent. The degradation is significantly suppressed when an *ompT* mutant lacking outer membrane protease OmpT is used as the host cell (Akiyama and Ito, 1990).

D.   Purifications of SecA, SecE, and SecY

The overproductions of components involved in the translocation of presecretory proteins have facilitated their purifications. Thus large-scale purifications of SecA (Kawasaki *et al.*, 1989; Cunningham *et al.*, 1989), SecE (Tokuda *et al.*, 1991), and SecY (Akimaru *et al.*, 1991) have been achieved. The purifications of SecD and SecF are in progress. As described in another chapter of this book, purification of a protein complex comprising SecE, SecY, and another protein has been reported (Brundage *et al.*, 1990).

1.   PURIFICATION OF SecA

SecA can easily be purified from a SecA-overproducing strain (Kawasaki *et al.*, 1989; Cunningham *et al.*, 1989). The following purification method was developed in our laboratory (Kawasaki *et al.*, 1989; Akita *et al.*, 1990).

Escherichia coli cells (RR1/pMAN400) grown in 4 liters of L broth containing 1 mM IPTG are disrupted through a French pressure cell (8000 psi, three times), and the supernatant obtained on centrifugation at 150,000 g for 2.5 hours is dialyzed against 10 mM Tris-acetate (pH 7.8) and 14 mM magnesium-acetate 60 mM potassium acetate and 1 mM dithiothreitol and is subjected to ammonium sulfate fractionation. The precipitate with 40–50% saturation is dialyzed against 10 mM sodium phosphate buffer (pH 7.2) and then applied to a hydroxylapatite column (2.5 × 20 cm) equilibrated with the same buffer. The column is washed with the buffer and then developed with a 300-ml linear gradient of 10–250 mM sodium phosphate (pH 7.2). Examination of individual fractions by SDS–polyacrylamide gel electrophoresis showed that SecA is eluted with about 180 mM sodium phosphate. Fractions containing SecA of the highest purity (more than 90%) are pooled and stored at − 80°C. The purified SecA is usually stable for several months at − 80°C. About 4 mg of the purified SecA can be obtained from the 4-liter culture.

2.   PURIFICATION OF SecE

The following method for purification of SecE was developed in our laboratory (Tokuda *et al.*, 1991). *Escherichia coli* W3110 M25 (*ompT*) cells (Sugimura, 1988) are transformed with pMAN809 carrying the *tac–secE* gene. The SecE-overproducing cells thus constructed are grown in 6 liters of L broth containing 1.5 mM IPTG. The cytoplasmic membrane fraction is isolated as described in Section III and is solubilized at a concentration of 1 mg/ml in 20 mM Tris-HCl (pH 7.5) containing 2.5 mg/ml of *E. coli* phospholipids and 2.5% *n*-octyl-β-D-glucopyranoside (octylglucoside) on ice for 10 minutes. The mixture is centrifuged at 50,000 rpm

(160,000 g) for 2 hours in a Beckman Ti70.1 rotor, and the supernatant fraction (20 ml), containing more than 95% of the total membrane proteins, is applied onto an anion exchange column, Mono Q (1 cm × 10 cm, Pharmacia), equilibrated with 20mM Tris-HCI (pH 7.5) and 2.5% octylglucoside. The column is developed with the same buffer at a flow rate of 4 ml/minute. Most of the SecE is not absorbed to the column, whereas most other proteins are retained. The SecE content of the pass-through fraction (6.3 mg protein) is usually about 30%. A portion of this fraction containing 1 mg of protein is concentrated and further fractionated by size-exclusion chromatography on a Superose 12HR column (1 cm × 30 cm, Pharmacia) equilibrated with 50 mM potassium phosphate (pH 6.95) containing 2.5% octylglucoside, 10% (w/v) glycerol, and 150 mM NaCI. The column is developed with the same buffer at a flow rate of 0.4 ml/min and fractions of 0.4 ml are collected. SecE is eluted as a single symmetrical peak with an apparent molecular mass of 40 kDa. The purity and yield of SecE after the Superose 12HR chromatography are usually about 80 and 70%, respectively.

3. Purification of SecY

E. coli W3110 M25 (*ompT⁻*) harboring both pMAN809 and pMAN510, which carries the *tac–secY* gene, was constructed for the simultaneous overproduction of SecE and SecY (Matsuyama *et al.*, 1990b). The overproduction is induced by growing the cells in 6 liters of L broth containing 1.5 mM IPTG. The cytoplasmic membrane fractions are prepared as described in Section III and solubilized at 1 mg protein/ml on ice with 2.5% octylglucoside containing 50 mM potassium phosphate (pH 6.95), 150 mM NaCl, 10% (w/v) glycerol, and 2.5 mg/ml *E. coli* phospholipids. After ultracentrifugation at 140, 000 g for 30 minutes in a Beckman TLA 100.3 rotor, the supernatant containing more than 95% of the total membrane proteins is applied on a Mono S, cation exchanger column (1 cm × 10 cm; Pharmacia), equilibrated with 2.5% octylglucoside containing 50 mM potassium phosphate (pH 6.95), 10% glycerol, and 150 mM NaCl. The column is then developed at the flow rate of 4 ml/minute with a linear gradient of NaCl in the same buffer. The amount of SecY in each fraction (2 ml) is determined by SDS–polyacrylamide gel electrophoresis followed by immunoblotting with anti-SecY antiserum. SecY is preferentially eluted immediately after the pass-through fraction, in which SecE and most of the other proteins appear. The fraction, which contains most of the SecY, is concentrated by means of membrane filtration. An aliquot (0.5 ml) of the concentrated fraction is further purified by size exclusion chromatography on a Superose 12 HR column (1 cm × 30cm; Pharmacia), equilibrated with 2.5% octylglucoside containing 50 mM potassium phosphate (pH 6.95), 10% glycerol, and 150 mM NaCl. The column is developed with the same buffer at the flow rate of 0.4 ml/minute. Size exclusion chromatography is performed ten to twenty times, and fractions containing SecY with a purity of about 70% are combined and concentrated.

VII.   Methods for Analysis of SecA Functions

SecA is a peripheral cytoplasmic membrane protein that is essential for the translocation of secretory proteins both *in vivo* (Oliver and Beckwith, 1981, 1982b) and *in vitro* (Kawasaki *et al.*, 1989; Cunningham *et al.*, 1989; Cabelli *et al.*, 1988). It is a translocation ATPase (Lill *et al.*, 1989) that interacts with both ATP and presecretory proteins (Cunningham *et al.*, 1989; Akita *et al.*, 1990; Matsuyama *et al.*, 1990a). Interactions between SecA and membrane vesicles/liposomes have also been desmonstrated (Cunningham *et al.*, 1989; Lill *et al.*, 1990). These interactions resulted in conformational change of the SecA molecule (Shinkai *et al.*, 1991). The possible interaction of SecA with SecY (Pandl *et al.*, 1988; Cunningham *et al.*, 1989) has also been suggested. Furthermore, evidence suggests the implication of $\Delta \bar{\mu} H^+$ in the function of SecA (Yamada *et al.*, 1989b). Thus, SecA most likely plays a central role in the translocation process, especially in an early step. Another prominent feature of SecA is that it is a large molecule, being a homodimer of a subunit protein with a molecular mass of 102 kDa (Akita *et el.*, 1991).

Several methods have been developed to study the function of this interesting large molecule. Here we describe these methods and stress those that we have developed. For further information the reader should also refer to other chapters dealing with SecA.

A.   SecA-Dependent *in Vitro* Translocation Assay

Both the everted membrane vesicles and the cytosolic fraction prepared by conventional methods contain substantial amounts of SecA, so the rate of translocation is usually not increased significantly by the addition of SecA. For studies of the role of SecA in the translocation reaction, therefore, the amount of SecA in the original reaction mixture should be reduced. Two methods have been used for this purpose. One is the use of a mutant defective in SecA synthesis to prepare both the membrane vesicles and the cytosolic fraction (Oliver and Beckwith, 1982b; Cabelli *et al.*, 1988), and the other is the removal of SecA from the membrane vesicles by means of urea washing (Kawasaki *et al.*, 1989; Cunningham *et al.*, 1989). An example of the latter method is described below.

SecA can be removed from everted membrane vesicles by urea treatment. Purified membrane vesicles (2 mg of protein) are treated with 0.5 ml of 50 mM Tris-acetate (pH 7.8), 1 mM dithiothreitol, and 2% glycerol containing various concentrations of urea for 30 minutes on ice, recovered by centrifugation on a Beckman TLA 100.3 rotor at 100,000 rpm for 30 minutes, and washed twice with 1 ml of the same buffer containing urea and then with 1 ml of 50 mM postassium phosphate buffer (pH 7.6). The pelleted membrane vesicles are suspended in 0.5 ml of the same buffer and stored at $-80°C$.

Western blotting analysis with anti-SecA IgG revealed that about 1/2, 3/4, and 6/7 of the SecA in the membrane vesicles were released upon treatment with 2.5, 4, and 6 M urea, respectively (Shinkai *et al.*, 1991). Essentially no translocation is observed when the 6 M urea-treated membrane vesicles are used unless a sufficient amount of SecA is supplemented. It should be noted that the rate of translocation into these membrane vesicles is appreciably slower than that into untreated ones, even when a sufficient amount of SecA is added.

B.   Assay of Translocation ATPase Activity of SecA

SecA exhibits ATPase activity that is significantly enhanced in the presence of both a presecretory protein and everted membrane vesicles (Lill *et al.*, 1989, 1990). The release of orthophosphate from ATP can be monitored using [γ-^{32}P]ATP or by photometric determination. An example of the ATPase assay is as follows (Lill *et al.*, 1990).

In a typical ATPase assay (50 μl), 20 μg/ml cytoplasmic membrane vesicles and 40 μg/ml proOmpA are used in buffer B [50 mM HEPES-KOH (pH 7.0), 30 mM KCl, 30 mM NH$_4$Cl, 1 mM dithiothreitol, and 5 mM magnesium acetate]. The components are mixed at 0°C, then samples are incubated at 40°C for 40 minutes before analysis for released inorganic phosphate by the photometric method of Lanzetta *et al.*, (1979). The reactions are stopped by the subsequent addition of 800 μl of color reagent (0.034% malachite green and 10.5 g/liter ammonium molybdate in 1 N HCl and 0.1% Triton X-100) and 100 μl of 34% citric acid. After 40 minutes at room temperature, absorption is measured at 660 nm and compared with a standard curve prepared with a phosphate salt.

SecA alone exhibits weak ATPase activity, but the activity is 100-fold higher in the presence of both proOmpA and membrane vesicles (Lill *et al.*, 1989). It is highly likely, therefore, that the observed ATPase activity is coupled to the translocation reaction.

C. Cross-Linking Studies

The interaction of SecA with ATP or a presecretory protein can be demonstrated by means of chemical cross-linking. In the case of ATP, the direct cross-linking of SecA with either [α-^{32}P]ATP or azido-ATP by means of photoaffinity labeling has been achieved (Matsuyama *et al.*, 1990a; Lill *et al.*, 1989). The cross-linking with presecretory proteins can be demonstrated with the aid of chemical cross-linkers (Akita *et al.*, 1990). Examples of such cross-linking experiments are described below.

1. PHOTOAFFINITY CROSS-LINKING OF SECA WITH [α-^{32}P]ATP

The general method of photoaffinity cross-linking of ATP has been described by Yue and Schimmel (1977). The reaction mixture for photo-affinity cross-linking of ATP with SecA (Matsuyama *et al.*, 1990a) consists of 50 mM Tris-acetate (pH 7.5), 100 mM potassium acetate, 2 mM magnesium acetate, 1 mM dithiothreitol, 0.1 μM [α-^{32}P]ATP (3000 Ci mmol^{-1}), and 30 pmol of SecA previously dialyzed against buffer of the same composition. The reaction mixture is incubated for 15 minutes at 0°C and then subjected to photoaffinity cross-linking for 40 minutes at 0°C using a 254-nm lamp (Transilluminator TS15 UVP Inc.) at a distance of 2 cm. UV irradiation under these conditions results in extensive inactivation of the translocation activity of SecA. Samples are then analyzed by SDS–polyacrylamide gel electrophoresis. The gels are dried and then exposed to an X-ray film, using an intensifying screen, for 12 hours at −80°C to detect the [^{32}P]ATP–SecA complex.

2. CROSS-LINKING OF SECA WITH PRESECRETORY PROTEINS

The direct cross-linking of SecA with presecretory proteins can be demonstrated by means of cross-linking (Akita *et al.*, 1990). Several cross-linkers may be used, but we obtained the best results with 1-ethyl-3-(3-dimethyl-aminopropyl)carbodiimide (EDAC). The cross-linking is signal peptide dependent and increases with increase in the number of positively charged amino acid residues at the amino-terminal region of the signal peptide, irrespective of the species of amino acid residues donating the charge. This positive charge effect is essentially the same as that for the entire reaction of protein translocation. The participation of the signal peptide region in SecA–presecretory protein interaction has also been demonstrated (Lill *et al.*, 1990).

Cross-Linking Procedures.   SecA (2.5 μg in 2 μl of 150 mM potassium phosphate, pH 7.2) is mixed with 2.5 μl of H$_2$O and 7.5 μl of gel-filtered translation mixture containing ^{35}S-labeled presecretory protein (1.5 × 10[5] dpm; proOmpF–Lpp or proapolipoprotein in this experiment), followed by incubation at 25°C for 10 minutes. Then 3 μl of 12.5 mM EDAC is added, and incubation is continued for a further 50 minutes for cross-linking. The cross-linking reaction is quenched by the addition of 200 mM Tris-HCl (pH 7.5) for 10 minutes, and cross-linked complexes are then analyzed by SDS–polyacrylamide gel electrophoresis, followed by fluorography.

Cross-lining is enhanced in the presence of ATP and competitively suppressed by other secretory proteins such as proOmpA, but not by its mature domains (Kimura *et al.*, 1991).

D.   Denaturation and Renaturation of SecA

Although SecA is a large molecule, being a homodimer (Akita *et al.*, 1991) of a protein subunit with a molecular mass of 102 kDa, it can be reversibly denatured and renature almost quantitatively (Shinkai *et al.*, 1990). This fact has been applied for reconstitution of SecA analogs from truncated fragments (Fig. 4) (Matsuyama *et al.*, 1990a), which have been

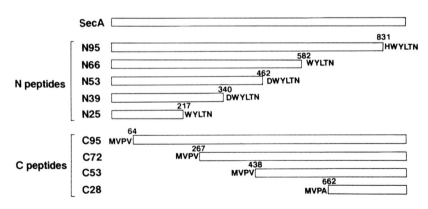

Fig. 4.   Structures of truncated SecA proteins. Open boxes indicate segments derived from SecA. Letters adjacent to open boxes denote amino acid residues derived from the vector sequence. The one-letter abbreviations for amino acids are: A, Ala; D, Asp; H, His; L, Leu; M, Met; N, Asn; P, Pro; T, Thr; V, Val; W, Trp; Y, Tyr. The amino acid residues at the truncation positions in SecA are numbered from the amino terminus. From Matsuyama *et al.* (1990a).

used to determine the sites in the SecA molecule that interact with ATP and presecretory proteins (Matsuyama *et al.*, 1990a; Kimura *et al.*, 1991).

RENATURATION OF SECA FROM GUANIDINE–HCl SOLUTION

The renaturation conditions are basically the same as those described previously for egg white lysozyme (Saxena and Wetlaufer, 1970). A solution of 1 nmol of SecA in 80 μl of denaturation buffer (6 M guanidine-HCl and 50 mM Tris-acetate, pH 7.5) is diluted with 1.2 ml of dilution buffer (50 mM Tris-acetate, pH 7.5, 100 mM potassium acetate, 3 mM reduced form of glutathione, and 0.3 mM oxidized form of glutathione). The solution is allowed to stand for 1 hour at 4°C and is then dialyzed against the dialysis buffer (50 mM Tris-acetate, pH 7.5, 100 mM potassium acetate, and 1 mM dithiothreitol). The dialyzed sample is concentrated by sprinkling Bio-Gel P-300 powder (BioRad) over the dialysis tubing.

E. Reconstitution of SecA Analogs from Truncated Fragments

As described previously, SecA is a large molecule that play a central role possibly in the early step of the translocation reaction by interacting with several components involved in the translocation reaction. For elucidation of the roles of submolecular domains of SecA in individual functions, several *secA* deletion mutants were constructed (Matsuyama *et al.*, 1990a). The *secA* gene was engineered so as to code for SecA fragments of different sizes, from either the amino or carboxyl terminus. SecA fragments encoded by these truncated genes were used in combination with the renaturation technique for determination of the SecA domains involved in interactions with ATP and presecretory proteins.

1. PREPARATION OF SECA FRAGMENTS

a. Construction of secA Deletions.   The *secA* gene is engineered so as to code for SecA fragments of different sizes, from either the amino terminus or the carboxyl terminus. The resulting genes are placed under the control of the *tac* promoter–operator. Details of the construction of these mutants are described elsewhere (Matsuyama *et al.*, 1990a). The structures of the SecA fragments encoded by these genes are shown in Fig. 4, with nomenclature.

b. Overproduction of SecA Fragments.   *Escherichia coli* RRl is transformed with a plasmid carrying a truncated *secA* gene and then grown in L broth in the presence of 1 mM IPTG. Usually, a 3-hour induction is sufficient for overproduction.

   c. Purification of SecA Fragments.   Two SecA fragments, N95 and C28, are purified from the cytosolic solution by ammonium sulfate fractionation and hydroxylapatite chromatography by the same method as for purification of SecA. Other SecA fragments, which are overproduced as inclusion bodies, can be quickly purified as follows. A cell lysate is prepared by sonication and centrifuged at 10,000 g at 4°C for 5 minutes. Inclusion bodies form larger aggregates during sonication and can be recovered by this low-speed centrifugation. The resultant precipitate is washed twice with 10 mM Tris-acetate, pH 7.5 and is then solubilized in 50 mM Tris-acetate, pH 7.5 and 6 M guanidine-HCl, and insoluble materials are removed by centrifugation at 350,000 g at 4°C for 30 minutes. The resultant supernatant was then used as a preparation of purified SecA fragments. In most cases, several milligrams of SecA fragments can be obtained from 100 ml of cell cultures with a purity of about 60–80%.

2.   Reconstitution of SecA Analogs

   The method used for reconstitution of SecA analogs from SecA fragments (N- and C-peptides) is the same as that for renaturation of intact SecA from the guanidine-HCl solution described in Section VII,D; namely, 1 nmol each of two SecA fragments, N- and C-peptides, is mixed in 80 μl of denaturation buffer (6 M guanidine-HCl and 50 mM Tris-acetate, pH 7.5), and diluted with 1.2 ml of dilution buffer (50 mM Tris-acetate, pH 7.5, 100 mM potassium acetate, 3 mM reduced form of glutathione, and 0.3 mM oxidized form of glutathione). The mixture is allowed to stand for 1 hour at 4°C and is then dialyzed against the dialysis buffer (50 mM Tris-acetate, pH 7.5, 100 mM potassium acetate, and 1 mM dithiothreitol). The dialyzed sample is concentrated by sprinkling Bio-Gel P-300 powder (BioRad) over the dialysis tubing.

3.   Use of Reconstituted SecA Analogs for Determination of Functional Domains in the SecA Molecule

   When reconstitution is performed with both N- and C-peptides that are large enough to complement each other structurally, the resultant SecA analogs exhibit some SecA functions, although they are nearly inactive in supporting the entire translocation reaction: namely, they are active as to cross-linking with ATP (Matsuyama *et al.*, 1990a) and with presecretory proteins (Kimura *et al.*, 1991). Thus we could determine the loci in the SecA molecule that interact with these compounds as described below.

a.   Determination of the ATP-Binding Site.   N- and C-peptides are mixed in different combinations under the reconstitution conditions. Photoaffinity cross-linking of $[\alpha\text{-}^{32}P]ATP$ is then carried out as described in Section VII,C,1, and the cross-linked samples are subjected to SDS–polyacrylamide gel electrophoresis to identify $[\alpha\text{-}^{32}P]ATP$-labeled peptides.

The N-peptides were always found to be cross-linked with ATP in the concomitant presence of a carboxyl-terminal fragment that was large enough to cover the region deleted from an N-peptide. From the results of a series of these experiments, we conclude that a domain within the first 217 amino acid residues from the amino-terminus of SecA is the region for ATP binding. This region contains a sequence (^{102}MetArgThrGlyGluGly-LysThr^{109}) that is somewhat homologous to the consensus sequence for ATP-binding segments (Gill *et al.*, 1986). The binding is specific in that (1) it takes place only at the amino-terminal region, (2) the site of interaction revealed by means of protease digestion was the same in both the intact SecA and reconstituted derivatives, and (3) the interaction with the amino-terminal fragments takes place only in the presence of particular carboxyl-terminal fragments. We conclude, therefore, that the interaction we observed is physiologically important.

b.   Determination of the Site of Interaction with Presecretory Proteins. Essentially the same procedures can be used to determine the site in SecA that interacts with presecretory proteins (Kimura *et al.*, 1991). ProOmpF–Lpp, a model presecretory protein, is used. The cross-linking of reconstituted SecA analogs with $[^{35}S]$proOmpF–Lpp is carried out using EDAC as a cross-linker as described in Section VII,C,2. The reconstituted SecA analogs are active in cross-linking with proOmpF–Lpp when the SecA fragments used are large enough to complement each other structurally. The cross-linking is enhanced in the presence of ATP, is signal peptide-dependent, and is suppressed in the presence of other presecretory proteins, indicating that the observed interaction is physiological.

The SecA fragments that cross-linked with proOmpF–Lpp are then analyzed on SDS–polyacrylamide gels. Cross-linking preferentially takes place on N-peptides, except N25. Weak cross-linking is also observed with sufficiently large C-peptides. From these results, the region responsible for cross-linking with presecretory proteins was deduced to be located between amino acid residues 267 and 340 from the amino-terminus of SecA.

It should be noted that the site of cross-linking does not necessarily mean the site of interaction in a general sense. In the present work, we used EDAC, which mediates peptide bond formation between the carboxyl group of one amino acid residue and the amino group of another as a cross-linker. EDAC-mediated cross-linking possibly takes place only between residues that are very close to each other.

VIII. Reconstitution of a Protein Translocation System

Several proteins present in the cytoplasmic membrane of *E. coli* have been suggested, mainly from the genetic studies, to be involved in protein translocation across the membrane. These are SecY (Ito *et al.*, 1983; Shiba *et al.*, 1984), SecE (Riggs *et al.*, 1988), and SecD and SecF (Gardel *et al.*, 1990). For elucidation of the overall mechanism of protein translocation, the membrane components necessary for protein translocation and characterization of their roles must be identified. Reconstitution of a protein translocation system is, therefore, essential.

Various membrane proteins have been reconstituted into liposomes by using the octylglucoside dilution method of Racker *et al.* (1979). The lactose carrier protein thus reconstituted has been analyzed at a molecular level (Kaback, 1988). The method is useful for simple and rapid examination of reconstituted samples. We have applied this method in studies on reconstitution of a protein translocation system (Tokuda *et al.*, 1990). A similar system has been developed by Driessen and Wickner (1990).

A. Materials

Cyctoplasmic membrane vesicles containing SecE/SecY at the normal or overproduced level are prepared from *E. coli* cells as described previously. SecE and SecY purified as described previously are also used for reconstitution. Phospholipids are prepared from *E. coli* MC4100 by the method of Viitanen *et al.* (1986). Acetone-insoluble/ether-soluble phospholipids are dried in a rotary evaporator and suspended in $2 \text{ m}M$ β-mercaptoethanol at a concentration of 100 mg/ml and are bath sonicated until a clear suspension is obtained. The liposomes thus formed are divided into small aliquots, flushed with N_2 gas, sealed in air-tight tubes, and kept at $-80°C$ until use. Frozen liposomes are thawed at room temperature and again sonicated to clarity before use for reconstitution. Octylglucoside is dissolved in $50 \text{ m}M$ potassium phosphate (pH 7.5) at a concentration of 15% (w/v) and is kept at $-20°C$.

B. Solubilization of Cytoplasmic Membrane Vesicles

The reported conditions for the solubilization of crude membrane fractions (Tokuda *et al.*, 1990) are slightly modified. The cytoplasmic membrane vesicles are solubilized at a concentration of 1 mg protein/ml in $50 \text{ m}M$ potassium phosphate (pH 7.5) containing $1 \text{ m}M$ dithiothreitol,

10% (w/v) glycerol, 150 mM NaCl, 2.5 mg/ml $E.$ $coli$ phospholipids, and 2.5% octylglucoside (Tokuda et $al.,$ 1991). Octylglucoside is added last and the mixture is kept on ice for 10 minutes. The mixture is then centrifuged at 50,000 rpm (140,000 g) for 30 minutes in a Beckman TLA 100.3 rotor, and the supernatant is used for reconstitution.

C. Reconstitution

1. RECONSTITUTION WITH UNFRACTIONATED MEMBRANE PROTEINS

When the octylglucoside supernatant prepared from membranes containing overproduced amounts of SecE and SecY is used, the supernatant may be diluted (H. Tokuda, J. Akimaru, K. Nishiyama, Y. Kabuyama, S. Matsuyama, and S. Mizushima, unpublished). Dilution should be made with the phospholipid-containing solution used for solubilization of membrane vesicles. For reconstitution, the octylglucoside supernatant is mixed with one-eighth volume of 100 mg/ml phospholipids and 1/40 volume of 15% octylglucoside, incubated on ice for 20 minutes, and then diluted 40-fold with 50 mM potassium phosphate (pH 7.5) containing 150 mM NaCl. The diluted sample is incubated at room temperature for 5 minutes. Reconstituted proteoliposomes thus formed are then collected by centrifugation at 50,000 rpm (160,000 g) for 2 hours at 4°C and are suspended at the original volume of supernatant in 50 mM potassium phosphate (pH 7.5) containing 150 mM NaCl. The proteoliposome suspension is then rapidly frozen in dry ice–ethanol and thawed at room temperature. The slightly turbid suspension thus obtained is sonicated briefly (less than 5 seconds) in a Branson bath-type sonicator Model B-12 to make it more transparent. The sonicator bath should contain 0.02% Triton X-100 and the level of water should be adjusted to achieve maximal agitation. The proteoliposomes thus obtained are promptly used for the translocation assay.

2. RECONSTITUTION WITH PURIFIED Sec PROTEINS

Purified SecE and SecY are mixed together at a final concentration of 10–20 μg each per ml in 100 μl of 50 mM potassium phosphate (pH 6.95) containing 150 mM NaCl, 10% glycerol, 1.25 mg $E.$ $coli$ phospholipids, and 2.5% octylglucoside (Akimaru et $al.,$ 1991). After 20 minutes incubation on ice, the proteoliposomes are reconstituted by using the octylglucoside dilution method as described previously.

D.   Protein Translocation into Reconstituted Proteoliposomes

Aliquots (15 μl) of the reconstituted proteoliposomes, from either un-fractionated proteins or purified Sec proteins, are mixed with 1 μl of 1.5 mg/ml SecA and preincubated at 37°C for 3 minutes. The reaction is started by the addition of 5 μl of 50 mM potassium phosphate (pH 7.5) containing 150 mM NaCl, 10 mM MgSO$_4$, 10 mM ATP, 50 mM creatine phosphate, and 6.25 μg (5 units) of creatine kinase (EC 2.7.3.2) and 4 μl of [^{35}S]methionine-labeled presecretory protein (2 × 10^5 dpm), which has been partially purified by gel filtration. Translocation at 37°C is terminated by transferring the reaction mixture into an ice-water bath, with simul-taneous addition of 5 μl of 5 mg/ml proteinase K. After 30 minutes of treatment with proteinase K on ice, the translocated protein can be de-tected as a proteinase K-resistant band on an SDS–polyacrylamide gel by fluorography as described in previous sections.

E.   Comments

Proteoliposomes reconstituted from the unfractionated proteins and those from purified SecE and SecY can take up more than 50% and 30%, respectively, of the input proOmpA D26 during incubation for 15 minutes at 37°C. Proteoliposomes are also appreciably active against the intact proOmpA. An ATP-generating system composed of creatine phosphate and creatine kinase considerably stimulates the rate of translocation (Tokuda *et al.,* 1990). Both ATP and SecA are obligatory for the trans-location reaction. In the case of reconstitution from purified proteins, no translocation activity appears when any one of SecE, SecY, and SecA is omitted, indicating that these are essential components of protein trans-location machinery. As described by Viitanen *et al.* (1986) as to reconstitu-tion of the lactose carrier protein, we noticed that freezing–thawing and subsequent brief sonication of proteoliposomes increased the reproducibil-ity of reconstitution.

IX.   Preparation of Fusion Proteins Composed of a Truncated SecA and a Component Involved in Protein Translocation—Possible Use for Preparation of Antisera

Many of the truncated SecAs, especially those that lack the amino-terminal region, form inclusion bodies in the cells upon overproduction, and hence can be easily purified (Matsuyama *et al.,* 1990a). Furthermore, the overproduction usually does not interfere with the cell growth. Taking advantage of these particular features, we have developed a rapid method for a large-scale preparation of membrane proteins, involved in protein translocation, in the form fused to the downstream of a truncated SecA (S. Matsuyama, and S. Mizushima, unpublished). Overproduction of a membrane protein alone often results in a severe growth inhibition.

A.   Construction of a Plasmid Vector for Expression of Proteins Fused to a Truncated SecA

Plasmid pMAN789-C95 (Matsuyama *et al.,* 1990a), carrying 95% of the carboxyl terminal of the coding region of the *tac–secA* gene, is digested with *Nco*I and *Sac*I. The 3′ terminus of the coding region corresponding to the five carboxyl-terminal amino acid residues is removed by this treatment. The *Nco*I–*Sac*I large fragment thus obtained is then ligated with a synthetic DNA linker [d(5′CATGGTCCGGTTGGTCCGGTTGG-TCCCGGGAGCT-3′) and d(3′-CAGGCCAACCAGGCCAACCAGG-GCCC-5′)] to construct pMAN811 (Fig. 5). On this plasmid, the three amino acid residues at the carboxyl terminus of the *tac–secA* gene is replaced by Pro-Val-Gly-Pro-Val-Gly-Pro-Gly-Ser (with no termination codon), which possesses a cleavage site for collagenase (Harper, 1970) and a *Sma*I site.

B.   Insertion of Foreign Genes into the 3′ Terminus of the Truncated *tac–secA* Gene to Overproduce Proteins Fused to SecA-C95

In general, foreign genes are fused in frame by blunt-end ligation to the *Sma*I site in the linker region (Fig. 5).

1.   SECA-C95–SECY FUSION

SecY alone cannot be overproduced, probably because of its rapid degradation, but it can be overproduced during simultaneous overproduction of SecE (Matsuyama *et al.,* 1990b). A *secY* gene-carrying plasmid is digested with *Dra*I and *Sal*I to isolate a DNA fragment containing the entire *secY* gene. The fragment is treated with T4 DNA polymerase and then inserted into the *Sma*I site of pMAN811 by blunt-end ligation to construct pMAN812. The resulting SecA-C95–SecY fusion product encoded by the hybrid gene possesses Lys-Phe at the fusion point. The fusion protein forms inclusion bodies as SecA-C95 does, when plasmid-containing cells are induced with IPTG. The protein is, therefore, easily purified on a

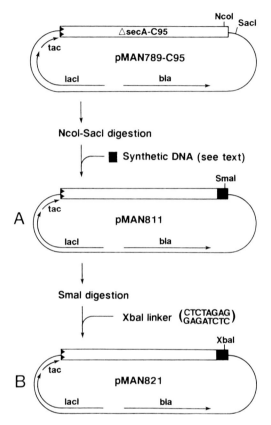

Fig. 5. Construction of (A) a plasmid vector (pMAN811) that can be used for construction of a *tac–secA-C95–X* gene used for overproduction and rapid purification of protein X fused to the carboxyl terminus of SecA-C95, an amino-terminal deletion of SecA, and (B) pMAN821, a SecA-C95-overproducing plasmid.

large scale as described in Section VII,E,2. About 5 mg of purified fusion protein is obtained from 100 ml of culture.

2. SecA-C95–SecD Fusion

SecD can be easily purified in large quantity as a form fused to SecA-C95. The *secD* gene possesses a *Bam*HI site at the coding region for Asp[32]. The *Bam*HI–*XMa*III fragment covering most of the coding region for SecD is isolated from pCG169 (Gardel *et al.*, 1987), treated with T4 DNA polymerase, and inserted into the *Sma*I site of pMAN811 to construct pMAN820D carrying the gene encoding the SecA-C95–SecD fusion protein. The fusion protein, which also forms inclusion bodies, can be purified by the same method as for SecA-C95–SecY in a yield of about 5 mg per 100 ml of culture.

3. SecA-C95–SecF Fusion

The *secF* gene has a unique *Xma*III site in the coding region for Gly[13]. The *Nsp*V–*Hin*dIII large fragment is isolated from pCG169, treated with T4 DNA polymerase, and ligated with the aid of the *Xba*I linker to construct pMAN813. The *Xma*III–*Xba*I fragment covering most of the coding region for SecF is then isolated from pMAN813, treated with T4 DNA polymerase, and inserted into the *Sma*I site of pMAN811 to construct pMAN820F carrying the gene coding for SecA-C95–SecF fusion, which is also overproduced as an inclusion body upon IPTG induction. About 5 mg of purified fusion protein can be obtained from 100 ml of culture. The method of purification is the same as described above.

C. Use of the Fusion Proteins as Antigens

Fusion proteins constructed and purified as described above can be used directly as antigens to raise antibodies against SecA, SecY, SecD, and SecF. It is advisable, however, to purify them further by SDS–polyacrylamide gel electrophoresis followed by electroelution.

For preparation of antibodies specific to either SecY, SecD, or SecF, anti-SecA antibodies raised simultaneously should be eliminated by passing the antisera through a SecA-immobilized column. For this purpose, we have constructed pMAN821 carrying the *tac–secA-C95* gene possessing a translation termination codon in the linker region at the 3' terminus as shown in Fig. 5. SecA-C95, which is overproduced upon IPTG induction, forms inclusion bodies and so can be purified rapidly.

Acknowledgments

We would like to thank Iyoko Sugihara for secretarial support. This work was supported by grants from the Ministry of Education, Science, and Culture of Japan (61060001, 02404013, and 02680153).

References

Akimaru, J., Matsuyama, S., Tokuda, H., and Mizushima, S. (1991). *Proc. Natl. Acad. Sci. U.S.A.* (in press).
Akita, M., Sasaki, S., Matsuyama, S., and Mizushima, S. (1990). *J. Biol. Chem.* **265,** 8164–8169.

Akita, M., Shinkai, A., Matsuyama, S., and Mizushima, S. (1991). *Biochem. Biophys. Res. Commun.* **174**, 211–216.

Akiyama, Y., and Ito, K. (1990). *Biochem. Biophys. Res. Commun.* **167**. 711–715.

Beckwith, J., and Ferro-Novick, S. (1986). *Curr. Top. Microbiol. Immunol.* **125**, 5–27.

Bieker, K. L., and Silhavy, T. J. (1990). *Cell (Cambridge, Mass.)* **61**, 833–842.

Bochkareva, E. S., Lissin, N. M., and Girshovich, A. S. (1988). *Nature (London)* **336**, 254–257.

Bosch, D., DeBoer, P., Bitter, W., and Tommassen, J. (1989). *Biochim. Biophys. Acta* **979**, 69–76.

Brundage, L., Hendrick, J. P., Schiebel, E., Driessen, A. J. M., and Wickner, W. (1990). *Cell (Cambridge, Mass.)* **62**, 649–657.

Cabelli, R. J., Chen, L., Tai, P. C., and Oliver, D. B. (1988). *Cell (Cambridge, Mass.)* **55**, 683–692.

Chen, L., and Tai, P. C. (1985). *Proc. Natl. Acad. Sci. U.S.A.* **82**, 4384–4388.

Crooke, E., Brundage, L., Rice, M., and Wickner, W. (1988a). *EMBO J.* **7**, 1831–1835.

Crooke, E., Guthrie, B., Lecker, S., Lill, R., and Wickner, W. (1988b). *Cell (Cambridge, Mass.)* **54**, 1003–1011.

Cunningham, K., Lill, R., Crooke, E., Rice, M., Moore, K., Wickner, W., and Oliver, D. (1989). *EMBO J.* **8**, 955–959.

Downing, W. L., Sullivan, S. L., Gottesman, M. E., and Dennis, P. P. (1990). *J. Bacteriol.* **172**, 1621–1627.

Driessen, A. J. M., and Wickner, W. (1990). *Proc. Natl. Acad. Sci. U.S.A.* **87**, 3107–3111.

Futai, M. (1974). *J. Membr. Biol.* **15**, 15–28.

Gardel, C., Benson, S., Hunt, J., Michaelis, S., and Beckwith, J. (1987). *J. Bacteriol.* **169**, 1286–1290.

Gardel, C., Johnson, K., Jacq. A., and Beckwith, J. (1990). *EMBO J.* **9**, 3209–3216.

Geller, B. L., and Green, H. M. (1989). *J. Biol. Chem.* **264**, 16465–16469.

Geller, B. L., Movva, N. R., and Wickner, W. (1986). *Proc. Natl. Acad. Sci. U.S.A.* **83**, 4219–4222.

Gill, D. R., Hatfull, G. F., and Salmond, G. P. C. (1986). *Mol. Gen. Genet.* **205**, 134–145.

Harper, E. (1970). *In* "Collagenase" (I. Mandel, ed.), pp. 19–25. Gordon & Breach, New York.

Hartl, F.-V., Lecker, S., Schiebel, E., Hendrick, J. P., and Wickner, W. (1990). *Cell (Cambridge, Mass.)* **63**, 269–279.

Hertzberg, E. L., and Hinkle, P. C. (1974). *Biochem. Biophys. Res. Commun.* **58**, 178–184.

Hirashima. A., Wu, H. C., Venkateswaren, P. S., and Inouye, M. (1973). *J. Biol. Chem.* **248**, 5654–5659.

Iino, T., Takahashi, M., and Sako, T. (1987). *J. Biol. Chem.* **262**, 7412–7417.

Inouye, S., Soberon, X., Franceschini, T., Nakamura, K., Itakura, K., and Inouye, M. (1982). *Proc. Natl. Acad. Sci. U.S.A.* **79**, 3438–3441.

Ito, K., Wittekind, M., Nomura, M., Shiba, K., Yura, T., Miura, A., and Nashimoto, H. (1983). *Cell (Cambridge, Mass.)* **32**, 789–797.

Kaback, H. R. (1988). *Annu. Rev. Physiol.* **50**, 243–256.

Kawasaki, H., Matsuyama, S., Sasaki, S., Akita, M., and Mizushima, S. (1989). *FEBS Lett.* **242**, 431–434.

Kimura, E., Akita, M., Matsuyama, S., and Mizushima, S. (1991). *J. Biol. Chem.* **266**, 6600–6606.

Kohara, Y., Akiyama, K., and Isono, K. (1987). *Cell (Cambridge, Mass.)* **50**, 495–508.

Kumamoto, C., and Beckwith, J. (1983). *J. Bacteriol.* **154**, 254–260.

Lanzetta, P. A., Alvarez, L. J., Reinach, P. S., and Candia, O. A. (1979). *Anal. Biochem.* **100**, 95–97.

Lill, R., Crooke, E., Guthrie, B., and Wickner, W. (1988). *Cell (Cambridge, Mass.)* **54,** 1013–1018.

Lill, R., Cunningham, K., Brundage, L. A., Ito, K., Oliver, D., and Wickner, W. (1989). *EMBO J.* **8,** 961–966.

Lill, R., Dowhan, W., and Wickner, W. (1990). *Cell (Cambridge, Mass.)* **60,** 271–280.

Lu, H. M., Yamada, H., and Mizushima, S. (1991). *J. Biol. Chem.* (in press).

Matsuyama, S., and Mizushima, S. (1989). *J. Biol. Chem.* **264,** 3583–3587.

Matsuyama, S., Kimura, E., and Mizushima, S. (1990a). *J. Biol. Chem.* **265,** 8760–8765.

Matsuyama, S., Akimaru, J., and Mizushima, S. (1990b). *FEBS Lett.* **269,** 96–100

Melton, D. A., Krieg, P. A., Rebagliati, M. R., Maniatis, T., Zinn, K., and Green, M. R. (1984). *Nucleic Acids Res.* **12,** 7035–7056

Mizushima, S., and Tokuda, H. (1990). *J. Bioenerg. Biomembr.* **22,** 389–399.

Mizushima, S., and Yamada, H. (1975). *Biochim. Biophys. Acta* **375,** 44–53.

Oliver, D. B., and Beckwith, J. (1981). *Cell (Cambridge, Mass.)* **25,** 765–772.

Oliver, D. B., and Beckwith, J. (1982a). *J. Bacteriol.* **150,** 686–691.

Oliver, D. B., and Beckwith, J. (1982b). *Cell (Cambridge, Mass.)* **30,** 311–319.

Pandl, J. P., Cabelli, R., Oliver, D., and Tai, P. C. (1988). *Proc. Natl. Acad. Sci. U.S.A.* **85,** 8953–8957.

Park, S., Liu, G., Topping, T. B., Cover, W. H., and Randall, L. L. (1988). *Science* **239,** 1033–1035.

Racker, E., Violand, B., O'Neal, S., Alfonzo, M., and Telford, J. (1979). *Arch. Biochem. Biophys.* **198,** 470–477.

Riggs, P. D., Derman, A. I., and Beckwith, J. (1988). *Genetics* **118,** 571–579.

Sasaki, S., Matsuyama, S., and Mizushima, S. (1990). *J. Biol. Chem.* **265,** 4358–4363.

Saxena, V. P., and Wetlaufer, D. B. (1970). *Biochemistry* **9,** 5015–5023.

Shiba, K., Ito, K., Yura, T., and Cerretti, D. P. (1984). *EMBO J.* **3,** 631–636.

Shibui, T., Uchida, M., and Teranishi, Y. (1988). *Agric. Biol. Chem.* **52,** 983–988.

Shinkai, A., Akita, M. Matsuyama, S., and Mizushima, S. (1990). *Biochem. Biophys. Res. Commun.* **172,** 1217–1223.

Shinkai, A., Lu, H. M., Tokuda, H., and Mizushima, S. (1991). *J. Biol. Chem.* **266,** 5827–5833.

Shiozuka, K., Tani, K., Mizushima, S., and Tokuda, H. (1990). *J. Biol. Chem.* **265,** 18843–18847.

Sugimura, K. (1988). *Biochem. Biophys. Res. Commun.* **153,** 753–759.

Tani, K., Shiozuka, K., Tokuda, H., and Mizushima, S. (1989). *J. Biol. Chem.* **264,** 18582–18588.

Tani, K., Tokuda, H., and Mizushima, S. (1990). *J. Biol. Chem.* **265,** 17341–17347.

Tokuda, H., Shiozuka, K., and Mizushima, S. (1990). *Eur. J. Biochem.* **192,** 583–589.

Tokuda, H., Akimaru, J., Matsuyama, S., Nishiyama, K., and Mizushima, S. (1991). FEBS Lett. **279,** 233–236.

Traub, P., Mizushima, S., Lowry, C. V., and Nomura, M. (1971). *In* "Methods in Enzymology" (K. Moldave and L. Grossman, eds.), Vol. 20, pp. 391–408. Academic Press, New York.

Viitanen, P., Newman, M. J., Foster, D. L., Wilson, T. H., and Kaback, H. R. (1986). *In* "Methods in Enzymology" (S. Fleischer and B. Fleischer, eds.), Vol.125, pp. 429–452. Academic Press, Orlando, Florida.

Vlasuk, G. P., Inouye, S., Ito, H., Itakura, K., and Inouye, M. (1983). *J. Biol. Chem.* **258,** 7141–7148.

von Heijne, G. (1985). *J. Mol. Biol.* **184,** 99–105.

Watanabe, M., and Blobel, G. (1989). *Cell (Cambridge, Mass.)* **58,** 695–705.

Weis, J. B., Ray, P. H., and Bassford, P. J., Jr. (1988). *Proc. Natl. Acad. Sci. U.S.A.* **85,**

8978–8982.

Yamada, H., Tokuda, H., and Mizushima, S. (1989a). *J. Biol. Chem.* **264**, 1723–1728.

Yamada, H., Matsuyama, S., Tokuda, H., and Mizushima, S. (1989b). *J. Biol. Chem.* **264**, 18577–18581.

Yamane, K., and Mizushima, S. (1988). *J. Biol. Chem.* **263**, 19690–19696.

Yamane, K., Ichihara, S., and Mizushima, S. (1987). *J. Biol. Chem.* **262**, 2358–2362.

Yamane, K., Matsuyama, S., and Mizushima, S. (1988). *J. Biol. Chem.* **263**, 5368–5372.

Yamato, I., Futai, M., Anraku, Y., and Nonomura, Y. (1978). *J. Biochem. (Tokyo)* **83**, 117–128.

Yasuda, S., and Takagi, T. (1983). *J. Bacteriol.* **154**, 1153–1161.

Yue, V. T., and Schimmel, P. R. (1977). *Biochemistry* **9**, 5015–5023.

Chapter 6

Preprotein Translocase of Escherichia coli: Solubilization, Purification, and Reconstitution of the Integral Membrane Subunits SecY/E

ARNOLD J. M. DRIESSEN,[1] LORNA BRUNDAGE,
JOSEPH P. HENDRICK, ELMAR SCHIEBEL, AND
WILLIAM WICKNER

Department of Biological Chemistry
Molecular Biology Institute
University of California, Los Angeles
Los Angeles, California 90024

Don't waste clean thinking on dirty enzymes
(Efraim Racker)

[1] Permanent address: Department of Microbiology, University of Groningen, 9751 NN Haren, The Netherlands.

147

I. Introduction

Escherichia coli proteins, which are exported to the cytoplasmic membrane, the periplasmic space, and the outer membrane, are made with N-terminal leader sequences that direct their targeting to the cytoplasmic membrane (Verner and Schatz, 1988). Once the protein has crossed the inner membrane, the leader sequence is removed by one of two endoproteases, lipoprotein signal peptidase (Yu *et al.*, 1984) or leader peptidase (Wolfe *et al.*, 1982). Most secretory proteins require the enzymes encoded by the *sec* (secretion) genes for export (Bieker *et al.*, 1990). These include *secA, secB, secD, secE* (or *PrlG*), *secF*, and *secY* (or *PrlA*). SecB is a "molecular chaperone" and forms stoichiometric complexes with precursor proteins by interacting with sites in the mature region of these proteins (Liu *et al.*, 1989; Lecker *et al.*, 1989, 1990; Kumamoto *et al.*, 1989) (Fig. 1). Complex formation inhibits side reactions such as aggregation and misfolding and promotes the productive binding of the precursor protein to the membrane surface. The SecA protein, a peripheral membrane protein (Oliver and Beckwith, 1982), is an ATPase (Lill *et al.*, 1989) and functions as the primary receptor for the SecB/precursor protein complex (Hartl *et al.*, 1990). SecA interacts with both the leader and the mature domain of the precursor protein (Cunningham and Wickner, 1989a; Lill *et al.*, 1990; Akita *et al.*, 1990) and requires acidic phospholipids (Lill *et al.*, 1990) and the SecY/E protein (Lill *et al.*, 1989) for activity. SecA promotes the ATP-dependent cycles of release and rebinding of these proteins (Schiebel *et al.*, 1991) and allows proton motive force (Δp)-dependent translocation of precursor proteins across the membrane (Geller *et al.*, 1986; Geller and Green, 1989; Tani *et al.*, 1989). SecY and SecE are integral membrane proteins and function together with the SecA protein as translocase enzymes (Brundage *et al.*, 1990). The SecY/E protein serves as the membrane receptor for SecA (Hartl *et al.*, 1990) and promotes its stability and activity (Lill *et al.*, 1989). SecD (Gardel *et al.*, 1987, 1990) and SecF (Gardel *et al.*, 1990) are integral membrane proteins with large periplasmic domains with unknown functions.

The role of the integral membrane enzymes in transmembrane transit of precursor proteins is largely unresolved. Their functions may be best approached by the study of a reconstituted protein translocation reaction with purified components. It has been difficult to reconstitute translocation-competent proteoliposomes from micellar detergent extracts of membranes. The SecY protein shows a strong tendency to aggregate when membranes are solubilized with *n*-octyl-β-D-glucopyranoside (Cunningham and Wickner, 1989b) or sodium cholate (Watanabe *et al.*, 1990).

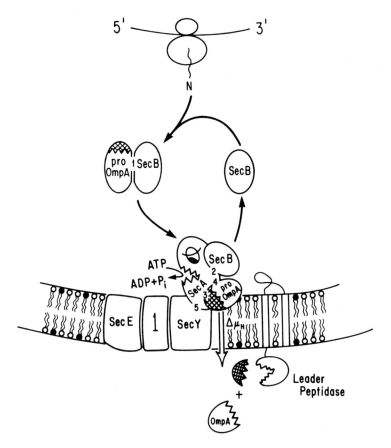

FIG. 1.   A current model of precursor protein translocation across the inner membrane of
Escherichia coli. The precursor of OmpA forms a one-to-one complex with the chaperone
SecB (1). High-affinity interactions of SecB (2) and of both the leader (3) and mature (4) parts
of proOmpA bind the proOmpA·SecB complex to SecA. SecA binds with high affinity to the
SecY/E protein (5). ATP hydrolysis by SecA drives the release of proOmpA from SecA and
allows translocation of the precursor. The proton motive force increases the rate of translocation. During membrane transit, the leader peptide is removed from the precursor by leader
peptidase. The spatial relationship and stoichiometry of the subunits of the translocase are
unknown. The hatched region of proOmpA symbolizes its leader peptide. Reprinted from
Hartl *et al.* (1990) with permission.

Under those conditions the protein is recovered in a particulate fraction.
SecY is soluble in strong ionic detergents such as sodium dedecyl sulfate
(Ito, 1984) or deoxycholate (L. Brundage and W. Wickner, unpublished
results). Proteoliposomes reconstituted from deoxycholate extracts stimulate the ATPase activity of SecA, i.e., "translocation ATPase," but are

unable to sequester precursor proteins in a proteinase-inaccessible form. We have shown that the use of glycerol and phospholipids with octylglucoside has a dramatic effect on the solubility of the SecY protein (Driessen and Wickner, 1990). Using these extraction conditions, we purified the SecY/E protein to homogeneity (Brundage *et al.*, 1990). The purified enzyme is a multisubunit integral membrane protein and has three subunits, two of which are the SecY and SecE polypeptides. Proteoliposomes reconstituted with the purified SecY/E protein support a SecA and ATP-dependent translocation reaction of the precursor of protein A, proOmpA. This chapter describes the procedures to solubilize and purify a functional SecY/E protein that, upon reconstitution, supports an authentic translocation reaction of precursor proteins.

About the Experimental Procedures.   The SecY/E protein is purified from octylglucoside-extracted membranes by the ability of the reconstituted enzyme to stimulate the ATP-hydrolyzing activity of the purified SecA protein in the presence of proOmpA. Sections II–IV describe in detail the reconstitution procedure and the various assays required to determine the activity of the reconstituted SecY/E protein. Protocols used for the isolation and solubilization of *E. coli* inner membranes and the purification of the SecY/E protein are described in Sections V–VII.

II.   Reconstitution of the SecY/E Protein into Liposomes

A.   Reconstitution

The SecY/E protein is reconstituted into proteoliposomes by a rapid detergent–dilution procedure (Racker, 1979). Proteoliposomes are fused by the addition of $CaCl_2$ (Papahadjopoulos *et al.*, 1975). Precipitates are collected by low-speed centrifugation and converted into large unilamellar proteoliposomes by the addition of EGTA. Proteoliposomes are subjected to a brief sonication step to disperse any remaining aggregates.

1.   MATERIALS

Buffer A: 50 mM Tris-HCl, pH 8.0, 50 mM KCl, and 1 mM dithiothreitol (DTT)

$CaCl_2$, 1 M

EGTA, 1 M, pH 8.0

Acetone/ether-washed *E. coli* phospholipids (Avanti, Pelham, AL).
   E. coli phospholipids dissolved in chloroform/methanol [9:1 (v/v)] are dried under a stream of nitrogen gas. Phospholipids are dried thoroughly as a film by rotary evaporation under vacuum. Dried

phospholipids are hydrated in 1 mM DTT at a concentration of 50 mg/ml and are dispersed by the use of a sonic bath. The phospholipid suspension is stored under nitrogen on ice, or frozen in liquid nitrogen and stored at $-80°C$.

Lyophilized purple membranes of *Halobacterium halobium* (Sigma, St. Louis, MO).

2. METHOD

Samples of 100 μl or less containing SecY/E protein are mixed with 7 μl of the phospholipid suspension. Samples are then diluted with 4 ml of buffer A to lower the octylglucoside concentration below its critical micellar concentration. Proteoliposomes are fused by the addition of 40 μl of 1 M CaCl$_2$. Diluted samples are kept on ice for at least 1 hour. Diluted samples are routinely stored on ice overnight without loss of activity. The white precipitate produced by the addition of CaCl$_2$ is collected by centrifugation at 41,000 g for 30 minutes at 4°C and is resuspended in 200 μl of buffer A containing 2.5 mM EGTA by stirring with a glass rod and by gently drawing the suspension up in a pipett several times. Samples are then briefly sonicated (8 seconds, using a bath sonicator); and stored on ice until use. Proteoliposomes are stable for at least 1 week.

B. Coreconstitution with Bacteriorhodopsin

The Δp is required for the translocation of most precursor proteins *in vivo* and stimulates *in vitro* translocation into cytoplasmic membrane vesicles. For the generation of a Δp, inside acid and positive, the SecY/E protein is coreconstituted with the light-driven proton pump bacteriorhodopsin (bR). bR is reconstituted into preformed liposomes by a sonication technique (van Dijck and van Dam, 1982). These liposomes predominantly contain bR inserted in an inside-out orientation, i.e., H$^+$ pumping from outside to inside (Driessen *et al.*, 1987a). bR liposomes are mixed with SecY/E protein in detergent solution, and proteoliposomes are formed by detergent dilution.

METHOD

Escherichia coli phospholipids are hydrated in buffer A at a final concentration of 25 mg/ml. The suspension is pulse sonicated for 10 minutes on ice using a Branson sonifier equipped with a microtip at a power setting of 2 and a duty cycle of 50%. Lyophilized purple membranes (1 mg of protein; 38.4 nmol of bR) are mixed with 400 μl of the preformed liposomes and then pulse sonicated for an additional 10 minutes as described above. Aliquots of 100 μl of the bR liposomes are frozen into liquid nitrogen and stored at − 80°C. For coreconstitution, bR liposomes are thawed at room temperature and briefly sonicated (10 seconds in a bath sonicator). bR liposomes (14 μl) are mixed with 100 μl or less of the (partial) purified SecY/E protein and are then diluted with 4 ml of buffer A. Further details on the reconstitution procedure are described in Section II,A.

III.   Assay of SecA Translocation ATPase

The ability of the proteoliposomes to stimulate the ATP-hydrolyzing activity of SecA in the presence of proOmpA is used as an assay for the isolation of a functional SecY/E protein. The amount of inorganic phosphorus released from ATP is determined by the procedure of Lanzetta *et al.* (1978).

1.   MATERIALS

10× buffer B: 0.5 M Tris-HCl, pH 8.0, 0.5 M KCl, 50 mM MgCl$_2$, and 10 mM DTT

Urea buffer: 6 M urea, 50 mM Tris-HCl, pH 8.0, and 1 mM DTT

Fatty acid-free bovine serum albumin (BSA) (Sigma), 100 mg/ml

ATP, 0.1 M, dissolved in 5 mM Tris-HCl, pH 8.0

proOmpA, 2 mg/ml, dissolved in urea buffer. proOmpA is purified from wild-type *E. coli* W3110 carrying the pTRC-Omp9 plasmid as described in Crooke *et al.* (1988)

SecA protein, 1 mg/ml, dissolved in buffer B containing 10% (v/v) glycerol. SecA protein is purified from the overproducing strain BL21 (λDE3/pT7-*secA*) (Cabelli *et al.*, 1988) as described in Cunningham *et al.* (1989)

34% (w/v) citric acid

20% (v/v) Triton X-100

Malachite green molybdate reagent: malachite green (Sigma) (340 mg) is dissolved in 75 ml of H$_2$O and mixed with a solution of ammonium molybdate (10.5 g) dissolved in 250 ml of 4 N HCl. The solution is made up to 1 liter in H$_2$O, allowed to clarify on ice for at least 1 hour, and then filtered through Whatman (No. 3) paper. This solution is stable at 4°C for at least 2 months. Malachite green molybdate reagent is freshly prepared by adding 250 μl of 20% (v/v) Triton X-100 to 50 ml of the above solution. This reagent is stored at 4°C and is renewed every 2–4 weeks. Old solutions may sometimes cause backgrounds. Plasticware is used to store and transfer the reagent

Phosphorus standard solution (Sigma) containing 645 nmoles of inorganic phosphorus per ml
Proteoliposomes, prepared as described in Section II

2.   METHOD

A stock solution (40 assays) is prepared at 0°C by mixing the following reagents: 200 μl of 10× buffer B, 1470 μl of H_2O, 10 μl of 100 mg/ml BSA, 20 μl of 0.1 M ATP, 20 μl of the SecA protein solution, and 40 μl of the proOmpA solution. Proteoliposomes (5 μl) are rapidly mixed with 45 μl of the above stock solution using plastic tubes. Control samples are mixed with a stock solution that received urea instead of proOmpA. Samples are incubated at 40°C and the reaction is terminated after 30 minutes by the addition of 800 μl of malachite green molybdate reagent. Citric acid (100 μl) is added after 5 minutes to facilitate the color development, and the tubes are kept at room temperature for 40 minutes. Absorbance is measured at 660 nm using plastic microcuvettes and are compared with a standard curve prepared with the phosphorus standard solution (0.5–20 nmol of phosphorus; 50 μl final volume). One OD_{660} unit corresponds to 12.5 nmol of phosphorus. Translocation ATPase is defined as the SecA-dependent, proOmpA-stimulated release of phosphorus from ATP. SecY/E/protein-independent ATPase activity of SecA, i.e., "SecA/lipid ATPase" (Lill et al., 1990) is suppressed in this assay by high concentrations of Mg^{2+} and lipid and a suboptimal concentration of proOmpA, i.e., 40 μg/ml. The concentration of proOmpA used is sufficient to saturate the translocation ATPase activity (Cunningham and Wickner, 1989a). One unit of translocation ATPase activity is defined as the hydrolysis of 1 μmol of ATP per minute under the assay conditions.

IV.   Assay of Translocation

Purified [^{35}S]proOmpA is renatured from 6 M urea by dilution into a reaction mixture containing proteoliposomes and purified SecA protein. ATP-dependent translocation of [^{35}S]proOmpA is assayed by its inaccessibility to added protease and is analyzed by SDS–PAGE and fluorography.

1.   MATERIALS

10 × buffer B: 0.5 M Tris-HCl, pH 8.0, 0.5 M KCl, 50 mM MgCl$_2$, and 10 mM DTT

Urea buffer: 6 M urea, 50 mM Tris-HCl, pH 8.0, and 1 mM DTT

BSA, 100 mg/ml

ATP, 0.1 M, dissolved in 5 mM Tris-HCl, pH 8.0

[^{35}S]proOmpA, dissolved in urea buffer. [^{35}S]proOmpA is synthesized in
a cell-free translation system using plasmid DNA containing the *omp9*
gene (Movva *et al.*, 1980) by the procedure of Gold and Schweiger
(1971) with the modification described in Bacallao *et al.* (1986).
[^{35}S]proOmpA is purified by an immunoaffinity and anion-exchange
column chromatography as described in Crooke and Wickner (1987)

SecA protein, 1 mg/ml, dissolved in buffer B containing 10% (v/v)
glycerol

SecB protein, 0.5 mg/ml, dissolved in 20 mM Tris-HCl, pH 7.4. SecB
was purified from the overproducing strain BL21 (λDE3)/pJW25 as
described in Weiss *et al.* (1988) with the modifications as in Lecker
et al. (1989)

Proteinase K (Boehringer, Mannheim, Germany) (EC 3.4.21.14),
10 mg/ml, dissolved in H_2O

25% (w/v) trichloroacetic acid

Acetone

2 × sample buffer: 4% (w/v) sodium dodecyl sulfate, 100 mM Tris-
HCl, pH 6.8, 20% (v/v) glycerol, 1.5% (v/v) β-mercaptoethanol, and
0.02% (w/v) bromophenolblue

2. METHOD

A stock solution (40 assays) is prepared at 0°C by mixing 200 μl of
10× buffer B, 1210 μl of H_2O, 10 μl of the BSA stock, and 20 μl each of
the SecA and SecB protein solutions. Proteoliopsomes (10 μl) are mixed
with 36.5 μl of the above solution using plastic tubes. Samples are supplemented with 2 μl of 0.1 M ATP, whereas controls receive 2 μl of H_2O.
Translocation is initiated by the addition of 1.5 μl of [^{35}S]proOmpA
(~ 50,000–70,000 cpm; 10,000 cpm/ng of proOmpA) from urea buffer,
and samples are incubated for 20 minutes at 40°C. Translocation is arrested
by rapidly chilling the samples in ice-cold water. Aliquots of 45 μl are
transferred to fresh tubes containing 5 μl of 10 mg/ml of proteinase K and
digestion is performed at 0°C for 15 minutes. Protein is precipitated by the
addition of 65 μl of ice-cold 25% (w/v) trichloroacetic acid and the samples are kept on ice for at least 1 hour. Trichloroacetic acid precipitation is
routinely performed overnight on ice. Precipitates are pelleted by centrifugation in a Brinkman microfuge for 10 minutes and are washed once with
1 ml of ice-cold acetone. Pellets are dried and solubilized in 30 μl of
2× sample buffer. Samples are analyzed by SDS–PAGE and fluorography
as described by Ito *et al.* (1980).

The effect of the Δp on the rate of precursor protein translocation can be studied with proteoliposomes containing both the SecY/E protein and bacteriorhodopsin. Translocation is assayed in the presence of ATP as described above. Uncapped tubes are illuminated with a saturating level of actinic light provided by a 150-W slide projector lamp via a glass fiber optic bundle (Oriel Co., Stratford, CT) using a fiber optic input adapter and a collimating beam probe to focus the light from the projector into the tube. Control samples are wrapped in aluminum foil to prevent stimulation by stray light. To collapse the Δp, the ionophores nigericin and valinomycin are added to the translocation mixture at a final concentration of 10 and 100 nM, respectively. Ionophores are added as ethanolic stock solutions such that the final ethanol concentration in the sample does not exceed 1% (v/v).

3.   REMARKS

SecB is not an essential component of the reconstituted translocation reaction. When diluted from urea, proOmpA aggregates slowly with time, thereby assuming a translocation-incompetent conformation (Lecker et al., 1990). SecB prevents aggregation of proOmpA by forming a iso-lable one-to-one complex (Lecker et al., 1989; Kumamoto et al., 1989). The shielding of apolar surfaces of the mature domain of proOmpA may account for the ability of SecB to suppress nonproductive membrane binding (Hartl et al., 1990). In the reconstituted translocation reaction, SecB improves the translocation efficiency, i.e., the level of translocation after 20 minutes, by approximately twofold.

ProOmpA assumes a protease-resistant conformation when it binds to liposomes, whereas it becomes protease inaccessible (translocated) when the liposomes bear the SecA and SecY/E protein (Schiebel et al., 1991; Brundage et al., 1990). A distinction between protease resistance and inaccessibility can only be made when high concentrations of proteinase K (up to 1 mg/ml) are used.

V.   Growth of Bacteria and Isolation of Cytoplasmic Membranes

A crude envelope preparation containing inner and outer membranes is obtained by sonication of the cells. Crude membranes are extracted with urea to remove and inactivate peripheral membrane proteins. Essentially, this treatment results in an inactivation of ATP-hydrolyzing activity associated with the membranes (Cunningham et al., 1989). Urea-treated membranes are used as a starting material for the purification of the SecY/E protein.

METHOD

Large-scale preparations of E. coli strain UT5600 (F$^-$; ara14, leuB6, azi6, lacY1, proC14, tsx67, Δ(ompT-fepC)266, entA403, λ$^-$, trpE38, rfbD1, rpsL109, xyl5, mtl1, thi1) (Elish et al., 1988) are grown to an OD$_{600}$ of 1.0 U at 37°C in L broth (Miller, 1972) in a 150-liter fermentor. Cells are harvested using a Sharpless centrifuge, homogenized in a Waring blender in an equal volume of 50 mM Tris-HCl, pH 7.5, containing 10% (w/v) sucrose, then frozen as small nuggets by rapid pipetting into liquid nitrogen, and stored frozen at − 80°C (Wickner et al., 1972).

Cells (80 g) are thawed in 200 ml of rapidly stirring buffer C [20% (v/v) glycerol, 50 mM HEPES/KOH, pH 7.0, 50 mM KCl, and 1 mM DTT] at room temperature. The suspension is sonicated at 0°C for 20 minutes using a Branson equipped with a standard probe at a power setting of 3 and a duty cycle of 50%. Lysates are centrifuged (10 minutes, 10,000 g, 0°C) to remove unbroken cells and then are centrifuged again (60 minutes, 150,000 g, 4°C) to collect the membranes. Pellets are suspended in 200 ml of buffer C containing 6 M urea at 0°C with a glass homogenizer and are centrifuged as before. Membranes are suspended in 200 ml of buffer C and centrifuged again. They are finally suspended in 18 ml of buffer C, frozen in liquid nitrogen, and stored at − 80°C. The procedure yields 80–100 mg of protein.

VI. Solubilization and Purification of the SecY/E Protein

A. Solubilization of Inner Membranes

Glycerol (Ambudkar and Maloney, 1986; Maloney and Ambudkar, 1989) and excess phospholipids (Newman and Wilson, 1980) present during the solubilization of the membranes with octylglucoside prevent the irreversible aggregation of SecY and minimize the loss of function upon reconstitution (Driessen and Wickner, 1990). This procedure has been used successfully to reconstitute many different transport systems (Ambudkar and Maloney, 1986; Ambudkar et al., 1986; Driessen et al., 1987b; D'Souza et al., 1987; Maloney and Ambudkar, 1989; Bishop et al., 1989;

Davidson and Nikaido, 1990). Membranes of an OmpT-deficient strain are solubilized in mixed micelles of octylglucoside and phospholipid in the presence of glycerol as described in Driessen and Wickner (1990). Nonextracted materials is then removed by ultracentrifugation. The SecY subunit is cleaved by OmpT during extraction of wild-type membranes (Akiyama and Ito, 1990), without affecting the activity of the purified SecY/E protein (Brundage *et al.*, 1990). During the extraction and further purification steps, the inhibitor *p*-aminobenzamidine (Sugimura and Nishihara, 1988) is included in the buffers to suppress residual proteolytic activity.

1. MATERIALS

$20\times$ Tris buffer: 0.33 M Tris-HCl, pH 7.9, and 0.1 M *p*-aminobenzamidine-HCl (Sigma)
12.5% (w/v) *n*-octyl-β-D-glucopyranoside (Calbiochem, San Diego, CA)
50% (v/v) glycerol
Escherichia coli phospholipids, 50 mg/ml, prepared as described previously
Crude membranes, prepared as described in Section V

2. METHOD

A freshly prepared solution of octylglucoside (8 ml) is mixed with 4 ml of $20\times$ Tris buffer, 40 ml of 50% (v/v) glycerol, and 5.4 ml of the phospholipid suspension. Crude membranes (400 mg of protein) are added and the solution is made up to 80 ml with H_2O. The suspension is stirred on a vortex mixer and incubated on ice for 30 minutes. The extract is then centrifuged at 250,000 g_{max} for 40 minutes at 4°C. The clear, brown supernatant is used for further fractionation. The SecY/E protein is unstable in the crude extract. It is therefore important to immediately proceed with the first chromatographic step (see below).

B. Purification of the SecY/E Protein

1. DEAE CHROMATOGRAPHY

DEAE–cellulose resin (Whatman DE-52) us swollen overnight at 4°C in 1 M Tris-HCl, pH 7.9. The slurry is allowed to settle, the top layer of clear solution is decanted. This procedure is repeated twice with 15 mM Tris-HCl, pH 7.9. A column (3.1 cm diameter by 10.4 cm height) is filled with the slurry and equilibrated with 15 mM Tris-HCl, pH 7.9, till the resin settles. Thereafter, the column is equilibrated with two to three times the column volume of 15 mM Tris-HCl, pH 7.9, containing 20% (v/v) glycerol. Finally, the column is equilibrated with 1.5 times the column volume of buffer D [15 mM Tris-HCl, pH 7.9, 5 mM *p*-aminobenzamidine-HCl, 20% (v/v) glycerol, 1.25% (w/v) octylglucoside, 0.5 mg/ml *E. coli* phospholipids, and 1 mM DTT]. Glycerol minimizes hydrophobic interactions with the resin and stabilizes the SecY/E protein (Driessen and Wickner, 1990; Brundage *et al.*, 1990). The column is operated by pumping buffer from the top using a flow rate of 8 ml/cm^2·hour. The octylglucoside extract is loaded and unabsorbed proteins are eluted with 80 ml of buffer D. The SecY/E protein is eluted from the column with 280 ml of a linear gradient of 0—100 mM KCl in buffer D. Fractions of 10 ml are collected and immediately mixed with 5 ml of glycerol and 0.5 ml of 12.5% (w/v) octylglucoside in order to stabilize the SecY/E protein. The complex elutes in the second half of the gradient, followed by yellowish fractions. Leader peptidase emerges just after the SecY/E protein (see also Wolfe *et al.*, 1982). Each fraction is reconstituted into proteoliposomes (see Section II) and assayed for translocation ATPase activity (see Section III). Active fractions are pooled (~ 80 ml) and stored at − 20°C. There is usually a two- to threefold increase in total units of translocation ATPase activity between the detergent extract and the DEAE-purified fraction. The partially purified SecY/E protein is stable for at least 1 month when stored at −20°C in mixed micellar solution in the presence of 40% (v/v) glycerol.

2.   Q SEPHAROSE FAST FLOW

Size-exclusion chromatography (Stellwagen, 1990) is used to exchange the buffer ion Tris in the DEAE pool for ethanolamine to be used for anion exchange. A column of Sephadex G-25 (Pharmacia, grade medium, 7 cm diameter by 6 cm height) is equilibrated with two times the column volume of buffer E [15 mM ethanolamine-HCl, pH 9.4, 5 mM *p*-amino-benzamidine-HCl, 40% (v/v) glycerol, 1.25% (w/v) octylglucoside, 0.5 mg/ml *E. coli* phospholipids, and 1 mM DTT] at a flow rate of 4 ml/cm^2·hour. Part of the DEAE pool (50 ml) is loaded on the column and then eluted with buffer E. Proteins eluting in the void volume are collected. Gel filtration results in a slight dilution of the sample without loss of translocation ATPase activity.

A slurry of Q Sepharose Fast Flow (Pharmacia) extensively washed with 15 mM ethanolamine-HCl, pH 9.4, is packed in a column (1 cm diameter by 7.5 cm height). The column is then equilibrated with two times the column volume of buffer E at 4°C, and the pool of the gel filtration run is

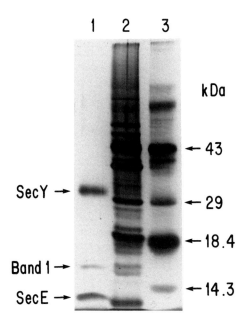

FIG. 2. Subunit composition of the purified SecY/E protein. Samples from the Fast Flow Q step (lane 1) or the DEAE step (lane 2) were analyzed on a 15% polyacrylamide gel by silver staining. Lane 3 contains molecular-weight markers. Reprinted from Brundage *et al.* (1990) with permission.

loaded on the column using a flow rate of 20 ml/cm$^2 \cdot$ hour. The unbound material is collected and the column is washed with 5 ml of buffer E followed by 100 ml of a linear gradient of 0–125 mM KCl in buffer E. Fractions of 8 ml are collected. The SecY/E protein elutes with the unbound material. Flow-through fractions containing translocation ATPase activity are collected, mixed with 0.02 volumes of 1 M Tris-HCl, pH 7.9, and stored at $-20°$C.

At this stage, the SecY/E protein is essentially pure (Fig. 2). A summary of a typical purification is shown in Table I. Although the presence of inhibitors in the crude extract interferes with a precise quantitation, the procedure results in a 85-fold purification with a 75% yield from the crude extract. The purified SecY/E protein is stable for at least 1 month when stored at $-20°$C in mixed micellar solution in the presence of 40% (v/v) glycerol.

In some of the Q Sepharose Fast Flow column runs we noted that a fraction of the SecY/E protein remained bound to the resin. The bound SecY/E protein is eluted upon the introduction of the KCl, and requires further purification by hydroxyapatite chromatography (Brundage *et al.*,

TABLE I

PURIFICATION OF SECY/E COMPLEX

Sample	Protein[a] (mg)	Total activity[b] (units)	Specific activity (units/mg of protein)	Recovery (%)
Extract	240	9.3	0.04	—
DEAE	15.9	13.3	0.84	100
Q Sepharose	2.02	7.0	3.45	52

[a] Protein was determined by the method of Bradford (1976).
[b] Translocation ATPase.

1990). Absorption on hydroxyapatite can also be used as a means to concentrate the SecY/E protein, although a considerable loss in activity has to be taken into account (for experimental details, see Brundage *et al.*, 1990).

VII. Functional and Biochemical Characterization of the Purified SecY/E Protein

The purified SecY/E protein contains three major polypeptide species. These have been identified by immunoblots with antisera to SecY and SecE and by N-terminal sequence analysis. The largest polypeptide of the purified protein reacts with antibodies to the SecY N-terminus (Watanabe and Blobel, 1989) and migrates on SDS–PAGE with an apparent molecular mass of 29 kDa. It has the sequence of AKQPGL..., which is the sequence of the SecY protein minus its amino-terminal methionine (Cerretti *et al.*, 1983). The next smaller polypeptide, labeled "band 1" in Fig. 2, has not been identified with respect to its amino acid sequence. Quantitative amino acid analysis suggest that it is present in substoichiometric amounts relative to the SecY and SecE subunits. Band 1 can be recognized by its blue appearance on silver-stained gels. The smallest polypeptide migrates on SDS–PAGE with an apparent molecular mass of 13.5 kDa, reacts with antibodies to SecE, and has the sequence SANTE..., which is identical to the N-terminal sequence of the SecE protein minus its initiating methionine (Schatz *et al.*, 1989). These polypeptides form a single multisubunit protein as demonstrated by immunoprecipitation with antibodies specific for the N-terminal domain of the SecY (Brundage *et al.*, 1990) and SecE (L. Brundage and W. Wickner, unpublished results) proteins. The quantitative subunit composition and their spatial relationship in the membrane is unknown.

The SecY/E protein is the major integral membrane protein needed for efficient translocation of proOmpA and translocation ATPase (Brundage *et al.*, 1990). The ratio of the translocation activity and translocation ATPase remains constant for each stage of purification and is very similar to the ratio found for inner membrane vesicles. Approximately twice the amount of purified SecY/E protein is required to achieve the same rate of translocation as observed with inner membrane vesicles, factors that may account for this difference are inactivation of the SecY/E protein during purification and loss of asymmetry after reconstitution.

It has been suggested that translocation may be reconstituted without the SecY protein (Watanabe *et al.*, 1990). Our results demonstrate that precursor protein translocation can be reconstituted with a defined set of purified components. In agreement with prior *in vivo* studies (Emr *et al.*, 1981; Ito *et al.*, 1983; Schatz et al., 1989), the reconstituted translocation reaction requires the SecY/E protein, SecA, and ATP (Brundage *et al.*, 1990). The rate of translocation in the reconstituted system is stimulated by the Δp (Driessen and Wickner, 1990; Brundage *et al.*, 1990) (Fig. 3). This demonstrates that the effect of Δp seen both *in vivo* (Zimmermann and Wickner, 1983) and *in vitro* (Geller *et al.*, 1986) reflects an intrinsic part of the translocation mechanism rather than an indirect effect of Δp on cellular physiology.

The availability of pure Sec/E protein and a reconstituted translocation reaction may now allow a clear definition of its functions. Two of the

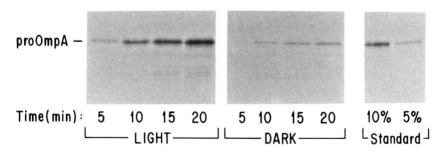

FIG. 3. The rate of translocation of proOmpA into SecY/E proteoliposomes is stimulated by a proton motive force. The purified SecY/E protein was coreconstituted with the light-driven proton pump bacteriorhodopsin. The ATP-dependent translocation of proOmpA was assayed in the presence and absence of light at the indicated times. Lanes labeled "standard" have the indicated percentage of total added proOmpA. Reprinted from Brundage *et al.* (1990) with permission.

biochemical functions of the SecY/E protein have been identified. In conjunction with acidic phospholipids (Hendrick and Wickner, 1991), the SecY/E protein serves as the high-affinity receptor for the SecA protein (Hartl *et al.*, 1990) and is essential to activate and stabilize the ATP hydrolytic activity of SecA (Lill *et al.*, 1989). Certain alleles of *secA*, *secY*, and *secE* can suppress leader sequence defects (Bieker *et al.*, 1990), suggesting that the SecY/E protein directly interacts with the leader sequence after its release from SecA. The SecY/E protein may act as a proton-conducting enzyme to couple Δp to the translocation of proteins across the membrane. Alternatively, Δp may increase the efficiency of the translocation process by changing the conformation of the SecY/E or the precursor protein. A major question is whether the integral membrane enzymes form a proteinaceous transport system to allow the passage of precursor proteins across the membrane, or whether these proteins cross through the bilayer *per se*.

Acknowledgments

We thank Marilyn Rice and Douglas Geissert for expert technical assistance. This work was supported by a grant from the NIH and by gifts from Biogen, Inc., and Merck. A.D. is supported a grant from the Netherlands Organization for Scientific Research (N. W. O.). L.B. is supported by USPHS National Institute Research Service Award CA-09056. J.H. is a fellow (PF 3432) of the American Cancer Society. E.S. is a fellow of the Deutsche Forschungsgemeinschaft.

References

Akita, M., Sasaki, S., Matsuyama, S., and Mizushima, S. (1990). SecA interacts with secretory proteins by recognizing the positive charge at the amino terminus of the signal peptide in *Escherichia coli. J. Biol. Chem.* **265**, 8164–8169.

Akiyama, Y., and Ito, K. (1990). SecY protein, a membrane-embedded secretion factor of *Escherichia coli*, is cleaved by the OmpT protein *in vitro. Biochem. Biophys. Res. Commun.* **167**, 711–715.

Ambudkar, S. V., and Maloney, P. C. (1986). Bacterial anion exchange. Use of osmolytes during solubilization and reconstitution of phosphate linked antiport from *Streptococcus lactis. J. Biol. Chem.* 10079–10086.

Ambudkar, S. V., Lynn, A. R., Maloney, P. C., and Rosen, B. P. (1986). Reconstitution of ATP-dependent calcium transport from streptococci. *J. Biol. Chem.* **261**, 15596–15600.

Bacallao, R., Crooke, E., Shiba, K., Wickner, W., and Ito, K. (1986). The SecY protein can act post-translocationally to promote bacterial protein export. *J. Biol. Chem.* **261**, 12907–12910.

Bieker, K. L., Phillips, G. J., and Silhavy, T. (1990). The *sec* and *prl* genes of *Escherichia coli. J. Bioenerg. Biomembr.* **22**, 291–310.

Bishop, L., Agbayani, R., Jr., Ambudkar, S. V., Maloney, P. C., and Ames, G. F. L. (1989). Reconstitution of a bacterial periplasmic permease in proteoliposomes and demonstration of ATP hydrolysis concomitant with transport. *Proc. Natl. Acad. Sci. U.S.A.* **86,** 6953–6957.

Bradford, M. M. (1976). A rapid and sensitive method for the quantitation of microgram quantities of protein utilizing the principle of protein dye binding. *Anal. Biochem.* **72,** 248–254.

Brundage, L., Hendrick, J. P., Schiebel, E., Driessen, A. J. M., and Wickner, W. (1990). The purified *Escherichia coli* integral membrane protein SecY/E is sufficient for reconstitution of SecA-dependent precursor protein translocation. *Cell (Cambridge, Mass.)* **61,** 649–657.

Cabelli, R. C., Chen, L., Tai, P. C., and Oliver, D. B. (1988). SecA protein is required for secretory protein translocation into *Escherichia coli* membrane vesicles. *Cell (Cambridge, Mass.)* **55,** 683–692.

Cerretti, D., Dean, D., Davis, G., Bedwell, D., and Nomura, M. (1983). The *spc* ribosomal operon of *Escherichia coli:* Sequence and contranscription of the ribosomal protein genes and a protein export gene. *Nucleic Acids Res.* **11,** 2599–2616.

Crooke, E., and Wickner, W. (1987). Trigger factor: A soluble protein which folds proOmpA into a membrane assembly competent form. *Proc. Natl. Acad. Sci. U.S.A.* **84,** 5216–5220.

Crooke, E., Guthrie, B., Lecker, S., Lill, R., and Wickner, W. (1988). ProOmpA is stabilized for membrane translocation by either purified *Escherichia coli* trigger factor or canine signal recognition particle. *Cell (Cambridge, Mass.)* **54,** 1003–1011.

Cunningham, K., and Wickner, W. (1989a). Specific recognition of the leader region of precursor proteins is required for the activation of translocation ATPase of *Escherichia coli. Proc. Natl. Acad. Sci. U.S.A.* **86,** 8630–8634.

Cunningham, K., and Wickner, W. (1989b). Detergent disruption of bacterial inner membranes and recovery of protein translocation activity. *Proc. Natl. Acad. Sci. U.S.A.* **86,** 8673–8677.

Cunningham, K., Lill, R., Crooke, E., Rice, M., Moore, K., Wickner, W., and Oliver, D. (1989). SecA protein, a peripheral protein of the *Escherichia coli* plasma membrane, is essential for the functional binding and translocation of proOmpA. *EMBO J.* **8,** 955–959.

Davidson, A. L., and Nikaido, H. (1990). Overproduction, solubilization, and reconstitution of the maltose transport system from *Escherichia coli. J. Biol. Chem.* **265,** 4254–4260.

Driessen, A. J. M., and Wickner, W. (1990). Solubilization and functional reconstitution of the protein translocation enzymes of *Escherichia coli. Proc. Natl. Acad. Sci. U.S.A.* **87,** 3107–3111.

Driessen, A. J. M., Helligwerf, K. J., and Konings, W. N. (1987a). The effect of trypsin treatment on the incorporation and energy transducing properties of bacteriorhodopsin in liposomes. *Biochim. Biophys. Acta* **891,** 165–176.

Driessen, A. J. M., Poolman, B., Kiewiet, R., and Konings, W. N. (1987b). Arginine transport in *Streptococcus lactis* is catalyzed by a cationic exchanger. *Proc. Natl. Acad. Sci. U.S.A.* **84,** 6093–6097.

D'Souza, M. P., Ambudkar, S. V., August, J. T., and Maloney, P. C. (1987). Reconstitution of the lysosomal proton pump. *Proc. Natl. Acad. Sci. U.S.A.* **84,** 6980–6984.

Elish, M. E., Pierce, J. R., and Earhart, C. F. (1988). Biochemical analysis of spontaneous *fepA* mutants of *Escherichia coli. J. Gen. Microbiol.* **134,** 1355–1364.

Emr, S. D., Hanley-Way, S., and Silhavy, T. J. (1981). Suppressor mutations that restore export of a protein with a defective signal sequence. *Cell (Cambridge, Mass.)* **23,** 79–88.

Gardel, C., Benson, S., Hunt, J., Michaelis, S., and Beckwith, J. (1987). *secD,* a new gene involved in protein export in *Escherichia coli. J. Bacteriol.* **169,** 1286–1290.

Gardel, C., Johnson, K., Jacq, A., and Beckwith, J. (1990). The *secD* locus of *Escherichia coli* codes for two membrane proteins required for protein export. *EMBO J.* **9,** 3209–3216.

Geller, B. L., and Green, H. M. (1989). Translocation of proOmpA across inner membrane vesicles of *Escherichia coli* occurs in two consecutive energetically distinct steps. *J. Biol. Chem.* **264,** 16465–16469.

Geller, B. L., Movva, N. R., and Wickner, W. (1986). Both ATP and electrochemical potential are required for optimal assembly of pro-OmpA into *Escherichia coli* inner membrane vesicles. *Proc. Natl. Acad. Sci. U.S.A.* **83,** 4219–4222.

Gold, L. M., and Schweiger, M. (1971). Synthesis of bacteriophage-specific enzymes directed by DNA *in vitro. In "Methods in Enzymology"* (K. Moldave and L. Grossman, eds.), Vol. 20, pp. 537–542. Academic Press, New York.

Hartl, F.-U., Lecker, S. H., Schiebel, E., Hendrick, J. P., and Wickner, W. (1990). The binding of SecB to SecA mediates preprotein targeting to the *Escherichia coli* membrane. *Cell (Cambridge, Mass.)* **63,** 269–279.

Hendrick, J. P., and Wickner, W. (1991). In preparation.

Ito, K. (1984). Identification of the *secY (prlA)* gene product involved in protein export in *Escherichia coli. Mol. Gen. Genet.* **197,** 204–208.

Ito, K., Date, T., and Wickner, W. (1980). Synthesis, assembly into the cytoplasmic membrane and proteolytic processing of the precursor of *Escherichia coli* phage M13 coat protein. *J. Biol. Chem.* **255,** 2123–2130.

Ito, K., Wittekind, M., Nomura, M., Shiba, K., Yura, T., Miura, A., and Nashimoto, H. (1983). A temperature-sensitive mutant of *Escherichia coli* exhibiting slow processing of exported proteins. *Cell (Cambridge, Mass.)* **32,** 789–797.

Kumamoto, C. A., Chen, L. L., Fandl, J., and Tai, P. C. (1989). Purification of the *Escherichia coli secB* gene product and demonstration of its activity in an *in vitro* protein translocation system. *J. Biol. Chem.* **264,** 2242–2249.

Lanzetta, P. A., Alvarez, L. J., Reinach, P. S., and Candia, O. A. (1978). An improved assay for nanomole amounts of inorganic phosphate. *Anal. Biochem.* **100,** 95–97.

Lecker, S. H., Lill, R., Ziegelhoffer, T., Georgopoulos, C., Bassford, P. J., Jr., Kumamoto, C. A., and Wickner, W. (1989). Three pure chaperone proteins of *Escherichia coli,* SecB, trigger factor, and GroEL, form soluble complexes with precursor proteins *in vitro. EMBO J.* **8,** 2703–2709.

Lecker, S. H., Driessen, A. J. M., and Wickner, W. (1990). ProOmpA contains secondary and tertiary structure prior to translocation and is shielded from aggregation by association with SecB protein. *EMBO J.* **9,** 2309–2314.

Lill, R., Cunningham, K., Brundage, L., Ito, K., Oliver, D., and Wickner, W. (1989). The SecA protein hydrolyzes ATP and is an essential component of the protein translocation ATPase of *Escherichia coli. EMBO J.* **8,** 961–966.

Lill, R., Dowhan, W., and Wickner, W. (1990). The ATPase activity of SecA is regulated by acidic phospholipids, SecY, and the leader and mature domains of precursor proteins. *Cell (Cambridge, Mass.)* **60,** 271–280.

Liu, G., Topping, T. B., and Randall, L. L. (1989). Physiological role during export for the retardation of folding by the leader peptide of maltose-binding protein. *Proc. Natl. Acad. Sci. U.S.A.* **86,** 9213–9217.

Maloney, P. C., and Ambudkar, S. V. (1989). Functional reconstitution of prokaryote and eukaryote membrane proteins. *Arch. Biochem. Biophys.* **269,** 1–10.

Miller, J. H. (1972). "Experiments in Molecular Genetics," p. 433. Cold Spring Harbor Lab., Cold Spring Harbor, New York.

Movva, N. R., Nakamura, K., and Inouye, M. (1980). Gene structure of the OmpA protein, a major surface protein of *Escherichia coli* required for cell–cell interaction. *J. Mol. Biol.* **143,** 317–328.

Newman, M. J., and Wilson, T. H. (1980). Solubilization and reconstitution of the lactose transport system from *Escherichia coli. J. Biol. Chem.* **255**, 10583–10586.

Oliver, D. B., and Beckwith, J. (1982). Regulation of a membrane component required for protein secretion in *Escherichia coli. Cell (Cambridge, Mass.)* **30**, 311–319.

Papahadjopoulos, D., Vail, W. J., Jacobson, K., and Poste, G. (1975). Coachleate lipid cylinders: Formation by fusion of unilamellar lipid vesicles. *Biochim. Biophys. Acta* **394**, 483–491

Racker, E. (1979). Reconstitution of membrane processes. *In* "Methods in Enzymology" (S. Fleischer and L. Packer, eds.), vol. 55, pp. 699–711. Academic Press, New York.

Schatz, P. J., Riggs, P. D., Annick, J., Fath, M. J., and Beckwith, J. (1989). The SecE protein encodes an integral membrane protein required for protein export in *Escherichia coli. Genes Dev.* **3**, 1035–1044.

Schiebel, E., Driessen, A. J. M., Hartl, F.-U., and Wickner, W. (1991). $\Delta\bar{\mu}_{H^+}$ and ATP function at different steps of the catalytic cycle of preprotein translocase. *Cell (Cambridge, Mass.)* **64**, 927–939.

Stellwagen E. (1990). Gel filtration. *In* "*Methods in Enzymology*" (M. P. Wentscher, ed.), Vol. 182. pp. 317–328. Academic Press, San Diego, California.

Sugimura, K., and Nishihara, T. (1988), Purification, characterization, and primary structure of *Escherichia coli* protease VII with specificity for parired basic residues: Identity of protease VII and OmpT. *J. Bacteriol.* **170**, 5625–5632.

Tani, K., Shiozuka, K., Tokuda, H., and Mizushima, S. (1989). *In vitro* analysis of the process of translocation of OmpA across the *Escherichia coli* cytoplasmic membrane. A translocation intermediate accumulates transiently in the absence of the proton motive force. *J. Biol. Chem.* **264**, 18582–18588.

van Dijck, P. W. M., and van Dam, K. (1982). Bacteriorhodopsin into proteoliposomes. *In* "*Methods in Enzymology*" (L. Packer, ed.), Vol. 88, pp. 17–25. Academic Press, New York.

Verner, K., and Schatz G. (1988). Protein translocation across membranes. *Science* **241**, 1307–1313.

Watanabe, M., and Blobel, G. (1989). Site-specific antibodies against PrlA (SecY) protein of *Escherichia coli* inhibit protein export by interfering with plasma membrane binding of preprotein. *Proc. Natl. Acad. Sci. U.S.A.* **86**, 1895–1899.

Watanabe, M., Nicchitta, C. V., and Blobel, G. (1990). Reconstitution of protein translocation from detergent-solubilized *Escherichia coli* inverted vesicles: PrlA protein-deficient vesicles efficiently translocate precursor proteins. *Proc. Natl. Acad. Sci. U.S.A.* **87**, 1960–1964.

Weiss, J. B., Ray, J. H., and Bassford, P. J., Jr. (1988). Purified SecB protein of *Escherichia coli* retards folding and promotes membrane translocation of the maltose binding protein *in vitro. Proc Natl. Acad. Sci. U.S.A.* **85**, 8978–8982.

Wickner, W., Brutlag, D., Schekman, R., and Kornberg, A. (1972). RNA synthesis initiates *in vitro* conversion of M13 DNA to its replicative form. *Proc. Natl. Acad. Sci. U.S.A.* **69**, 965–969.

Wolfe, P. B., Silver, O., and Wickner, W. (1982). The isolation of homogeneous leader peptidase from a strain of *Escherichia coli* which overproduces the enzyme. *J. Biol. Chem.* **257**, 7898–7902.

Yu, F., Yamada, K., Daishima, K., and Mizushima, S. (1984). Nucleotide sequence of the IspA gene, the structural gene for lipoprotein signal peptidase of *Escherichia coli. FEBS Lett.* **173**, 264–268.

Zimmermann, R., and Wickner, W. (1983). Energetics and intermediates of the assembly of protein OmpA into the outer membrane of *Escherichia coli. J. Biol. Chem.* **258**, 3920–3925.

Chapter 7

In Vitro Protein Translocation into Escherichia coli Inverted Membrane Vesicles

PHANG C. TAI, GUOLING TIAN, HAODA XU, JIAN P. LIAN,
AND JACK N. YU

Department of Cell and Molecular Biology
Boston Biomedical Research Institute
Boston, Massachusetts 02114
and
Department of Microbiology and Molecular Genetics
Harvard Medical School
Boston, Massachusetts 02115

METHODS IN CELL BIOLOGY, VOL. 34

I. Introduction

In bacteria, as in all living cells, secretory proteins are synthesized in the cytoplasm and are transferred across membranes. Studies on the mechanisms of bacterial protein secretion and export have lagged behind those in eukaryotic cells (see Davis and Tai, 1980). Work on the mechanism of eukaryotic protein translocation has been much influenced by the signal hypothesis originally proposed by Blobel and Dobberstein (1975) and later refined by several laboratories on the basis of *in vitro* studies with endoplasmic reticulum microsomes. According to this view, an N-terminal signal peptide emerging from a ribosome interacts with a multisubunit signal recognition particle, and translation is arrested until the particle binds to a receptor docking protein in the membrane; subsequent translocation in a membrane proteinaceous channel and cleavage of the signal are presumed to be coupled to protein synthesis. The tight coupling between translation and translocation and firm attachment of the ribosomes to the membrane have led to the view that energy derived from protein synthesis is sufficient to drive proteins across membranes.

In the past few years, genetic and biochemical studies with bacteria have contributed significantly to the understanding of the mechanism of protein secretion and transport (see a collection of reviews in Tai, 1990a). In particular, isolation of mutants has defined the function of the signal sequence and identified many gene products that are involved in protein export (see Bieker *et al.*, 1990). The realization that protein synthesis is not the source of energy for protein translocation across the bacterial membrane is also important in understanding the mechanism of this process. Thus, neither the attachment of ribosomes to membranes nor the tight coupling of translocation to translation has been found in bacteria. Indeed, *in vitro* systems have been established for protein translocation across *Escherichia coli* membrane vesicles in which posttranslational translocation is almost as efficient as cotranslational translocation, unlike the endoplasmic reticulum system (Rhoads *et al.*, 1984; Muller and Blobel, 1984; Chen *et al.*, 1985). Metabolic energy, derived from ATP hydrolysis and also contributed by proton motive force, is required for both co- and posttranslational translocation (Chen and Tai, 1985, 1987b; Geller *et al.*, 1986). Moreover, the *in vitro* systems have also characterized biochemically the functions of gene products identified genetically to be involved in protein translocation (Fandl and Tai, 1987; Fandl *et al.*, 1988; Cabelli *et al.*, 1988; Collier *et al.*, 1988; Kumamoto *et al.*, 1989; Watanabe and Blobel, 1989; Weiss *et al.*, 1988; Cunningham *et al.*, 1989). The convergence of both genetic and biochemical approaches provides a more confident list of the protein components that participate in the bacterial protein translocation (see Fandl and Tai, 1990).

The *in vitro* protein translocation system is basically composed of a soluble extract (S30) capable of synthesizing proteins when programmed with mRNA coded for secretory precursors, and a membrane fraction of inverted cytoplasmic membrane vesicles (Rhoads *et al.*, 1984; Muller and Blobel, 1984). In some cases, purified precursor molecules are used (Crooke and Wickner, 1987; Crooke *et al.*, 1988; Yamane *et al.*, 1988; Tian *et al.*, 1989). In these systems, proteins are considered translocated across membranes if processing of the signal sequences coincides with proteins resistant to proteolytic digestion and the processed proteins cosediment with membrane vesicles. This is taken as an indication that the proteins are translocated into the lumen of the vesicles, although it is possible that some of the protein is embedded in the membranes. Protein translocation can be examined cotranslationally by synthesizing proteins in the presence of membrane vesicles, or posttranslationally by blocking further synthesis prior to the addition of membrane vesicles (Chen *et al.*, 1985).

II.   Materials, Buffers, and Various Preparations

A.   Media and Growth of Cells

Standard preparations of membrane vesicles and S30 extracts are prepared from an RNase-free *E. coli* K12 strain D10 (RNase I$^-$, Met$^-$) originally isolated by Gesteland (1966). In assessing the role of proton motive force, its derivative, deleted of H$^+$-ATPase ($\Delta atpB$-C) is used for preparing membranes (Chen and Tai, 1986a; Geller *et al.*, 1986). Cells are grown with vigorous aeration at 37°C in L broth supplemented with A salts (LinA medium), 0.5% glucose, and 20 μg/ml methionine. The LinA medium is as follows: K$_2$HPO$_4\cdot$3H$_2$O, 9.2 g; KH$_2$PO$_4$, 3 g; Bacto-tryptone (Difco), 10 g; yeast extracts (Difco), 2 g; (NH$_4$)$_2$SO$_4$, 1 g; sodium citrate, 0.5 g; NaCl, 5 g. This is made up with deionized water to 1 liter in a 6-liter flask, adjusted to pH 7.2 with 2 pellets of NaOH, and autoclaved at 121°C for 30 minutes.

B.   Membrane Preparation

The important steps for preparing active inverted membrane vesicles are to prepare and lyse spheroplasts by the lysozyme/EDTA methods of Witholt *et al.*, 1976), to invert the membrane vesicles using a French press cell, and to subfractionate the inner cytoplasmic membrane by sucrose gradients with modification of the procedures of Osborn and Munson (1974). More than 95% of the inner membrane vesicles are inverted, and the lighter subfractions of cytoplasmic membranes are active in protein translocation.

To prepare membrane vesicles for protein translocation, D10 or H^+-ATPase deletion-derivative cells are grown exponentially in LinA media to about 1.5×10^9 cells/ml (300 Klett units at 540-nm brown filter; reading by one to five dilutions). The culture is poured over ice, centrifuged, and the cells are converted to spheroplasts as described previously (Rhoads *et al.*, 1984). Typically, cell pellets from 2.5 liters are washed with 0.2 M Tris, pH 8.0, and are quickly dispersed mechanically in 150 ml of 0.2 M Tris-HCl (pH 8.0) containing 0.5 M sucrose and 50 μg/ml chloramphenicol at room temperature. The following additions are then made sequentially, each followed by gentle swirling to ensure complete mixing: 1.5 ml of 50 mM trisodium EDTA (pH 8.0), 1.5 ml of 4 mg/ml of freshly prepared lysozyme (egg white, from Sigma), and 150 ml of water. When spheroplast formation, monitored by phasecontrast microscopy, is about 80–90% complete (usually 8–15 minutes; cells may not round completely, but may only shoren and fatten), the suspension is chilled (all the subsequent steps are at 4°C) and the spheroplasts are collected by centrifugation (7000 g, 5 minutes): this step is omitted if the spheroplasts are lysed. The pellets are thoroughly suspended in 60 ml DE_{20} solution (1 mM DTT and 20 mM EDTA, pH 7.2) and are partially lysed. The suspension is passed through a French press cell at 10,000 psi to complete the lysis, to shear the released DNA (addition of DNase usually not necessary), to disrupt membrane aggregrates, and to invert membrane vesicles. Large debris is removed by centrifugation (7000 g, 5 minutes). The supernatant lysate is layered over two-step block gradients (1 ml each 0.5 M and 1.4 M sucrose in DE_{20}) and is centrifuged in a Beckman SW41 rotor at 37,000 rpm for 3 hours; much of the outer membrane sediments through the cushion. The visible interface band, consisting mostly of crude inner membranes, is collected with a Pasteur pipet, diluted four- to fivefold with DE_{20}, and layered over a four-step sucrose gradient (3 ml sample, 2 ml each of 0.8, 1.0, 1.2, and 1.4M sucrose in DE_{20}) in a Beckman SW41 rotor. After centrifugation at 37,000 rpm for 17–18 hours, the gradient yields two visible bands at an interace of 1.0–1.2 M (fraction II) and 1.2–1.4 M sucrose (fraction III) cushion, and some diffuse lighter membranes (fraction I). (Note: for larger samples, these step sucrose gradients may also be run in a Beckman Ti60 rotor with the corresponding increase in volume with the same speed and duration of centrifugation.) These fractions are collected, diluted at least fourfold with DTK buffer (1 mM DTT, 10 mM Tris-HCl, pH 7.6, and

50 mM KCl), and membranes are collected by centrifugation (75 minutes at 37,000 rpm in a Beckman Ti60 rotor). The pelleted membranes are resuspended thoroughly in minimal amounts of DTK buffer with a motorized mixer. Membrane concentration is usually measured by absorbance at 280 nm in 2% sodium dodecyl sulfate (SDS), adjusted to 20 OD_{280} U/ml, and stored at −76°C in small aliquots. Membranes can be frozen and thawed several times without loss of activity. This protocol usually yields about 50–100 OD_{280} units each of three membrane factions. The protein patterns are relatively similar (Fig. 1), with increasing amounts of outer membrane proteins in the denser fraction (fraction III). The lighter fractions contain some ribosomal proteins and stable mRNA (Rhoads *et al.*, 1984). These membranes are isolated not based on density alone, but are similar in activity to membranes isolated by equilibrium centrifugation (Table I). The denser brownish fraction III is usually not active in protein translocation for OmpA, but is still active for lipoprotein (see Table I).

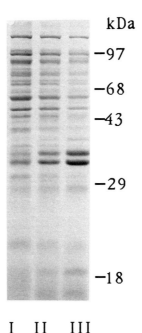

I   II   III

FIG. 1. Protein compositions of membrane fractions. The 50-μg proteins of membrane fractions I, II, and III were run on SDS–page (12% acrylamide) and stained with Coomassie blue. The molecular weight standards are phosphorylase B (97 kDa), bovine serum albumin (68 kDa), ovalbumin (43 kDa), carbonic anhydrase (29 kDa), and β-lactoglobulin (18 kDa).

TABLE I

TRANSLOCATION ACTIVITY OF MEMBRANE VESICLE PREPARATIONS[a]

	Membrane translocation activity (%)				
	Fraction II		Fraction III		
Protein	Block	Equilibrium	Block	Equilibrium	Flotation equilibrium
OmpA	100	109	5	7	5
Lipoprotein	100	92	119	101	138

[a] Membranes were fractionated with block gradients into fraction II and fraction III as described for regular procedures. These membranes were then centrifuged to equilibrium density in 1.0–1.5 M sucrose gradients in a Beckman SW41 rotor at 37,000 rpm for 70 hr. For flotation equilibrium, the membranes were mixed with 1.6 M sucrose and loaded at the bottom of centrifuge tubes before centrifugation. Each membrane preparation (at 0.1 A_{280}) was assayed for translocation of prolipoprotein and proOmpA protein, respectively. The ratios of translocated precursors to mature proteins were about the same in all cases; and the translocation activities of the regular membranes (fraction II) based on protease-resistant precursor and mature forms of OmpA or lipoprotein are taken as 100%.

C.   Preparation of S30 Extracts

In preparing extracts for *in vitro* protein synthesis, it is important that there be no ribonucleases. Thus the ribonuclease I-free D10 strain is used, and all the containers, tubes, tips, and buffers are autoclaved when possible. It is also essential to prepare extracts from fast-growing cells and to have no endogenous mRNA and small membrane vesicles.

Culture of exponentially grown D10 cells (0.8×10^8 cells/ml; Klett reading at 150) at 37°C is shifted to a waterbath at 15°C and stirred for 15 minutes to run off the endogenous messenger RNA. Cells are harvested and washed with DTKM buffer [1 mM DTT, 10 mM Tris-HCL, pH 7.6, 50 mM KCl, and 10 mM Mg(OAC)$_2$] and are pelleted solidly to remove excess liquid (30,000 g for 15 minutes in a Sorvall SS34 rotor). Cell pellets are quickly frozen, weighed (to determine the amount of aluminum and buffer needed), and kept at −76°C until use.

The frozen cell pellets are ground with 1.5 times their cell weight of abrasive aluminum (North Co., Worcester, MA) in the cold, until cells lyse and the paste becomes viscous (avoid excessive grinding). DTKM buffer (twice the weight of the cell pellets) and DNase I (2 μg/ml final concentration) are added. The mixture is centrifuged (10,000 g, 10 minutes) to remove aluminum, unbroken cells, and large debris. The supernatant is transferred to a narrow, long centrifuge tune (1 × 10 cm) and centrifuged at 30,000 g for 30 minutes to minimize the contaminating membranes. The upper two-thirds of the supernatant is carefully removed, incubated at 4°C for 15 minutes with 0.5% octylglucoside (Calbiochem Co.) to solubilize any remaining membranes, and dialyzed against a 50-fold volume of DTKM buffer for 3 hours at 4°C with one change of buffer. After centrifugation at 30,000 g for 10 minutes, the supernatant is removed (its OD_{260} is determined), divided into samll portions (the S30 is used only once after thawing), frozen rapidly, and stored at -76°C. The S30 is most active when lysis yields about 200–300 OD_{260} U/ml, and is stable for years.

D.   Preparation of Messenger RNA

The mRNAs for synthesizing secretory protein precursors may be prepared *in vitro* from plasmids encoding the gene, or *in vivo* from cells harboring such plasmids. We prefer the latter, because it is less expensive for making large quantities and is more reproducible for synthesizing proteins once it is made. The synthesis of marker m RNA (we normally used alkaline phosphatase and OmpA) is induced under proper conditions, and total RNA is extracted. The stable mRNAs, such as outer membrane proteins OmpA and lipoprotein, remain and provide other marker precursors for translocation. All materials, and containers are autoclaved before use, when possible.

1.   PHOA AND OMPA mRNAs

Total RNA containing mRNa was prepared as described (Rhoads *et al.*, 1984; Chen *et al.*, 1985) with modifications adapted from Legault-Demare and Chambliss (1974). MC1000/pHI-5, which contains the *PhoA* gene in its plasmid (Inouye *et al.*, 1981), produces about 10 times as much alkaline phosphatas as did the wild type when phosphate is limited, and is used as the source for alkaline phosphatase mRNA and the more stable OmpA mRNA.

MC1000/pHI-5 cells in 2.5 liters are grown with forced oxygenation in phosphate-limited medium P2, containing 0.1 M Tris-maleate (pH 7.0), 10 mM KCl, 0.5 mM MgSO$_4$, 2 mM KH$_2$PO$_4$, 1 μM each of FeSO$_4$ and ZnSO$_4$, 1 mM Na citrate, 10 μg/ml thiamine, 1% glucose, 0.5% low-phosphate casamino acids (ICN Pharmaceuticals), 40 μg/ml leucine, and 10 μg/ml tetracycline. The culture is maintained near neutrality by addition of 2 M Tris base. At 1–2 hours after depletion of the phosphatase in the medium, the cells begin to induce alkaline phosphatase (qualitatively assayed by *p*-nitrophenyl phosphate hydrolysis); prewarmed chloramphenicol (50 ml, 2.5 mg/ml) is added and the culture is quickly poured over an equal volume of ice. The cells are harvested, washed, and suspended in

70 ml of Tris-HCl (pH 7.6) containing 10 mM each of KCl, MgCl$_2$, and NaN$_3$. SDS is added to 1% and the suspension is mixed immediately with an equal volume of redistilled m-cresol. The mixture is quickly passed through a French press at 10,000 lb/in^2 and centrifuged at 3650 g for 20 minutes. The upper aqueous phase is removed and saved. The lower phase is mixed with one-third volume each of the same Tris-buffer containing 1% SDS and m-cresol, and is passed through the French press again. After centrifugation, the aqueous phase is combined with the first aqueous solution and mixed with an equal volume of 4 M NaCl. The RNA is precipitated by the addition of two volumes of ethanol at −20°C for at least 2 hours or −76°C for 1 hour (or overnight). After centrifugation, the pellet is dissolved in water and is shaken with an equal volume of redistilled phenol at room temperature, and brought to 1 M NaCl, and ethanol is added to precipitate RNA as above. The pellet is dissolved in 20 ml TKM buffer; the solution is treated with 60 μg of DNase and then mixed drop by drop with 20 ml of 4M LiCl and set on ice for 30 minutes to precipitate large RNAs. Pellets are collected by centrifugation, dissolved, and precipitated with LiCl. The pellet is again dissolved and precipitated with ethanol as above. Portions are lyophilized and stored at −76°C. The optimal amount of mRNA for protein synthesis for each preparation is determined experimentally.

2. OmpA mRNA

To prepare the stable OmpA mRNA, the following simplified procedures are used. Cells of HJM114 containing plasmids pOmp9 and *pCI857* [the *ompA* is under λ promoter P$_L$ control with a temperature-sensitive repressor; obtained from Geller *et al.* (1986)], are grown at 30°C in LinA medium containing 100 μg/ml ampicillin and 40 μg/ml kanamycin until the Klett reading reaches 180. The culture is shifted to 42°C for 2 hours to induce OmpA mRNA, and the cells are harvested over ice and resuspended in the same buffer as above. Proteinase K (7 mg) and 4 ml of 20% SDS are added, and the mixture is shaken for 10 minutes then 30 ml of warm saturated phenol is added for another 15 minutes. After centrifugation at room temperature, the upper aqueous-phase solution is saved, and the phenol is back-extracted with 10 ml of the buffer. The combined aqueous upper "supernatant" (75 ml) is mixed with 75 ml of 4 M NaCl, and three volumes of cold ethanol is added. After storage at −20 or −76°C for 2 hours, the pellet is collected by centrifugation and resuspended in 20 ml of water. Phenol (20 ml) and 40 ml of chloroform are added and the mixture is shaken thoroughly. The upper aqueous phase (28 ml) from the first extract and back-extract is mixed with 9 ml of 4 M NaCL and three volumes of cold ethanol and kept at −20°C for 2 hours or overnight. The pellets are collected by centrifugation, resuspended in 20 ml of TKM buffer, and mixed with 20 ml of 4 M LiCl at 4°C for 30 minutes. The pellet, after centrifugation, is again suspended in 15 ml of water, then 5 ml of 4 M NaCl and 40 ml of cold ethanol as above. The pellet after centrifugation is thoroughly dried and resuspended in 10 ml of water and lyophilized in small portions, as above.

III. Protein Synthesis and Protein Translocation

A. Protein Synthesis and SDS–Gel Electrophoresis

Protein synthesis mixtures (in 1 ml) contain 50 mM Tris-HCl (pH 7.6), 40 mM KCl, 20 mM NH$_4$Cl, 0.1 ml of an ATP-regenerating stock (10 mM Tris-ATP, 0.2 mM GTP-Tris, 50 mM phosphoenol pyruvate-Tris, 30 μg pyruvate kinase), 6.5 mM Mg(OAc)$_2$, 1 mM spermidine-HCl, 8 mM putrescine-HCl, 0.1 ml of optimal amounts (usually 20 OD$_{260}$ U) of activated S30 (0.1 ml of S30, 2 μl of 2M Tris-HCl, pH 8.0, and 10 μl of ATP-regenerating stock, 37°C for 7 minutes), 2 mM dithiotheritol, and optimal amounts of mRNA (normally 200 μg of total RNA), 200 μCi of [^{35}S]methionine (1000 Ci/mmol), and 19 other amino acids at 0.05 mM each. Protein synthesis is carried out at 40°C for 10 to 15 minutes. [Note: synthesis of alkaline phosphatase is several fold higher at 40°C than at 37°C—see Rhoads *et al.* (1984)]. Samples (5–10 μl) are removed for analyzing precursor markers in SDS–gel electrophoresis.

SDS–PAGE (with 10% acrylamide) is carried out according to Laemmli (1970). To visualize the radioactive polypeptides, gels are treated with Autofluor (National Diagnostics), dried and exposed to Kodak XR-5 film at −76°C.

B. Cotranslational Translocation

The cotranslational translocation is carried out with the addition of inverted membrane vesicles during protein synthesis. The following procedures are adapted from Chen *et al.* (1985) and Chen and Tai (1987b), where more detailed discussion may be found. The protein synthesis mixture (95 μl, supplemented with additional 5 mM phosphoenol pyruvate to enhance translocation) is initiated at 40°C for 3 minutes (to minimize the inhibition of protein synthesis by membranes), then 5 μl of membrane vesicle (0.1 OD$_{280}$ U) is added and incubated for an additional 15 minutes to complete the translocation.

To analyze translocation into membrane vesicles, samples are exposed to proteinase K (100–500 μg/ml, which is about 10- to 50-fold more than necessary) at 0°C for 15 minutes, diluted with 1 ml of 10 mM Tris-HCl buffer, pH 7.6, containing 2 mM PMSF. The membranes are pelleted in a Beckman TLA 100.2 rotor at 90,000 rpm for 20 minutes or Ti50.3 rotor for 90 minutes. Carrier membranes (0.1 OD$_{260}$ U of inactive fraction III) is sometimes added just before centrifugation to improve the recovery and shorten the centrifugation time. The pellet is dissolved and analyzed in SDS–PAGE.

C. Posttranslational Translocation

Protein synthesis is carried out without membranes at 40°C for 15 minutes. Chloramphenicol (50–100 μg/ml) is added to stop further protein synthesis. This translational mixture (90 μl) may be used directly for posttranslational translocation with the addition of membrane vesicles (5 μl containing 0.1 OD$_{280}$ U) and 5 μl of additional ATP-regenerating stock and incubated at 40°C for 15 minutes. The translocated products in the membranes after proteinase K treatment are isolated and analyzed as above (see Fig. 2). Because precursors lose competence for translocation rapidly at 40°C (Chen et al., 1985), it is important that protein synthesis is not carried out too long. In fact, synthesis for only 10 minutes, though sacrificing some total synthesis, allows more efficient translocation in terms of percentage translocated products to total marker proteins. Translocation efficiency is usually 40–80% for OmpA and 10–25% for alkaline phosphatase.

In some instances, e.g., using different membranes, it may be desirable to remove precursor aggregates, ribosomes, and any contaminating membranes. The mixtures after translation are centrifuged in a Beckman TLA 100.2 rotor at 90,000 rpm for 20 minutes or 100,000 rpm for 10 minutes; the supernatant is then used for translocation with addition of desired membranes as above. In this case, the treatment of S30 with octylgucoside may be omitted.

To remove small molecules such as ATP and salts, or to exchange buffer (Chen and Tai, 1985), the mixtures are centrifuged through a Sephadex column. Three ml of Sephadex G-50–150 (Sigma) is packed in a 5-ml plastic syringe, equilibrated in 50 mM potassium phosphate (pH 7.6) (or any desired buffer), and placed in a conical plastic tube. Just before use, 1 ml of 0.5% bovine serum albumin in the buffer is layered onto the column and centrifuged in an International Equipment Model CL clinical centrifuge with swinging bucket at top speed for 2.5 minutes. A portion (0.3 ml) of mixtures after translocation is then layered onto the prespun Sephadex column, with a new conical plastic tube as a receptacle, and centrifuged for 2.5 minutes. If necessary, the centrifuged eluate can be centrifuged in a new Sephadex column once more to remove residual small

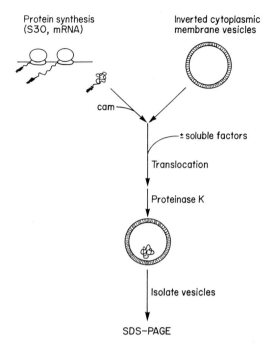

FIG. 2. Schematic illustration of posttranslational protein translocation assay. For co-translational translocation, chloramphenicol (cam) is omitted and membrane vesicles are added during protein synthesis.

molecules. The eluates are pooled and used for translocation. Translocation can then be carried out: mixtures, in 0.1 ml, contain 80 μl of centrifuged precursors in 50 mM phosphate (pH 7.6) buffer, 0.2 mM Mg(OAc)$_2$, 1 mM spermidine-phosphate, 8 mM putrescine-phosphate, 3 mM ATP, and 0.1 OD$_{280}$ U of membrane vesicles. [Note: the phosphate buffer may be changed to Tris-HCl buffer. As in many physiological systems, the translocation is more efficient with an ATP-regenerating system, instead of ATP alone (Chen and Tai, 1985). Thus, 10 μl of ATP-regenerating stock may replace 3 mM ATP in the translocation. Moreover, dithiothreitol is not necessary in this protocol for the translocation of OmpA and alkaline phosphatase precursor, but 1–2 mM DTT may be necessary in other translocation systems.]

If desirable, precursors may be purified away from other soluble factors by antibody affinity columns. After neutralization, the eluted precursors, such as OmpA or lipoprotein precursors, may be used directly for translocation. These affinity-purified precursors are sometimes more efficient for translocation.

178 PHANG C. TAI *et al.*

The limitations and advantages of *in vitro* translocation have been discussed (Tai, 1986). Some requirements and inhibitors of protein translocation have been reported (Rhoads *et al.,* 1984; Chen and Tai, 1985; 1986a,b, 1987a,b). The additional requirement of proton motive force in protein translocation is still controversial (see Chen and Tai, 1986a; Geller *et al.,* 1986; Tai, 1990b).

IV. Translocation of Lipoproteins

The export of the major lipoprotein (Lpp) into the *E. coli* outer membrane is unique in several respects. Lipoprotein is a small protein; its precursor undergoes modification and is cleaved by a specific prolipoprotein signal peptidase, signal peptidase II (Tokunaga *et al.,* 1982). Requirements for the translocation of lipoprotein are similar in someways but unique in others, as compared to requirements for nonlipoproteins such as alkaline phosphatase and OmpA (see Wu and Tokunaga, 1986; Tian *et al.,* 1989).

A. Preparation of S30 Containing Endogenous Lpp mRNA

Because of the stable nature of Lpp mRNA (Hirashima *et al.,* 1973), the S30 extracts prepared from cells upon the induction of a cloned *lpp* gene contain sufficient lipoprotein mRNA for *in vitro* protein synthesis. Cells of strain D10 carrying pYM140, which contains the cloned *lpp* gene (Coleman *et al.,* 1984), are grown in LinA medium supplemented with 0.5% glucose, 20 μg/ml methionine, and 40 μg/ml ampicillin, they are induced with 1 m*M* isopropyl-β-*D*-thiogalactopyranoside at Klett 100 units for one doubling. The culture is cold run-off at 15°C for 10 minutes, the cells are harvested, and the cell extracts are prepared as in Section II,C, except that octylglucoside treatment is omitted because the translocation is carried out posttranslationally, due to the small size of the lipoprotein.

B. Translation and Translocation

1. TRANSLATION

Synthesis of prolipoprotein is carried out as for the OmpA precursor (Section II,B) except that S30 extracts are not preactivated, 9 m*M*

$Mg(OAc)_2$ is used, and polyamines and exogenous mRNA are omitted. Protein synthesis is at 40°C for 10 minutes, chloramphenicol (50 $\mu g/ml$) is added to stop further synthesis, and the translation mixtures are centrifuged in a Beckman TLA 100.2 rotor at 90,000 rpm for 20 minutes or 100,000 rpm for 10 minutes to remove ribosomes, aggregates, and endogenous membrane vesicles. The supernatants are used directly for translocation or the precursors are purified.

2. REMOVAL OF NUCLEOTIDES AND SMALL MOLECULES

The supernatant (0.5 ml) of the translation mixture is applied to a Sephadex G-25–80 column (1 × 40 cm) that is equilibrated with 1 mM DTT and 50 mM Tris-HCl, pH 7.6 (or any desired buffer), and eluted with the same buffer. Fractions are collected and those containing prolipoprotein near the void volume are pooled and used for determining the energy source and other requirements.

3. PURIFICATION OF PROLIPOPROTEIN

The supernatant (0.4 ml) of the translation mixture is applied to a column of antilipoprotein IgG coupled to Sepharose (Dev *et al.*, 1985) that has been equilibrated with DTNE buffer containing 1 mM DTT, 20 mM Tris-HCl (pH 8.0), 150 mM NaCl, and 5 mM EDTA. After binding, the column is first washed extensively with the same buffer and then with 20 mM Tris-HCl (pH 8.0)–1 mM DTT. Prolipoprotein is eluted either with 0.2M glucine-HCl (pH 2.2)–1 mM DTT and adjusted with Tris base to pH 7.6 or eluted with 0.1 M triethylamine, pH 11.5–1 mM DTT and adjusted to pH 7.6 with 1 M KH_2PO_4 (pH 4.7) (900 μl of eluant with 45 μl of the neutralizing buffer) and used immediately for translocation.

To prepare prolipoprotein that is free of other potential complexed proteins, the translational products are treated with 1% SDS at 100°C for 3 minutes, cooled, and diluted 20-fold with DTNE buffer containing 1% Triton X-100. This mixture is applied to the lipoprotein antibody column, washed successively with DTNE buffer containing 1% Triton X-100, DTNE buffer, and 20 mM Tris-HCl (pH 8.0)–1 mM DTT. The prolipoprotein is eluted and treated as above.

4. TRANSLOCATION

Crude translation mixture supernatant (90 μl) may be used directly for translocation with near 100% efficiency by the addition of 5 μl of membrane vesicles (0.1 OD_{280} U) and 5 μl of ATP-regenerating stock. For purified prolipoprotein, the translocation mixtures (0.1 ml) contain the precursors in the respective buffer, membrane vesicles (0.1 OD_{280} U), 1 mM spermidine-HCl, 8 mM putrescine-HCl, 1 mM DTT, 0.2 mM Mg(OAc)$_2$, 3 mM ATP, or 5 μl of ATP-regenerating stock. After incubation at 37°C for 10 minutes, the translocation mixtures are cooled, treated with 15 μg/ml of proteinase K at 0°C for 15 minutes, and diluted with 1 ml of 10 mM Tris-HCl (pH 7.6) buffer containing 2 mM PMSF. Membrane vesicles containing translocated products are pelleted in a Beckman TLA 100.2 rotor at 90,000 rpm for 20 minutes and dissolved for electrophoresis.

Unlike nonlipoproteins, the processing of lipoprotein is inhibited by globomycin (Inukai *et al.*, 1978), giving rise to the unprocessed modified form (Tian *et al.*, 1989). Moreover, the denser membrane fraction III, which is normally not active for nonlipoproteins, is often active for the translocation of lipoprotein (Table I).

5. POLYACRYLAMIDE GEL ELECTROPHORESIS

Because of the small size and modifications (Tian *et al.*, 1989), various forms of lipoproteins are separated on 15% acrylamide–0.4% bisacrylamide gels in 375 mM Tris-HCl (pH 8.8) buffer containing 1% SDS and 7.5 M urea at 35 mA for 5 hours (the stacking gel and the running buffer are the same as in Laemmli gel). The gels are fixed with 10% trichloroacetic acid–30% methanol, and treated with Autofluor for fluorograms as in Section II,A.

The translocation (insertion) of lipoprotein is a temperature-dependent, spontaneous process, occurring with negatively charged liposomes in the absence of an energy source. However, the overall process, including modification and cleavage of precursors, requires ATP, SecA, and other membrane proteins, but not SecB (Tian *et al.*, 1989; Yu and Tai, 1991; Lian and Tai, 1991).

V. *In Vitro* Characterization of Genetically Defined Translocation Components

As in any *in vitro* system to study a physiological process, it is important to correlate the *in vitro* findings with *in vivo* functions, and vice versa. The concurrent genetic and biochemical approaches in protein secretion in *E. coli* make it possible to utilize the *in vitro* system to characterize the genetically identified translocation components (see Fandl and Tai, 1990); fractionation of the *in vitro* system facilitates defining the components of the translocation machinery and their interactions. To examine the effects of mutations on the functions of the gene products in the *in vitro* system, either the soluble protein fraction or the membrane fraction from mutant cells grown under restrictive conditions is used in combination with wild-type fractions. Alternatively, the role of a genetically defined component can be examined by immunochemically removing it from the soluble fraction, or by inactivating the component *in vitro*. Both these approaches work well; it is especially powerful when used in combination (see Cabelli *et al.*, 1988), because mutants grown at nonpermissive conditions may have an unusual cell physiology (see de Cock *et al.*, 1989).

A. SecA

The *secA* gene encodes a 102-kDa soluble protein (SecA) that is also found peripherally bound to the cytoplasmic membrane (Oliver and Beckwith, 1982; Schmidt *et al.*, 1988). SecA is essential for cell growth, and mutations in *secA* cause pleiotropic defects in protein translocation (Oliver and Beckwith, 1981) or in the suppression of signal sequence defects (*prlD* alleles) (Bankaitis and Bassford, 1985; Fikes and Bassford, 1989; Stader *et al.*, 1989). The essential requirement of SecA for *in vitro* protein translocation has recently been demonstrated (Cabelli *et al.*, 1988). Membranes with greatly reduced levels of SecA due to a conditional *secA* amber mutation, *secA13(am)*, are inactive for protein translocation only if the soluble component of the *in vitro* system is also depleted in SecA, due either to the conditional amber mutation or immunochemical depletion of the soluble fraction; the translocation activity can be restored with purified SecA (Cabelli *et al.*, 1988). SecA possesses ATPase activity and may be related to an ATP requirement (Lill *et al.*, 1989).

To prepare S30 and membranes free of SecA, *E. coli* strain BL15.5 (F$^-$, *rna19 relA1[λ] secA13 (am) supF^{ts}*) is grown in LinA medium supplemented with 0.5% glucose at 30°C. At a cell density of 30–35 Klett units, the cultures are rapidly shifted to 40°C and continued until growth ceases (to 250–300 Klett units, usually about 2–3 hours). The cells are harvested and prepared for membranes and S30 as in Section II. Under proper growth dilution, little SecA (determined immunologically) remains in S30, whereas membranes retain some inactive SecA. Translocation activity in S30 or the membranes can be restored by addition of purified SecA (Cabelli *et al.*, 1988).

B. SecY

The *secY* gene (also known as the *prlA* gene) encodes a 49-kDa integral membrane protein (SecY) with 10 potential transmembrane domains

(Akiyama and Ito, 1985, 1987). Mutations in *secY* result in a pleiotropic accumulation of precursors of secretory proteins (Shiba *et al.*, 1984) or in suppressing signal sequence defects (Emr *et al.*, 1981). The translocation defect of a temperature-sensitive mutation (*secY24*) has been demonstrated in membranes prepared from cells grown at a restrictive temperature (Bacallo *et al.*, 1986; Fandl and Tai, 1987), or, more specifically, from membranes that are functionally inactivated *in vitro* (Fandl and Tai, 1987). The translocation defect of SecY24 membranes can be significantly compensated by a soluble fraction, including SecA and SecB (Fandl *et al.*, 1988; Kumamoto *et al.*, 1989).

To prepare SecY24 membranes inactivated at a restrictive temperature, IQ85 cultures (MC4100, *secY24*), isolated by Shiba *et al.* (1984), in LinA medium supplemented with 0.5% glucose, are grown at 32°C to a cell density of 150 Klett units and are shifted to 42°C for 2.5 hours (to about 350 Klett units). The membranes are prepared according to standard procedures. To prepare SecY24 membranes inactivated *in vitro*, the functional membranes are prepared from IQ85 cells grown at 32°C and incubated at 40°C [1 OD_{280} U SecY24 membrane per 0.1 ml of 10 mM Tris-HCl (pH 7.6), 50 mM KCl]; almost 90% of subsequent translocation activity is lost after a 30-minute incubation (Fig. 3).

SecY can also be depleted from cells harboring an upstream polar amber mutation resulting in the impaired expression of the *secY* gene (Ito *et al.*, 1984). Cultures of KI200 [MC4100, *rplOam215 Tn10* (*ϕ80sus2supF*[ts])] grown exponentially in LinA medium supplemented with 0.5% glucose at 32°C are diluted to a cell density of 10 Klett units in the same medium prewarmed at 42°C. Cells are harvested when growth ceases, at about 360 Klett units, and the membranes are prepared as in standard procedures. Such membranes contain less than 3% of SecY as normal membranes, determined immunologically, but are about 50% as active for translocating OmpA precursor, and even more for lipoprotein precursors (Lian and Tai, 1991), suggesting that under certain conditions, translocation can be quite efficient with little SecY.

C.   SecB

The *secB* gene encodes a 16-kDa soluble protein (SecB) that is not essential for cell growth and affects the translocation of a subset of secretory proteins (Kumamoto and Beckwith, 1983). The primary role of SecB appears to be targeting and modulation of some precursor molecules, and its functions in *in vitro* translocation have been shown (Weiss *et al.*, 1988; Kumamoto *et al.*, 1989; Watanabe and Blobel, 1989; Lecer *et al.*, 1989). Because SecB is not essential for cell growth, the deprivation of SecB in

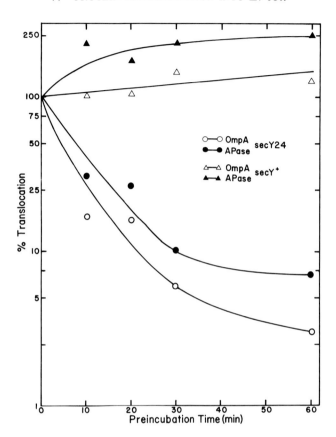

FIG. 3. *In vitro* heat inactivation of SecY24 membrane protein translocation activity. Membranes from strain IQ85 (*secY24*) and from IQ86 (SecY⁺) grown at 32°C were incubated in 10 m*M* Tris-HCl (pH 7.5) per 50 m*M* KCl at 40°C for the times indicated, then were added to translocation mixtures and incubated at 30°C for 25 minutes. Posttranslational transloca-tion of OmpA protein and alkaline phosphatase (APase) are expressed as a percentage of the activity of nonheated membranes. From Fandl and Tai (1987).

cell extracts can be easily achieved by growing the SecB null strain MM152 (MC4100 *zhe::Tn10 malT^c secB::Tn5*) in minimal medium supplemented with 0.5% glycerol (Kumamoto and Beckwith, 1983), or by removing it immunologically. Purified SecB then can be used to restore translocation activity.

D. Other *sec* Gene Products

Genetic studies indicated that the gene products of *secE(prlG)*, *secD*, and *secF* are also involved in protein translocation (Gardel *et al.*, 1987;

Riggs *et al.*, 1988; Schatz *et al.*, 1989; Stader *et al.*, 1989; Bieker and Silhavy, 1990; Gardel *et al.*, 1990). *In vitro* studies have shown that SecE may be required (Brundage *et al.*, 1990), but further experiments are necessary to establish the roles of SecE and other Sec proteins. Involvement of GroEL in protein translocation has also been shown (Bochkareva *et al.*, 1988; Philips and Silhavy, 1990).

ACKNOWLEDGMENTS

We appreciate the contribution of D. Rhoads, L. Chen, and J. Fandl in the early development and improvement of the *in vitro* translocation system. We also thank J. Fandl for Fig. 2. The experimental work in our studies has been supported by NIH Grants GM 34766 and GM 41845.

REFERENCES

Akiyama, Y., and Ito, K. (1985). The SecY membrane component of the bacterial protein export machinery: Analysis by new electrophoretic methods for integral membrane proteins. *EMBO J.* **4,** 3351–3356.

Akiyama, Y., and Ito, K. (1987). Topology analysis of the SecY protein, an integral membrane protein involved in protein export in *Escherichia coli. EMBO J.* **6,** 3465–3470.

Bacallo, R., Crooke, E., Shiba, K., Wickner, W., and Ito, K. (1986). The SecY protein can act post-translationally to promote bacterial protein export . *J. Biol. Chem.* **261,** 12907–12910.

Bankaitis, V. A., and Bassford, P. J., Jr. (1985). Proper interaction between at least two components is required for efficient export of proteins to the *Escherichia coli* cell envelope. *J. Bacteriol.* **161,** 169–178.

Bieker, K. L., and Silhavy, T. J. (1990). PrlA(Sec Y) and PrlG(SecE) interact directly and function sequentially during protein translocation in *E. coli. Cell (Cambridge, Mass.)* **61,** 833–842.

Bieker, K. L., Phillips, G. J., and Silhavy, T. J. (1990). The *sec* and *prl* genes of *Escherichia coli. J. Bioenerg. Biomembr.* **22,** 291–310.

Blobel, G., and Dobberstein, B. (1975). Transfer of protein across membranes. I. Presence of proteolytically processed and unprocessed nascent immunoglobulin light chains on membrane-bound ribosomes of murine myeloma. *J. Cell Biol.* **67,** 835–851.

Bochkareva, E. S., Lissin, N. M., and Girshovich, A. S. (1988). Transient association of newly synthesized unfolded proteins with the heat-shock GroEL protein. *Nature (London)* **336,** 254–257.

Brundage, L., Hendrick, J. P., Schiebel, E., Arnold, J. M., Driessen, A., and Wickner, W. (1990). The purified *E. coli* integral membrane protein SecY/E is sufficient for reconstitution of SecA-dependent precursor protein translocation. *Cell (Cambridge, Mass)* **62,** 649–657.

Cabelli, R. J., Chen, L. L., Tai, P. C., and Oliver, D. B. (1988). SecA protein is required for secretory protein translocation into *E. coli* membrane vesicles, *Cell (Cambridge, Mass.)* **55,** 683–692.

Chen, L. L., and Tai, P. C. (1985). ATP is essential for protein translocation into *Escherichia coli* membrane vesicles. *Proc. Natl. Acad. Sci. U. S. A.* **82,** 4384–4388.

Chen, L. L., and Tai, P. C. (1986a). Roles of H^+-ATPase and proton motive force in ATP-dependent protein translocation *in vitro. J. Bacteriol.* **167**, 389–392.

Chen, L. L., and Tai, P. C. (1986b). Effects of nucleotides on ATP-dependent protein translocation into *Escherichia coli* membrane vesicles. *J. Bacteriol.* **168**, 828–843.

Chen, L. L., and Tai, P. C. (1987a). Effects of antibiotics and other inhibitors on ATP-dependent protein translocation into membrane vesicles. *J. Bacteriol.* **169**, 2373–2379.

Chen, L. L., and Tai, P. C. (1987b). Evidence for the involvement of ATP in cotranslational protein translocation. *Nature (London)* **328**, 164–166.

Chen, L. L., Rhoads, D., and Tai, P. C. (1985). Alkaline phosphatase and OmpA protein can be translocated posttranslationally into membrane vesicles of *Escherichia coli. J. Bacteriol.* **161**, 973–980.

Coleman, J., Green, P. J., and Inouye, J. (1984). The use of RNAs complementary to specific mRNAs to regulate the expression of individual bacterial genes. *Cell (Cambridge, Mass.)* **37**, 429–436.

Collier, D. N., Bankaitis, V. A., Weiss, J. B., and Bassford, P. J. (1988). The antifolding activity of SecB promotes the export of the *E. coli* maltose-binding protein. *Cell (Cambridge, Mass.)* **33**, 273–283.

Crooke, E., and Wickner, W. (1987). Trigger factor: A soluble protein that folds pro-OmpA into a membrane-assembly competent form. *Proc. Natl. Acad. Sci. U.S.A.* **84**, 5216–5220.

Crooke, E., Brundage, L., Rice, M., and Wickner, W. (1988). ProOmpA spontaneously folds in a membrane assembly competent state which trigger factor stabilizes. *EMBO J.* **1**, 1831–1835.

Cunningham, K., Lill, R., Crooke, E., Rice, M., Moore, K., Wickner, W., and Oliver, D. (1989). SecA protein, a peripheral protein of the *Escherichia coli* plasma membrane, is essential for the functional binding and translocation of proOmpA. *EMBO J.* **8**, 955–959.

Davis, B. D., and Tai, P. C. (1980). The mechanism of protein secretion across membranes. *Nature (London)* **283**, 433–438.

de Cock, H., Meeldijk, J., Overduin, P., Verkleij, A., and Tommassen, J. (1989). Membrane biogenesis in *Escherichia coli:* Effects of *secA* mutation. *Biochim. Biophys. Acta* **985**, 313–319.

Dev, I. K., Harvey, R. J., and Ray, P. H. (1985). Inhibition of prolipoprotein signal peptidase by globomycin. *J. Biol. Chem.* **260**, 5891–5894.

Emr, S. D. Hanley-Way, S., and Silhavy, T. J. (1981). Suppressor mutations that restore export of a protein with a defective signal sequence. *Cell (Cambridge, Mass.)* **23**, 79–88.

Fandl, J. P., and Tai, P. C. (1987). Biochemical evidence for the *secY24* defect in *Escherichia coli* protein translocation and its suppression by soluble cytoplasmic factors. *Proc. Natl. Acad. Sci. U.S.A.* **84**, 7448–7452.

Fandl, J. P., and Tai, P. C. (1990). Protein translocation *in vitro:* Biochemical characterization of genetically defined translocation components. *J. Bioenerg. Biomembr.* **22**, 369–387.

Fandl, J. P., Cabelli, R., Oliver, D., and Tai, P. C. (1988). SecA suppresses the temperature-sensitive SecY24 defect in protein translocation in *Escherichia coli* membrane vesicles. *Proc. Natl. Acad. Sci. U.S.A.* **84**, 7448–7452.

Fikes, J. D., and Bassford, P. J., Jr. (1989). Novel *secA* alleles improve export of maltose-binding protein synthesized with a defective signal peptide. *J. Bacteriol.* **171**, 402–409.

Gardel, C., Benson, S., Hunt, J., Michaelis, S., and Beckwith, J. (1987). *secD*, a new gene involved in protein export in *Escherichia coii. J. Bacteriol.* **169**, 1286–1290.

Gardel, C., Johnson, K., Jacq, A., and Beckwith, J. (1990). The *secD* locus of *E. coli* codes for two membrane proteins required for protein export. *EMBO J.* **9**, 3209–3216.

Geller, B. L., Moova, N. R., and Wickner, V. (1986). Both ATP and the electrochemical potential are required for optimal assembly of pro-OmpA into *Escherichia coli* inner membrane vesicles. *Proc. Natl. Acad. Sci. U.S.A.* **83**, 4219–4222.

Gesteland, R (1966). Isolation and characterization of ribonuclease I mutants of *Escherichia coli. J. Mol. Biol.* **16**, 67–84.

Hirashima, A., Childs, G., and Inouye, M. (1973). Differential inhibitory effects of antibiotics on the biosynthesis of envelope proteins of *Escherichia coli. J. Mol. Biol.* **79**, 373–379.

Inouye, H., Michaelis, S., Wright, A., and Beckwith, J. (1981). Cloning and restriction mapping of the alkaline phosphatase structural gene *phoA* of *Escherichia coli* and generation of deletion mutants *in vitro. J. Bacteriol.* **146**, 668–675.

Inukai, M., Takeuchi, M. Shimizu, K., and Arai, M. (1978). Mechanism of action of globomycin. *J. Antibiot.* **31**, 1203–1205.

Ito, K., Cerretti, D., Nashimoto, H., and Nomura, M. (1984). Characterization of an amber mutation in the structural gene for ribosomal protein, L15, which impairs the expression of the protein export gene, SecY, in Escherichia coli. EMBO J. 3, 2319–2324.

Kumamoto, C. A., and Beckwith, J. (1983). Mutations in a new gene, secB, cause defective protein localization in *Escherichia coli. J. Bacteriol.* **154**, 253–260.

Kumamoto, C. A., Chen L. L., Fandl, J. P., and Tai, P. C. (1989). Purification of the *Escherichia coli secB* gene product and demonstration of its activity in an *in vitro* protein translocation system. *J. Biol. Chem.* **264**, 2242–2249.

Laemmli, U. K. (1970). Cleavage of structural proteins during the assembly of the head of bacteriophage T4. *Nature (London)* **227**, 680–685.

Lecker, S., Lill, R., Ziegelhoffer, T., Georgopoulos, C., Bassford, P., Kumamoto, C., and Wickner, W. (1989). Three pure chaperone proteins of *Escherichia coli*—SecB, trigger factor, GroEL—form soluble complexes with precursor proteins *in vitro. EMBO J.* **8**, 2703–2709.

Legault-Demare, L., and Chambliss, G. (1974). Natural messenger ribonucleic acid-directed cell-free protein synthesizing system of *Bacillus subtilis. J. Bacteriol.* **120**, 1300–1307.

Lill, R., Cunningham, K., Brundage, L., Ito, K., Oliver, D., and Wickner, W. (1989). SecA protein hydrolyzes ATP and is an essential component of the protein translocation ATPase of *Escherichia coli. EMBO J.* **8**, 961–966.

Muller, M., and Blobel, G. (1984). *In vitro* translocation of bacterial proteins across the plasma membrane of *Escherichia coli. Proc. Natl. Acad. Sci. U.S.A.* **81**, 7421–7425.

Oliver, D. B., and Beckwith, J. (1981). *E. coli* mutant pleiotropically defective in the export of secreted proteins. *Cell (Cambridge, Mass.)* **25**, 765–772.

Oliver, D. B., and Beckwith, J. (1982). Regulation of a membrane component required for protein secretion in *E. coli. Cell (Cambridge, Mass.)* **30**, 311–319.

Osborn, M. J., and Munson, R. (1974). Separation of the inner (cytoplasmic) and outer membranes of gram-negative bacteria. *In* "Methods in Enzymology" (S. Fleischer and L. Pucker, eds.), Vol. 32, pp. 642–650. Academic Press, New York.

Phillips, G. J., and Silhavy, T. J. (1990). Heat-shock proteins DnaK and GroEL facilitate export of LacZ hybrid proteins in *E. coli. Nature (London)* **344**, 882–884.

Rhoads, D., Tai, P. C., and Davis, B. D. (1984). Energy-requiring translocation of the OmpA protein and alkaline phosphatase of *Escherichia coli* into inner membrane vesicles. *J. Bacteriol.* **159**, 63–70.

Riggs, P. D., Derman, A. I., and Beckwith, J. (1988). A mutation affecting the regulation of a *secA–lacZ* fusion defines a new *sec* gene. *Genetics* **118**, 571–579.

Schatz, P. J., Riggs, P. D., Jacq, A., Fath, M. J., and Beckwith, J. (1989). The *secE* gene encodes an integral membrane protein required for protein export in *E. coli. Genes Dev.* **3**, 1035–1044.

Schmidt, M., Rollo, E., Grodberg, J., and Oliver, D. (1988). Nucleotide sequence of the *secA* gene and *secA* temperature-sensitive mutations preventing protein export in *Escherichia coli. J. Bacteriol.* **170**, 3404–3414.

Shiba, K., Ito, K., Yura, T., and Cerretti, D. (1984). A defined mutation in the protein export gene with *spc* ribosomal protein operon of *Escherichia coli:* Isolation and characterization of a new temperature-sensitive *secY* mutant. *EMBO J.* **3**, 631–635.

Stader, J., Gansheroff, L., and Silhavy, T. J. (1989). New suppressors of signal-sequence mutations, *prlG*, are linked tightly to the *secE* gene of *Escherichia coli*. *Genes Dev.* **3**, 1045–1052.

Tai, P. C. (1986). Biochemical studies of bacterial protein export. *Curr. Top. Microbiol. Immunol.* **125**, 43–58.

Tai, P. C. (1990a). Protein export in bacteria: Structure, function and molecular genetics. *J. Bioenerg Biomembr.* **22**, 209–491.

Tai, P. C. (1990b). Energetic aspects of protein insertion and translocation into or across membranes *In* "Bacterial Energetics" (T. Krulwich, ed.), "The Bacteria" Vol. 12, pp. 393–416. Academic Press, San Diego, California.

Tian, G., Wu, H. C., Ray, P. H., and Tai, P. C. (1989). Temperature-dependent insertion of prolipoprotein into *Escherichia coli* membrane vesicles and subsequent requirements of ATP, soluble factors and functional SecY protein for the overall translocation process. *J. Bacteriol.* **171**, 1987–1997.

Tokunaga, M., Loranger, J. M., Wolfe, P. B., and Wu, H. C. (1982). Prolipoprotein signal peptidase in *Escherichia coli* is distinct from the M13 procoat protein signal peptidase. *J. Biol. Chem.* **257**, 9922–9925.

Watanabe, M., and Blobel, G. (1989). Cytosolic factor purified from *Escherichia coli* is necessary and sufficient for the export of a preprotein and is a homotetramer of SecB. *Proc. Natl. Acad. Sci. U.S.A.* **86**, 2728–2732.

Weiss, J. B., Ray, P. H., and Bassford, P. J. (1988). Purified SecB protein of *Escherichia coli* retards folding and promotes membrane translocation of the maltose-binding protein *in vitro*. *Proc. Natl. Acad. Sci. U.S.A.* **85**, 8978–8982.

Witholt, B., Boekhout, M., Brock, M., Kingman, J., van Heerikhuizen, H., and de Leij, L. (1976). An efficient and reproducible procedure for the formation of spheroplasts from variously grown *Escherichia coli*. *Anal. Biochem.* **74**, 160–170.

Wu, H. C., and Tokunaga, M. (1986). Biogenesis of lipoproteins in bacteria. *Curr. Top. Microbiol. Immunol.* **125**, 127–157.

Yamane, K., Matsuyama, S. I., and Mizushima, S. (1988). Efficient *in vitro* translocation into *Escherichia coli* membrane vesicles of a protein carrying an uncleavable signal peptide. *J. Biol. Chem.* **263**, 5368–5372.

Yu, J., and Tai, P. C. (1991). In preparation.

Chapter 8

Membrane Components of the Protein Secretion Machinery

KOREAKI ITO AND YOSHINORI AKIYAMA

Institute for Virus Research
Kyoto University
Kyoto 606, Japan

I. Introduction

Protein translocation across the biological membrane, a key process in protein secretion and membrane protein biogenesis, involves a series of protein–protein interactions. In other words, it is a process facilitated by cellular proteinaceous factors. The secretion factors of *Escherichia coli* have been defined genetically by isolation of the *sec* and *prl* mutations (see

189

Bieker *et al.*, 1990, for review). Some of them are hydrophilic proteins in the cytoplasm (e.g., SecB) or on the cytoplasmic periphery of the plasma membrane (cytoplasmic membrane) (e.g., SecA), whereas other secretion factors are hydrophobic proteins integrally associated with the membrane (e.g., SecY). In addition to SecY (see Ito, 1990, for review), membrane factors now include SecE, SecD, and SecF, although this assignment for the latter three proteins is based only on analyses of the deduced amino acid sequences and the PhoA fusion proteins (Schatz *et al.*, 1989; C. Gardel, K. Johnson, A. Jacq, and J. Beckwith, personal communication).

Obviously, integral membrane components, which may catalyze trans-bilayer movement of the translocating polypeptide chain, are important to study, but they are often difficult to handle because of their hydrophobic character. We have been characterizing the *secY(prlA)* gene product, the first multipath membrane protein identified as a secretion factor, and this chapter describes some technical tips that we learned from the studies of the SecY protein. First we summarize what is known about the SecY protein from the methodological viewpoints, and then we describe some methods useful for enrichment, detection, and identification of hydrophobic membrane proteins.

II.   SecY as an Integral Membrane Protein

A.   Overproduction of SecY

The gene *secY* is located within a multicistronic operon encoding 11 ribosomal and one membrane (SecY) proteins (Cerretti *et al.*, 1983). The expression level of *secY* is much lower than those of the ribosomal proteins (Ueguchi *et al.*, 1989), presumably due to inefficient translation initiation. In addition, translation of *secY* is coupled to the continued translation of the upstream genes (Ito *et al.*, 1984). To ensure a highlevel overproduction of SecY from a cloned *secY* gene, it is important not only that *secY* is fused to an efficient promoter, but also that it is preceded by an efficient translation initiation region such that the upstream *rplO* gene fragment is actively translated (Akiyama and Ito, 1985). Ideally, *secY* should be fused directly to the initiation codon of a highly expressed gene by site-directed DNA manipulations. To maintain a SecY-overexpressible plasmid, the promoter should be strictly controllable, because a moderate overproduction of SecY is deleterious to the cell. The best overproducer we constructed is pKY120 (Y. Akiyama and K. Ito, unpublished results), in which *secY* is placed under the *lac* promoter on pHSG399 and fused in frame to the initiation codon of *lacZ*. pHSG399 is a pUC-derived plasmid containing the chloramphenicol-resistance gene instead of the ampicillin-resistance gene (Takeshita *et al.*, 1987); we avoid the presence of a vector gene whose product interacts in some way with the membrane. To maintain such a SecY overexpressible plasmid, the host cell should contain both the *lac* repressor-overproducing mutation (*lacI*^{Q}) and an adenylate cyclase-deficient mutation (*cya*). Induction is achieved by addition of 1 m*M* iso-propyl-β-D-thiogalactoside (IPTG) and 5 m*M* cyclic AMP (Akiyama and Ito, 1985).

Although the synthesis rate of SecY can be accelerated markedly by an overexpression plasmid, amount of the SecY accumulation is limited, because overproduced SecY molecules are rapidly degraded *in vivo* (Akiyama and Ito, 1986). The proteolytic system responsible for this degradation has not been identified, but it works somewhat less at lower temperature. On the other hand, the copy number of a pUC plasmid increases at higher temperature. Thus, by amplifying pKY120 or pNO1573 (Akiyama and Ito, 1986) at 42°C under the uninduced conditions and then allowing induction of SecY at 30°C, accumulation of SecY in amounts 10- to 20-fold the normal level can be achieved (Akiyama and Ito, 1986, also unpublished results).

The instability of the overproduced SecY may be due to the lack of simultaneous overproduction of some other component(s) that is normally complexed with SecY. One such candidate is the SecE protein (Schatz *et al.*, 1989), because more SecY accumulates when both SecY and SecE are oversynthesized than when only SecY is oversynthesized (Matsuyama *et al.*, 1990).

B. Localization of SecY

Evidence for the SecY protein being an integral component of the cytoplasmic membrane comes from the following observations. (1) It fractionates with the cytoplasmic membrane in the standard method (Osborn and Munson, 1974) of membrane fractionation, including isopycnic sucrose gradient centrifugation. This is true for both the amplified and unamplified SecY (Akiyama and Ito, 1985). (2) It remains rapidly sedimenting after treatment with alkaline solution (see below). This is true for the unamplified SecY protein, but not necessarily for the amplified SecY protein. (3) It contains 10 hydrophobic segments of sufficient hydrophobicity and length to span the membrane (Cerretti *et al.*, 1983; Akiyama and Ito, 1987). (4) It partitions to the detergent-containing gel during detergent blotting (see below). (5) It is partly protected from trypsin digestion unless membrane is solubilized (Akiyama and Ito, 1985). (6) Finally, the existence of multiple transmembrane segments in SecY is suggested by the analysis of a series of SecY–PhoA (alkaline phosphatase) fusion proteins (see below).

C. Analysis of SecY–PhoA Fusion Proteins

Fusions of the alkaline phosphatase mature sequence to various regions of SecY yield SecY–PhoA fusion proteins whose PhoA portions are either periplasmically exposed or internal to the cytoplasm, depending upon the site of fusion joint along the SecY sequence (Akiyama and Ito, 1987). The PhoA fusion technique is described in Chapter 3 by Manoil. Our application of this method to the analysis of SecY protein topology confirmed that it is a powerful way of analyzing the topological disposition of a complex membrane protein. The plasmid used for fusion construction contained a *lac* promoter-controlled *secY* fragment, but its translation depended only on its own initiation signal; the expression level of *secY* was low and just enough to express the biological activity (to complement the *secY* mutation).

Localization of the PhoA moiety of the SecY–PhoA fusion proteins was assigned not only by the standard method of phosphatase activity measurement, but also by a trypsin sensitivity assay of the PhoA domain in detergent extracts of the cell (Akiyama and Ito, 1989). Because this assay has not been used in other studies of PhoA fusions, we briefly decribe the details. Labeled cells (100–200 μl) are centrifuged, washed with 10 mM Tris-HCl (pH 8.1), and resuspended in 100 μl of 20% (w/v) sucrose–30 mM Tris-HCl (pH 8.1), to which 10 μl of 1 mg/ml lysozyme freshly dissolved in 0.1 M EDTA (pH 7.5) is added. After incubation at 0°C for 30 minutes, a 45-μl portion is mixed with 2.5 μl of 20% Triton X-100 and subsequently with 2.5 μl of trysin (500 μg/ml in 10 mM Tris-HCl, pH 8.1), followed by incubation at 0°C for 30 minutes. The reaction is stopped by 1 mM (final concentration) of phenymethylsulfonyl fluoride, 3 mM tosyl-L-lysine chloromethylketone, and 100 μg/ml of soybean trypsin inhibitor. The mixture is then treated with trichloroacetic acid (5% final concentration). The acid-denatured proteins are collected by centrifugation, washed with 5% trichloroacetic acid and then with acetone, and dissolved in 1% SDS–50 mM Tris-HCl (pH 8.1)–1 mM EDTA, containing the inhibitors of trypsin as specified above, by incubation at 37°C for 5 minutes. This is then diluted with Triton buffer (Ito *et al.*, 1981) or Lubrol buffer (see Section III,D) supplemented with the trypsin inhibitors for immunoprecipitation with antialkaline phosphatase (Ito *et al.,* 1981). It is important that all the reagents after the trysin treatment contain the inhibitors of trypsin. The fusion protein whose PhoA domain is exported to the periplasm yields a trypsin-resistant (correctly folded) PhoA fragment of about 50,000 DA, whereas the internalized PhoA polypeptide is completely digested under these condition.

The results with the SecY–PhoA fusion proteins, together with the other observations described in the preceding section, support the notion that SecY contains five periplasmic regions, 10 transmembrane segments, and six cytoplasmic domains, including both termini (Akiyama and Ito, 1987, 1989).

D. *In Vitro* Proteolytic Cleavage of SecY and Its Prevention

Proteolytic degradation causes problems not only *in vivo* for the over-produced SecY, but also *in vitro* after cell disruption. Significant proportions of SecY in isolated membrane vesicle preparations can be in a cleaved form, in which the cleavage occurs at the central cytoplasmic region into the amino- and carboxy-terminal fragments (Akiyama and Ito, 1990). The extent of this cleavage varies depending upon how carefully the sample was maintained at low temperatures during membrane isolation, the length of storage, and how many times the membrane preparation experienced freezing and thawing. This cleavage is catalyzed by an outer membrane-located protease, OmpT, and it is generally difficult to completely remove outer membrane fragments from a cytoplasmic membrane preparation. The OmpT-catalyzed cleavage of SecY is enhanced after solubilization of the membranes with a detergent. Thus, it is important to use strains that do not contain the OmpT protease as starting material for *in vitro* studies of protein translocation, including solubilization–reconstitution of the membrane components.

We constructed an insertion mutation in the OmpT structural gene. This mutation, *ompt::kan*, can easily be introduced into other specific *E. coli* strains by P1 phage-mediated transduction, selecting for kanamycin (10 μg/ml)-resistant transductants. In the cell extract of the *ompT::kan* mutant, SecY remains intact even after incubation at 37°C for up to 20 hours (Akiyama and Ito, 1990).

Because the OmpT protease has specificity toward paired basic amino acids, and the cytoplasmic domains of integral membrane proteins are often enriched for basic residues, membrane proteins other than SecY can also be a substrate for this protease.

III. Some Methods Useful for Enrichment and Identification of Integral Membrane Proteins

A. Gel Electrophoresis of Membrane Proteins

1. SDS–POLYACRYLAMIDE GEL ELECTROPHORESIS IN ONE DIMENSION

A class of hydrophobic membrane proteins, including SecY, aggregates when heated at 100°C after solubilization with SDS. Instead of this standard procedure of sample preparation in SDS–PAGE, incubation in SDS–sample buffer at 37°C for 5 minutes should be used for electrophoresis of samples containing membrane proteins (Akiyama and Ito, 1985). We observe that membrane proteins such as SecY and LacY (lactose permease) form sharper bands when electrophoresed through a gel of a modified composition than that of the original composition of Laemmli (1970). This modification (Table I) includes a lowered proportion of N,N'-methylenebis-acrylamide to acrylamide and inclusion of a low concentration of NaCl in the gel. Interestingly, the precursor form of the maltose-binding protein isolated in chemical amounts can only be electrophoresed as a sharp band in this modified system (Ito, 1982; Ohno-Iwashita *et al.*, 1984), although the trace amount of the precursor protein pulse labeled *in vivo* can be resolved in the regular Laemmli gel.

Care should be taken in estimating molecular weights of integral membrane proteins by their mobility relative to standard proteins. Frequently, hydrophobic membrane proteins migrate faster than expected based on their true molecular weight (Ito, 1984).

2. TWO-DIMENSIONAL ELECTROPHORESIS

Gel electrophoresis in two dimension offers an effective way of resolving and identifying individual proteins in complex biological samples. The most powerful and widely used system is that developed by O'Farrell *et al.*, (1977). However, the isoelectric-focusing step, especially its nonequilibrium version (O'Farrell *et al.*, 1977), used in the first dimension of this system is not always appropriate for hydrophobic membrane proteins. For instance, SecY forms a horizontal line in the nonequilibrium O'Farrell two-dimensional gel (the basic character of SecY requires the nonequilibrium system). This appears to be due to interaction between SecY and Nonidet P-40 (NP-40), a nonionic detergent present in the first-dimension step; presumably they form mixed micelles whose migration through polyacrylamide gel is restricted.

TABLE I

SDS–POLYACRYLAMIDE GEL USED FOR SEPARATION OF MEMBRANE PROTEINS

Stock solution	Volume (ml)	Final concentration
1. 30% Acrylamide	10.0	15% (acrylamide)
2. 1% N,N'-Methylenebisacrylamide	2.4	1.2% (bisacrylamide)
3. 1 M Tris-HCl (pH 8.7) and 0.04 M NaCl	6.7	0.335 M (Tris-HCl), 0.013 M (NaCl)
4. 10% SDS	0.2	0.1% (SDS)
5. H_2O	0.7	—

In SDS–PAGE, SecY, LacY, and other hydrophobic membrane proteins migrate faster than hydrophilic proteins of the same corresponding molecular weights (Ito, 1984). This deviation is more pronounced in gels with lower concentrations and lower extents of cross-linking. Thus, this class of proteins can be separated in two dimensions by successive SDS–PAGE through two gels of different molecular sieving effects ("SDS–SDS two-dimensional gel electrophoresis") (Akiyama and Ito, 1985). The first-dimensional electrophoresis is carried out through a low-cross-linked gel described in the previous section, with the exception that the concentration of methylenebisacrylamide is further reduced to 0.08%. After electrophoresis, each lane was cut out by pressing the edge of a thin metal plate, placed horizontally on the top of the stacking gel (regular Laemmli composition) of the second dimension gel, such that two gels are in tight contact without bubbles. The placement of the first-dimension gel on the second-dimension gel is easier when the second-dimension gel is slightly thicker than the first-dimension one. We achieve this by simply assembling the first-dimension glass plates with stronger clips than are used on the second-dimension glass plates, when pouring the gel solutions. Alternatively, one or two sheets of Parafilm strips might be placed on the gel spacer when pouring the second-dimension gel solution.

One typical example for the composition of the second-dimension separation gel is shown in Table II. This gel is cross-linked with both N,N'-methylenebisacrylamide and acrylaide (product of FMC BioProducts), and can be dried down for autoradiography without cracking despite that it is highly cross-linked. The running time for the second dimension should be determined empirically, but sufficient time should be allowed. In this two-dimensional gel, almost all soluble proteins are electrophoresed along the diagonal line, whereas many membrane proteins are separated as spots off the diagonal (Akiyama and Ito, 1985) (Fig. 1). Using this two-dimensional system, we can separate more than 100 proteins from

TABLE II

Second-Dimension Gel for SDS–SDS Two-Dimensional Electrophoresis

Stock solution	Volume (ml)	Final concentration
1. 60% Acrylamide–2.4% N,N'-methylenebisacrylamide	6.7	20% (acrylamide), 0.8% (bisacrylamide)
2. 2% Acrylaide	6.0	0.6% (Acrylaide)
3. 1.5 M Tris-HCl (pH 8.8) and 0.4% SDS	5.0	0.375 M (Tris-HCl)
4. H_2O	2.3	0.1% (SDS)

the *E. coli* cytoplasmic membrane, among which SecY can be identified without overproduction (Fig. 1A). This system can also be applied to eukaryotic membrane proteins as illustrated in Fig. 1B, wherein dog pancreas microsomes were electrophoresed.

B.   Enhanced Detection of Hydrophobic Proteins by Detergent Blotting

The inability of the NP-40-containing gel to separate SecY suggested a possible use of an NP-40-containing gel to selectively trap hydrophobic proteins during their electrophoretic movement. In fact, electroblotting can be performed with a polyacrylamide gel that contains NP-40 sandwiched between the original SDS–gel and a membrane filter (Fig. 2). Membrane proteins, including SecY, LacY, and the melibiose carrier protein, are retained on the NP-40-containing gel. By this variation of protein blotting, which we call "detergent blotting," detection of these membrane proteins is greatly facilitated (Ito and Akiyama, 1985).

After SDS–PAGE, the gel is soaked briefly (about 5 minutes) at room temperature in 2.5 mM Tris–19.2 mM glycine buffer. The dye front part is removed to avoid spread of radioactive low-molecular-weight materials. The gel is laid down on a filter paper (preequilibrated with the Tris–glycine buffer), and covered successively with the NP-40-containing gel (10% acrylamide, 0.27% N,N'-methylenebisacrylamide, 2.5 mM Tris, 19.2 mM glycine, and 2% NP40) that had been cut into an appropriate size, a nylon membrane filter (Zeta probe, obtained from BioRad) preequilibrated with the Tris–glycine buffer, and then with another buffer-equilibrated filter paper. This complex is sandwiched between porous plastic sheets that are supported by plastic frames and is placed into an electroblotting apparatus. Electroblotting is performed at 20 V/cm for 20–24 hours with 4°C water circulating through the cooling

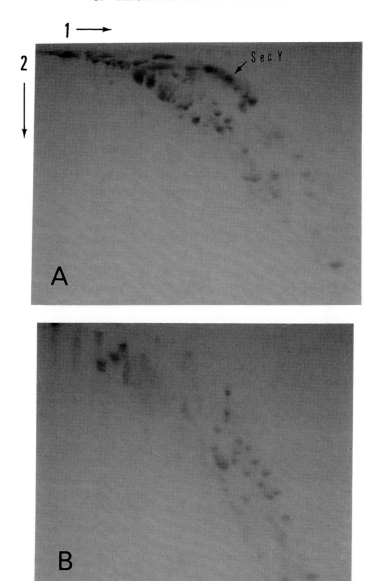

FIG. 1.   SDS–SDS two-dimensional gel electrophoresis of membrane proteins. Samples of *E. coli* cytoplasmic membranes (A) and dog pancreas microsomes (B) were subjected to SDS–PAGE in two dimensions, through gels of different acrylamide concentrations and different extents of cross-linking. The arrows numbered 1 and 2 indicate the direction of the first- and second-dimension electrophoresis, respectively. Gels were stained with Commassie brilliant blue, and the spot of SecY protein is indicated.

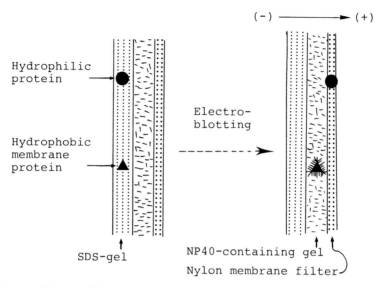

FIG. 2. Schematic illustration of detergent blotting. Behaviors of typical hydrophilic (circle) and hydrophobic (triangle) proteins are shown before (left) and after (right) electro-blotting.

jacket. For autoradiographic detection, the surface of the NP-40 gel that has been in contact with the original gel is placed on X-ray film.

When a gel after SDS–SDS two-dimensional electrophoresis is subjected to detergent blotting, the proteins off the diagonal line are completely retained in the NP-40-containing intermediate gel, whereas proteins on the diagonal line largely pass through it (Akiyama and Ito, 1985). This combination of two-dimensional gel and detergent blotting, which we called "three-dimensional gel (Akiyama and Ito, 1985)" may be of use for further enrichment of membrane proteins.

C. Enrichment of Integral Membrane Proteins by Alkali Fractionation of the Cell

The bilayer structure of membranes is largely preserved at high pH, and consequently, the proteins directly anchored in the lipid bilayer are not extractable by alkaline solutions such as 0.1 N NaOH. In contrast, proteins peripherally bound to the membrane surface are solubilized by alkali along with normally soluble proteins. Thus, centrifugation following treatment with a strongly alkaline agent selectively pellets down the integral membrane proteins (Steck and Yu, 1973; Fujiki *et al.*, 1982). Russel and Model (1982) first applied this technique to fractionate *E. coli* proteins. In their main protocols, the bacterial culture was directly treated with 0.1 N NaOH. However, such direct treatment of the intact cells (growing culture) with NaOH cannot be used, because it releases only a fraction of the total proteins, whose effective solubilization is possible only when cells have been disrupted before exposure to alkali. A simple way of disrupting the cell is to digest the peptidoglycan layer with lysozyme and subject the cell suspension to freezing and thawing. After this pretreatment, the alkali treatment then solubilizes a vast majority of cellular proteins in the supernatant fraction, leaving putative integral membrane proteins in the pellets.

Alkali treatment of *E. coli* cells is done in the following manner. Labeled cultures (usually 0.1 ml) are chilled with crushed ice, mixed with chloramphenicol (100 μg/ml) and NaN_3 (200 μg/ml), followed by centrifugation in a microfuge for 4 minutes at 4°C. Cell pellets are then suspended in 100 μl of 10 mM Tris-HCl (pH 8.0)–5 mM EDTA–5 mM β-mercaptoethanol–100 μg/ml chloramphenicol–200 μg/ml NaN_3, mixed with 10 μl of 10 mg/ml of a solution of lysozyme, incubated for 5 minutes at 0°C, and subjected to three cycles of freezing (with dry ice–ethanol) and thawing (with vigorous vortexing at room temperature). After leaving the suspension at 0°C for an additional 20 minutes, an 80-μl portion was mixed with an equal volume of 0.2 N NaOH, vortexed for about 10 seconds, and centrifuged in a microfuge for 15 minutes at 4°C. The supernatant was carefully removed into another microtube and mixed with 1/10 volume of 100% (w/v) trichloroacetic acid, whereas the precipitates (usually not visible) were suspended in 160 μl of 5% trichloroacetic acid. After standing at 0°C for 15 minutes, samples were centrifuged for 2 minutes, the supernatant was discarded, and the precipitates were washed with 1 ml of acetone by vortexing and centrifuging again. The final protein precipitates were solubilized in 50 μl of 1% SDS–50 mM Tris-HCl (pH 8.0)–1 mM EDTA by vortexing vigorously for 30 minutes at room temperature, using a multisample microtube mixer (product of Tomy, Tokyo, Japan), followed by incubation at 37°C for 5 minutes. This careful solubilization step was essential for a good recovery of membrane proteins after the alkali–acid treatments.

Alkali treatment of fractionated soluble and membrane preparations demonstrated that, unlike the proteins in the soluble compartments, which remained soluble at high pH, roughly 50% of the proteins in the membranes were in the alkali pellets. A large proportion of the alkali-insoluble membrane proteins are retained in the NP-40-containing gel when detergent blotted, whereas the alkali-soluble proteins mostly pass through it.

Although detergent blotting probably concerns the hydrophobicity of the SDS-separated polypeptides, alkali fractionation concerns whether a protein is anchored to the membrane and could be used to follow the process of membrane protein integration (K. Ito and Y. Akiyama, unpublished results).

Alkali fractionation of lysozyme-treated cells (see above) yields integral membrane proteins in the pellet fraction; the proteins identified include SecY and LacY (Fig. 3), cytochrome o subunit b, and leader peptidase (K. Ito and Y. Akiyama, unpublished results). This method of enriching and assigning integral membrane proteins is simple and particularly

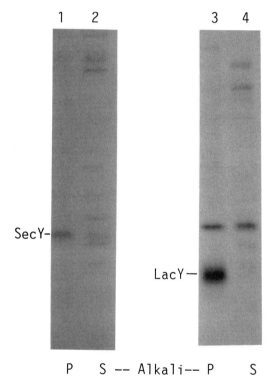

FIG. 3. Alkali fractionation of *E. coli* cells and immunoprecipitation of integral membrane proteins. Cells of *E. coli* strains MC4100 (lanes 1 and 2) and MC4100 carrying F'*lac* factor (lanes 3 and 4) were labeled with [^{35}S]methionine. The latter strain had been induced with IPTG. The labeled cells were alkali fractionated as described in the text. The pellet (lanes 1 and 3) and supernatant (lanes 2 and 4) fractions were immunoprecipitated with anti-SecY (lanes 1 and 2) or anti-LacY (lanes 3 and 4) sera. After SDS–PAGE, labeled proteins were visualized by fluorography.

suited for small-scale or labeled samples. In applying this method for studies of *E. coli* protesins, it should be noted that outer membrane proteins often behave anomalously; they do not contain hydrophobic stretches typically found in the cytoplasmic membrane proteins, and their modes of association with the outer membrane appear different from those for the typical integral membrane proteins of the cytoplasmic membrane.

D. Immunoprecipation of Integral Membrane Proteins

Although immunoblotting after gel electrophoresis can be applied to membrane proteins without major problems [note, however, that some proteins including SecY and SecE cannot effectively be transferred to a membrane filter if the gel has been equiliberated with the blotting buffer, a procedure generally recommended in protein blotting (K. Ito and Y. Akiyama, unpublished results)], some experiments require pulse labeling and immunoprecipitation, rather than blotting and staining. This is the case for experiments addressing the *in vivo* synthesis rate and metabolic stability of a protein. There has been no established method for immuno-precipitation of integral membrane proteins. Seckler (1986) reported an immunoprecipitation procedure for the LacY protein that had been solu-bilized from the isolated membranes with a nonionic detergent. In the method of Seckler, however, background levels are very high, such that overproduction of the protein and previous isolation of the membrane are essential to obtain a reasonable signal-to-noise ratio.

An alternative procedure of immunoprecipitation (described below) gives a much higher signal-to-noise ratio and enables detection of the unamplified level of SecY, LacY, and cytochrome *o* subunit *b* with-out previous cell fractionation or after alkali fractionation (Fig. 3) of the cell.

Protein samples are first solubilized in 1% SDS. A sample can be prepared by subjecting cells to the lysozyme-freezing and -thawing proce-dure described in the previous section followed by addition of 1/10 volume of 10% SDS and incubation at 37°C for 5 minutes. Alternatively, pulse-labeled culture (0.1 ml) is directly treated with an equal volume of 10% trichloroacetic acid. After standing at 0°C for 15 minutes or longer, the sample is centrifuged in a microfuge for 2 minutes, and the supernatant is discarded. The precipitates are washed with 1 ml of acetone by vortexing and centrifuging again and are dissolved in 30 μl of 1% SDS–50 mM Tris-HCl (pH 8.0)–1 mM EDTA by vortexing for 30 minutes at room temperature, followed by incubation at 37°C for 5 minutes. The SDS-solubilized protein samples after the alkali fractionation (see above) can also be a starting material for immunoprecipitation (Fig. 3).

Samples thus solubilized in 1% SDS are diluted 33-fold with Lubrol buffer consisting of 50 mM Tris-HCl (pH 8.0), 0.15 M NaCl, 0.1 mM EDTA, and 0.1% Lubrol (Type-PX), clarified of nonspecific precipitates by centrifugation in a microfuge at 4°C for 10 minutes; the supernatant is carefully removed into another tube containing an appropriate volume of antiserum. In the last step, some 20 μl should be left at the bottom of the tube to avoid contamination with nonspecific precipitates. After incubation at 4°C overnight, antigen–antibody complexes were isolated with the protein A immunoadsorbent. The mixture is incubated at 4°C for 1 hour with a 10% suspension of fixed *Staphylococcus aureus* cells (Kessler, 1975) (20-fold the antiserum volume is used; obtained from The Enzyme Center), with continuous mixing using a rotating mixer. Precipitates were collected by centrifugation in a microfuge for 1 minute, washed with 1 ml of Lubrol buffer, centrifuged again, washed with 1 ml of 10 mM Tris-HCl, centrifuged, and finally suspended in SDS–sample buffer. The final protein solubilization is at 37°C for 5 minutes. In this method, the concentration of the detergent Lubrol is critical to minimize the background.

ACKNOWLEDGMENTS

The work in the authors' laboratory was supported by grants from the Ministry of Education, Science, and Culture, Japan, and from the Naito Foundation. We thank W. Wickner and H. R. Kaback for antisera against some integral membrane proteins.

REFERENCES

Akiyama, Y., and Ito, K. (1985). *EMBO J.* **4**, 3351–3356.
Akiyama, Y., and Ito, K. (1986). *Eur. J. Biochem.* **159**, 263–266.
Akiyama, Y., and Ito, K. (1987). *EMBO J.* **6**, 3465–3470.
Akiyama, Y., and Ito, K. (1989). *J. Biol. Chem.* **264**, 437–442.
Akiyama, Y., and Ito, K. (1990). *Biochem. Biophys. Res. Commun.* **167**, 711–715.
Bieker, K. L., Phillips, G. J., and Silhavy, T. J. (1990). *J. Bioenerg. Biomembr.* **22**, 291–310.
Cerretti, D. P., Dean, D., Davis, G. R., Bedwell, D. M., and Nomura, M. (1983). *Nucleic Acids Res.* **11**, 2599–2619.
Fujiki, Y., Hubbard, A. L., Fowler, S. and Lazarow, P. B. (1982). *J. Cell Biol.* **93**, 97–102.
Ito, K. (1982). *J. Biol. Chem.* **257**, 9895–9897.
Ito, K. (1984). *Mol. Gen. Genet.* **197**, 204–208.
Ito, K. (1990). *J. Bioenerg. Biomembr.* **22**, 353–367.
Ito, K., and Akiyama, Y. (1985). *Biochem. Biophy. Res. Commun.* **133**, 214–221.
Ito, K., Bassford, P. J. and Beckwith, J. (1981). *Cell (Cambridge, Mass.)* **24**, 707–717.
Ito, K., Cerretti, D. P., Nashimoto, H., and Nomura, M. (1984). *EMBO J.* **3**, 2319–2324.
Kessler S. W. (1975). *J. Immunol.* **115**, 1617–1624.
Laemmli, U. K. (1970). *Nature (London)* **227**, 680–685.
Matsuyama, S., Akimura, J., and Mizushima, S. (1990). *FEBS Lett.* **269**, 96–100.
O'Farrell, P. Z., Goodman, H. M., and O'Farrell, P. H. (1977). *Cell (Cambridge, Mass.)* **12**, 1133–1142.

Ohno-Iwashita, Y., Wolfe, P., Ito, K., and Wickner, W. (1984). *Biochemistry* **23,** 6178–6184.
Osborn, M. J., and Munson, R. (1974). *In* "Methods in Enzymology" (S. Fleischer and L. Packer, eds.), Vol. 31, pp. 642–653. Academic Press, New York.
Russel, M., and Model, P. (1982). *Cell (Cambridge, Mass.)* **28,** 177–184.
Schatz, P. J., Riggs, P. D., Jacq, A., Fath, M. J., and Beckwith, J. (1989). *Genes Dev.* **3,** 1035–1044.
Seckler, R. (1986). *Biochem. Biophys. Res. Commun.* **134,** 975–981.
Steck, T. L., and Yu, J. (1973). *J. Supramol. Struct.* **1,** 220–232.
Takeshita, S., Sato, M., Toba, M., Masahashi, W., and Hashimoto-Gotoh, T. (1987). *Gene* **61,** 53–74.
Ueguchi, C., Wittekind, M., Nomura, M., Akiyama, Y., and Ito, K. (1989). *Mol. Gen. Genet.* **217,** 1–5.

Chapter 9

Signal Sequence-Independent Protein Secretion in Gram-Negative Bacteria: Colicin V and Microcin B17

RACHEL C. SKVIRSKY,[1] LYNNE GILSON,
AND ROBERTO KOLTER

Department of Microbiology and Molecular Genetics
Harvard Medical School
Boston, Massachusetts 02115

I. Introduction

In Gram-negative bacteria, most periplasmic and outer membrane proteins are exported from the cytoplasm by a signal sequence-dependent process (Randall *et al.*, 1987). Specifically, these proteins are synthesized

---

[1] Present address: Biology Department, University of Massachusetts at Boston, Harbor Campus, Boston, Massachusetts 02125.

METHODS IN CELL BIOLOGY, VOL. 34

as a precursor from which an amino-terminal leader peptide is cleaved during the translocation process. In order to excrete proteins to the extracellular medium, Gram-negative bacteria need to overcome the additional permeability barrier of the outer membrane. Despite this obstacle, Gram-negative bacteria, especially *Pseudomonas, Serratia, Aeromonas,* and *Erwinia* species, excrete a wide array of proteins, principally toxins and degradative enzymes (Pugsley *et al.,* 1990; Pugsley and Schwartz, 1985).

To transport extracellular proteins across both membranes, the Gram-negative cell requires additional strategies besides those used for exporting cell envelope proteins. Although the mechanisms whereby these proteins traverse the Gram-negative cell envelope are understood incompletely, it is clear that these pathways for excretion are diverse. In some instances, excretion appears to occur by a two-step process, in which the protein is first translocated to the periplasm by a signal sequence-mediated pathway, then is transported across the outer membrane via an additional mechanism (Pugsley *et al.,* 1990; Pugsley and Reyss, 1990). In other instances, entirely different mechanisms involving "direct" export across both membranes appear to be operative. Many of these alternative mechanisms are signal sequence independent.

An unusual signal sequence-independent mechanism, for example, mediates the cellular release of many colicins. Colicins are the best characterized of the bacteriocins, or extracellular toxins that target bacteria similar to the producing strain. Colicins A, E1, E2, N, and D13 are released from the cell by selective leakage of the envelope, caused by the actions of a lysis protein and a phospholipase (Pugsley and Schwartz, 1985).

Other signal sequence-independent pathways are highly specific and result in protein excretion without cell lysis or leakage. These include the colicin V (ColV) export pathway, which requires two transport proteins comprising a dedicated export apparatus (Gilson *et al.,* 1987). The predicted amino acid sequence of one of these proteins, CvaB, reveals that it belongs to a multidrug-resistance (MDR)-like subfamily of export proteins (Blight and Holland, 1990; Gilson *et al.,* 1990). This group consists of integral membrane proteins with similar hydropathy profiles and highly conserved ATP-binding domains. Other bacterial proteins excreted by transporters in this subfamily include the *Escherichia coli* hemolysin (Felmlee *et al.,* 1985; Hartlein *et al.,* 1983), the *Bordetella pertussis* cyclolysin (Glaser *et al.,* 1988), and the protease B produced by *Erwinia chrysanthemi* (Letoffe *et al.,* 1990). All of these proteins lack typical N-terminal signal peptides, and all are secreted into the medium by dedicated export mechanisms.

Although its release from the cell is not mediated by an MDR-like protein, the bacteriocin microcin B17 (MccB17) is also excreted by a signal sequence-independent pathway in *E. coli* (Garrido *et al.*, 1988). Furthermore this pathway is highly specific, requiring two dedicated export proteins.

II. Summary of Colicin V and Microcin B17 Export Systems

In our laboratory, investigations of signal sequence-independent transport processes have focused on two systems—the export of ColV and that of MccB17. For this reason, these systems are described in more detail, and they form the basis for the experimental protocols that are presented. These procedures, however, are intended to be applicable to a variety of dedicated export pathways.

A. Colicin V

Produced by diverse strains of enteric bacteria, ColV is a small protein toxin (approximately 4000 D) with antibacterial activity (Gilson *et al.*, 1987). It is thought to inhibit cells by destroying the membrane potential (Yang and Konisky, 1984). Determinants for ColV activity, export, and immunity are encoded on large low-copy plasmids; such ColV plasmids are often associated with *E. coli* pathogenicity (Hardy, 1975).

Our laboratory has identified three genes required for ColV production: *cvaC*, which is the structural gene for the toxin, and *cvaA* and *cvaB*, which are required for ColV export (Gilson *et al.*, 1987). A fourth gene, *cvi*, confers immunity to the producing cell. We have determined the complete nucleotide sequence of all four genes, and have shown that the *cvaA* and *cvaB* genes form one operon, and that *cvaC* and *cvi* form another (Gilson *et al.*, 1990).

The export gene *cvaB* encodes a putative integral inner membrane protein containing six potential transmembrane regions and a C-terminal nucleotide-binding domain. As described above, CvaB belongs to the MDR-related subfamily of proteins and, in conjunction with CvaA, forms a dedicated export apparatus. In addition, a third component involved in the export of ColV is the outer membrane protein TolC (Gilson *et al.*, 1990). This protein is also involved in the extracellular secretion of hemolysin (Wandersman and Delepelaire, 1990).

The pathway leading to ColV secretion presents distinct advantages as a model for investigating transport processes in the MDR subfamily. In particular, the system is readily amenable to genetic analysis, because the export pathway is specific and because ColV production and secretion are nonessential functions. A key reason for developing the ColV system as a model is that it is feasible to select and/or screen for both forward and reverse variants in the export functions. It is possible to select for mutations that destroy export function because ColV-producing cells with defective immunity (Cvi⁻) can survive only if CvaA or CvaB function is abolished. Conversely, it is possible to screen for extracellular production of ColV by the progeny of 1 in 10^7 cells on one agar plate. Finally, the ColV toxin can be assayed easily, using ColV antibodies as well as a bioassay. Thus, unlike many other transport processes in the MDR subgroup, the ColV system is simple and easily manipulable. Dissection of this system can provide a powerful strategy for elucidating the transport mechanisms shared by homologous proteins in this group.

B.   Microcin B17

MccB17 is a 3200-D peptide antibiotic that is active against most species of enteric bacteria (Davagnino *et al.*, 1986). It is produced by *E. coli* strains harboring pMccB17, a 70-kb single-copy conjugative plasmid. Production of the toxin is maximal during stationary phase (Connell *et al.*, 1987; Hernandez-Chico *et al.*, 1986). The antibiotic acts on sensitive bacteria by inhibiting DNA replication, with its target believed to be DNA gyrase (Herrero and Moreno, 1986). Cells harboring pMccB17 are immune to the lethal action of MccB17 (Herrero *et al.*, 1986).

Seven plasmid-encoded genes are required for MccB17 production, transport, and immunity (Garrido *et al.*, 1988; San-Millan *et al.*, 1985). The *mcbA* structural gene encodes an inactive form of MccB17, which is processed by functions encoded by *mcbBCD*. The *mcbE* and *mcbF* genes encode proteins that act as an inner membrane complex to mediate the specific translocation of MccB17. Finally, the McbG product confers immunity to the producing cell.

III.   Methodology

Many of the genetic and biochemical approaches used to study signal sequence-mediated transport of proteins can be applied to the analysis of signal sequence-independent transport as well. Accordingly, several chapters in this volume describe methods that are relevant to the systems and procedures discussed here. See, for example, Chapter 3 by Manoil.

A. Bioassay of Bacteriocin Activity

The bacteriocin produced from a given strain can be detected using a bioassay. This bioassay exploits the capacity of the bacteriocin to inhibit growth of a lawn of sensitive cells. The bioassay can detect either excreted or intracellular bacteriocin.

In either case, the bioassay plate, made of M63 minimal salts with 0.2% glucose (M63-glu) (Miller, 1972), is overlaid with a lawn of sensitive indicator cells. The choice of indicator strain depends on the bacteriocin activity being tested. Most *E. coli* laboratory strains are sensitive to ColV. We have found that strain 71-18 (Yanisch-Perron *et al.*, 1985) gives reproducible results. Given the mode of action of MccB17, strains deficient in the induction of the SOS repair pathway are markedly more sensitive to this microcin. Thus we routinely use MC4100*recA56* (ZK4) (Gilson *et al.*, 1987) as an indicator strain for MccB17 activity. The lawn is prepared by suspending several colonies of sensitive cells in $1\times$ M63 buffer to produce a slightly cloudy suspension (approximately 1×10^8 cells/ml). After adding 50 μl of this suspension to 3 ml of molten 0.7% H_2O agar, it is then poured onto the assay plate.

To test a strain's ability to produce extracellular bacteriocin, it is grown on solid medium, and a fresh colony is stabbed with a toothpick into the assay plate. After overnight growth at 37°C, excreted bacteriocin is indicated by a zone of growth inhibition of the lawn of indicator cells (halo), surrounding the stabbed colony. The area of this circular zone of inhibition surrounding the producer colony can be calculated to compare relative levels of bacteriocin export. (We have found that when purified bacteriocin is added to a lawn of sensitive cells, the area of the zone of growth inhibition is directly proportional to the amount of bacteriocin added.) The minimum measurable area is limited by the size of the toothpick hole, which has an area of approximately 0.8 mm^2. This represents 1.0% the area of a typical wild-type ColV or MccB17 halo (approximately 80 mm^2) produced from strains with bacteriocin genes on multicopy plasmids. Thus the minimum level of detection for this assay is approximately 1.0% the level of extracellular bacteriocin produced from cells carrying wild-type ColV or MccB17 genes on multicopy plasmids.

A more sensitive test resulting in larger zones of inhibition can be performed by pregrowing the cells to be assayed on an M63-glu plate before overlaying them with a lawn of sensitive cells. After pregrowing the producer cells overnight, the plate is inverted over several drops of chloroform on a flat surface. The cells are killed by 2–5 minutes of exposure to the chloroform vapors. The plates are then left open for 5 minutes to release the chloroform vapors, overlaid with a suspension of indicator cells as described above, and incubated overnight at 37°C.

Under both sets of assay conditions (co-incubation versus pre-growth), the ratios of halo areas between strains remain approximately constant. Thus halo area shows a linear relationship with levels of exported bacteriocin within the range found in these assays.

To assay cells for intracellular bacteriocin, the cells are grown under conditions that maximize expression of the bacteriocin. ColV-producing cells are grown overnight in liquid TB medium (Miller, 1972), diluted 1/10 into the same medium, and regrown 3 hours. A sample of culture (5 ml) is centrifuged 5 minutes at 5000 g. The cell pellet is resuspended in 0.2 ml of cold 25% sucrose, 50 mM Tris-HCl, pH 8.0, and is transferred to a 1.5-ml microcentrifuge tube. To this is added 40 µl of 5 mg/ml lysozyme (freshly made in 50 mM Tris-HCl, pH 8.0), and tubes are kept on ice for 5 minutes; 60 µl of 0.25M EDTA, pH 8.0, is added, and the mix is incubated another 5 minutes on ice, with occasional mixing. Lysis is effected by the quick addition of 300 µl of Lysis Mix (1% Brij 58, 0.4% sodium deoxycholate, 62.5 mM EDTA, and 50 mM Tris-HCl, pH 8.0) with rapid mixing. Tubes are kept on ice 2–5 minutes until the solution clears. Lysates are then centrifuged 10 minutes in a microcentrifuge at 4°C (Clewell and Helinski, 1969). Lysate supernatants are diluted serially and 10 µl is spotted onto lawns of sensitive cells as described above. Bacteriocin activities from these lysates and from liquid growth media can be quantitated by the critical dilution method (Mayr-Harting *et al.*, 1972). According to this method, twofold serial dilutions are assayed, with the final dilution that exhibits activity defined as the critical dilution.

B. Localization of Secreted Products

1. CELL FRACTIONATION

To analyze the pathway of export through the cellular compartments, it is useful to fractionate the producer cells and to assay the fractions for bacteriocin. Such an analysis can be performed on export-competent cells as well as export-deficient mutants, and with wild-type bacteriocin as well as fusion proteins. Final measurements of bacteriocin are performed, as appropriate, using immunological (Section III,D) or bioassay (Section III,A) techniques.

Cell fractionation is performed by the osmotic shock method (Boyd *et al.*, 1987b; Manoil and Beckwith, 1986). All of the fractionation steps are performed in 1.5-ml microcentrifuge tubes, and all centrifugation steps (except the sucrose gradients for separating membranes) are performed in a microcentrifuge at 4°C. According to this procedure, 1.5 ml of cells is resuspended in 150 μl of cold spheroplast buffer (0.1M Tris-HCl, pH 8, 0.5 mM EDTA, and 0.5M sucrose) with 20 μg/ml PMSF (phenylmethylsulfonyl fluoride, a serine protease inhibitor, Sigma), then is incubated 5 minutes on ice. Of this suspension, 50 μl is withdrawn and saved as the whole-cell fraction. Cells are pelleted, warmed to room temperature, then osmotically shocked by resuspension in 100 μl of cold H_2O. Cells are pipetted vigorously for 30 seconds, left 15 seconds on ice, and 5 μl of 20 mM $MgCl_2$ is then added. This osmotic shock releases the periplasmic contents. After centrifugation for 3 minutes, the supernatant, containing the periplasmic fraction, is collected and saved. The pellet, containing membranes plus cytoplasm, is resuspended in 150 μl of cold spheroplast buffer with PMSF. Lysozyme (15 μl) at 2 mg/ml is added, followed by 150 μl cold H_2O, and the samples are incubated 5 minutes on ice. This treatment digests cell wall material, and the spheroplasts are pelleted by a 3-minute centrifugation. The spheroplasts are then resuspended in 600 μl of 10 mM Tris-HCl, pH 8, and 20 μg/ml PMSF, and lysed by freeze–thaw treatment (three cycles of dry ice/28°C). To the lysed spheroplasts are added 20 μl of 1M $MgCl_2$ (to stabilize the outer membrane) and 6 μl of 1 mg/ml DNase I. Membranes are then pelleted (25-minute centrifugation at 4°C) and the supernatant is saved as the cytoplasmic fraction. The membrane pellet is resuspended in 100 μl of Tris-HCl, pH 8. Separation of inner and outer membrane components is carried out as described by Osborn and Munson (1974). According to this method, the membrane fraction is extracted with nonionic detergents, which solubilize differentially inner and outer membrane components.

To verify that the subcellular compartments have been separated and identified correctly, each fraction can be assayed for marker proteins. For example, the periplasmic enzyme alkaline phosphatase is detectable by enzyme assay (Michaelis et al., 1983) or Western blots. Other marker enzyme assays that are easily assayed include NADH oxidase (Osborn et al., 1972), associated with the inner membrane, and glucose-6-phosphate dehydrogenase (Malamy and Horecker, 1964), a cytoplasmic enzyme. Finally, the major outer membrane protein OmpA can be detected in the outer membrane fraction using Western blots.

2. ASSAY OF CULTURE SUPERNATANTS FOR SECRETED PRODUCTS

When analyzing the excretion of a specific protein from the cell via a dedicated export system, it is important to determine whether there is nonspecific leakage of other proteins, concomitant with release of the protein under study. For this purpose, the culture supernatants are assayed carefully for total proteins.

To perform this analysis, cells are grown in 10-ml cultures under conditions that maximize expression and export of the protein under study. Appropriate control strains are grown in parallel, including, for example, strains that produce but do not export the bacteriocin, as well as strains that bear the vector without the bacteriocin genes. Cultures are centrifuged at 10,000 g for 20 minutes, and the resulting supernatants recentrifuged at 100,000 g for 30 minutes. To ensure removal of all cells, supernatants are then filtered through membranes that minimize protein binding (0.2 μm Uniflo filter units; Schleicher and Schuell #46-02330). To demonstrate the absence of viable cells, 40 μl of filtrate is spread on LB agar plates and incubated at 37°C. The filtrates can then be assayed for the protein of interest by appropriate methods (for example, by bioassay or Western blot). In addition, they can be assayed for a cytoplasmic enzyme and a periplasmic enzyme to determine the extent of cell lysis. Chloramphenicol acetyl transferase (Shaw, 1976) and β-lactamase (Ross and O'Callahan, 1976) are convenient enzymes to assay if the strain harbors the appropriate plasmids. SDS–PAGE can then be used to analyze total proteins present in the culture supernatants.

Filtered supernatants are dialyzed against 100 volumes of 30 mM ammonium bicarbonate, using Spectra-Por Membrane #7 with 1000 MW cut-off (Spectrum Medical Industries, Inc.). Dialyzed samples are lyophilyzed to dryness, and each sample is dissolved in 60 μl of H_2O. To separate total proteins, the 20-μl sample plus 20-μl Laemmli loading buffer (Laemmli, 1970) are boiled and loaded onto a 15% SDS–PAGE gel. Proteins are visualized best by silver staining (Giulian et al., 1983).

C. Isolation of Export-Deficient Mutants and Extragenic Suppressors

To understand the mechanism of signal sequence-independent export at the level of protein interaction, it is useful to investigate which amino acids within the transported protein are essential for recognition by the export machinery. First, one can generate within the structural gene missense mutations that produce an active peptide that is deficient in export function. Analysis of such mutations can identify export signals within the transported substrate. One can then isolate, in the genes encoding transport functions, suppressors of the original mutations, by screening for restored export function. Using this approach, the domains of the transport protein(s) that recognize the transported substrate during export can be identified. All of the relevant strains and plasmids for this mutational analysis, as it applies to the ColV and MccB17 systems, are listed in Table I.

To perform this analysis with the ColV system, a collection of missense mutations in *cvaC* that abolish ColV export is first isolated. To this end, plasmids carrying *cvaCcvi* are mutagenized, either *in vitro* by chemical or

TABLE I

STRAINS AND PLASMIDS USEFUL FOR ANALYSIS OF ColV AND MccB17 SYSTEMS

Strain/ Plasmid	Relevant feature	Usefulness	Source or reference
Strain			
71-18	ColV/M13 sensitive	Indicator lawns to test ColV production and to plate M13 derivatives	Yanisch-Perron *et al.* (1985)
GC4415	*sfi::lacZ* fusion	Indicator strain to detect increases in the intracellular concentration of MccB17	Herrero and Moreno (1986)
ZK4	*recA56*	Indicator strain to detect MccB17	Gilson *et al.* (1987)
LE30	*mutD5*	*In vivo* mutagenesis	Fowler *et al.* (1974)
Plasmid			
pHK11	ColV$^+$ Ap^r	Source of all *cva* genes	Gilson *et al.* (1987)
pHK22	ColV$^+$ Cm^r	Source of all *cva* genes	Gilson *et al.* (1987)
pLY21	*cvaCcvi* Cm^r	Source of immunity and structural genes of ColV	Gilson *et al.* (1990)
pLY11	*cvaAB* Ap^r	Source of the genes encoding the ColV exporter	Gilson *et al.* (1990)
M13-ColV	*cvaCcvi*	Source of ColV structural and immunity genes in an M13 vector for the plaque assay of ColV production	Kolter lab stocks
M13McbA	*mcbA*	Source of the MccB17 structural gene independent of all other *mcb* genes	Kolter lab stocks
pPY113	*mcbABCDEFG*	Source of all MccB17 production genes	Kolter lab stocks
pPY117	*mcbBCD*	Source of the three MccB17 maturation genes	Kolter lab stocks
pPY121	*mcbBCDEFG*	Source of all MccB17 maturation and export genes, without structural gene	Kolter lab stocks
pSWF1b	*phoAΔss*	Source of a *phoA* gene without a signal sequence suitable for *in vitro* construction of fusions	Gilson *et al.* (1990)

oligonucleotide-directed mutagenesis, or *in vivo* in a strain with the *mutD* mutator allele (Fowler *et al.*, 1974). Mutagenized *cvaCcvi* plasmids are then used to transform a host strain harboring a compatible plasmid carrying the transport genes *cvaAcvaB*. Export-deficient transformants can be isolated by several methods. First, since such mutants fail to release biologically active ColV into the medium, they can be identified by their failure to produce zones of inhibition in the ColV bioassay (Section III,A). To verify an export defect and eliminate that class of mutants with abolished ColV activity, cell lysates of the mutants are then assayed for intracellular toxin activity (Section III,A).

A second strategy for isolating export mutants is based on the observation that ColV-producing cells with defective immunity (cvi⁻) can survive only if they are export defective. Thus, to isolate *cvaC* mutations that perturb ColV export, CvaC⁺ Cvi⁻ plasmids are mutagenized, then used to transform a host carrying *cvaAcvaB*. Transformants receiving an intact *cvaC* gene are killed due to the cell's defective immunity, whereas *cvaC* mutants with abolished export and/or toxin activity can survive and be selected. As in the first scheme, mutants with defective export are distinguished from those lacking toxin activity by assaying cell lysates for intracellular ColV activity.

In a third approach for isolating export-deficient mutants, M13 phage-carrying *cvaCcvi* are mutagenized using chemical or oligonucleotide-directed mutagenesis. The mutagenized DNA is then used to transfect an M13-sensitive strain carrying a plasmid with *cvaAcvaB*, and transfected cells are plated to produce phage plaques. Cells transfected with wild-type *cvaC* produce extracellular ColV, which kills adjacent cells in the sensitive lawn (cells lacking immunity, because Cvi is absent). The result is a large, clear "plaque," with a colony of M13 (*cvaCcvi*)-infected cells at its center. On the other hand, an incoming M13 carrying a mutant *cvaC* that prevents ColV release produces a typical small, turbid M13 plaque. Such small M13 plaques can be picked and these cells can be screened for intracellular ColV activity as described in Section III,A.

cvaC mutants considered to have export defects are then sequenced, using a primer from a site near *cvaC*. These sequences identify the amino acids within ColV that are important for export. Once export-deficient mutants have been identified and characterized, representatives are chosen for suppressor isolation. Extragenic mutations that can potentially suppress these *cvaC* mutations are then generated by mutagenizing *in vitro* a plasmid carrying *cvaAcvaB*. These mutagenized plasmids are used to transform a host harboring the original *cvaC* mutation as well as an intact immunity gene. Pseudorevertants of the *cvaC* mutation can be identified by screening for halo formation in the ColV bioassay.

This screen for suppressor mutations is performed as follows: Transformants are plated on agar at a density of 10^6 to 10^7 cells per plate. After growth, the resulting cell lawn is lifted onto a filter paper master and kept separately. The original plate is treated with chloroform to kill the remaining cells, then overlaid with ColV-sensitive cells. After growth of the overlay, areas on the plates where extracellular ColV had been produced by cells during the initial incubation appear as halos of growth inhibition. To isolate the export-proficient suppressor strains, cells are retrieved from the corresponding positions on the master and are rescreened for ColV production using the same bioassay procedure. This method is an effective screen in which low-frequency events can be detected; in reconstruction experiments using this procedure, ColV-producing cells can be recovered after being mixed with nonproducers in a ratio of 1 in 10^6 or 10^7.

In the analysis of MccB17 export, several different strategies can be used to isolate export-deficient mutants. First, as in the colicin V system, mutants deficient in MccB17 export fail to produce halos in the plate bioassay. Thus *mcbA* mutants lacking export function can be screened by this method. In addition, MccB17 export-deficient mutants have other specific properties that can be exploited in mutant isolations. In particular, when active microcin accumulates inside the cell, this intracellular microcin overwhelms the immunity conferred by McbG, causing an unhealthy phenotype. The endogenous microcin induces expression of the SOS repair system, including the *sfiA* (*sulA*) gene (Garrido et al., 1988; Herrero and Moreno, 1986). This induction provides a basis for screening export defects. Specifically, a plasmid carrying the mutagenized structural gene (*mcbA*) and another carrying *mcbBCDEFG* are used to transform a host strain bearing a *sfiA::lacZ* operon fusion. Due to accumulation of intracellular microcin, export-deficient mutants show elevated expression of SfiA, resulting in increased LacZ activity. These mutants are thus identifiable by their color on X-Gal or MacConkey lactose indicator plates (Miller, 1972). To verify an export defect, whole cells and cell lysates are then assayed for microcin activity.

The unhealthy phenotype of MccB17 export-deficient cells can also be exploited in the isolation of suppressor mutations that restore export function. Specifically, export mutants which also bear a *recA⁻* mutation grow poorly on rich medium and fail to grow on minimal medium (Garrido et al., 1988). Thus plasmids carrying mutagenized export genes can be used to transform a RecA⁻ host harboring the original *mcbA* mutation, plus *mcbBCD*. Pseudorevertants of the *mcbA* mutation with restored export function can be isolated by selecting for survival on minimal medium. To verify reversion of the export defect, the strains are then tested for halo formation in the microcin bioassay.

The collection of extragenic revertants isolated by the above procedures is then analyzed to determine the location of the suppressor mutation. The

gene that contains the suppressor mutation can be identified by complementation analysis and the exact site of mutation can be determined
by DNA sequencing.

D. Immunological Identification of Bacteriocins

1. GENERATION OF ANTIBODIES

Standard techniques for raising antibodies (Harlow and Lane, 1988)
have been found to be successful for bacteriocin-specific antibodies. The
predicted amino acid sequence of the bacteriocin can be used to identify
appropriate antigenic domains—in particular, domains that are hydrophilic and that exhibit secondary structure. Peptides of about 20 residues
having the sequence of such domains can be synthesized, conjugated to a
carrier such as BSA or keyhole limpet hemocyanin, and can be used to
raise antibodies (Harlow and Lane, 1988). Several companies, such as East
Acres Biologicals, Southbridge, MA, specialize in antibody production.

2. PREPARATION OF CELL LYSATES FOR WESTERN ANALYSIS

Once antibodies have been raised, active or inactive bacteriocin can be
detected in cell lysates by Western analysis. To prepare cell lysates, 1 ml of
an overnight culture (grown to maximize bacteriocin expression) is pelleted and the cells are washed in cold 50 mM Tris-HCl, pH 7.4. The cells
are resuspended in 100 μl Laemmli loading buffer and lysed as follows:
vortex 20 seconds, boil 5 minutes, vortex 20 seconds, freeze at $-70°C$ 15
minutes, thaw and vortex 20 seconds, boil 1 minute, and vortex. Approximately 50 μl is loaded onto a 15 or 20% SDS–PAGE gel (Laemmli, 1970).
In our experience, some small bacteriocins such as MccB17 are difficult to
resolve in the standard Laemmli SDS–PAGE gels. Much better resolution
of these small peptides can be achieved by simple modifications of the
Laemmli recipe (Thomas and Kornberg, 1978). Briefly, the Tris concentration of the running gel is doubled to 0.75 M, the Tris concentration of the
running buffer is doubled to 50 mM (384 mM glycine), and the stock
solution of acrylamide:bisacrylamide is prepared with an acrylamide:bisac-
rylamide ratio of 30:0.15. Western blots (Harlow and Lane, 1988) are
prepared and probed with bacteriocin-specific antibodies.

3. PULSE LABELING AND IMMUNOPRECIPITATION

Antibodies can also be used to immunoprecipitate bacteriocins either
from cell lysates or from culture supernatants. Cells are first pulse labeled

with [^{35}S]methionine or [^{35}S]cysteine by the following procedure. Cells are grown under conditions that maximize bacteriocin production. For MccB17 production, cells are grown to stationary phase, with aeration, in 10 ml of M63-glu in a 300-ml flask. In contrast, ColV-producing cells are grown overnight in M63-glu, diluted 1/10 into the same medium, and regrown 3 hours. For [^{35}S]methionine labeling, 0.5 ml of cells are pulse labeled with 100 μCi L-[^{35}S]methionine. After continued aeration for 1–2 minutes, the 0.5 ml of cells is added to 0.5 ml of Stop Mix (0.08% NaN$_3$ and 0.2 mg/ml Spectinomycin) on ice. Samples are then centrifuged (in 1.5-ml microcentrifuge tubes) 5 minutes at 4°C, and the cell pellet and culture medium supernatant are processed separately.

To extract the cell pellet, cells are washed with 0.5 ml of 150 mM NaCl, resuspended in 50 μl STEB (1% SDS, 10 mM Tris-HCl, pH8, 1 mM EDTA, and 5% β-mercaptoethanol), and boiled 2 minutes. After samples are cooled on ice, 800 μl of chilled 50 mM Tris-HCl, pH 8.0, 0.15 M NaCl, 2% Triton X-100, and 1 mM EDTA is added, and samples are vortexed and centrifuged 5 minutes in a microcentrifuge at room temperature. The extracted cell supernatants are transferred to fresh tubes for immunoprecipitation. A 10-μl aliquot is saved separately; this serves as a sample of total cell-associated labeled protein for SDS–PAGE.

To immunoprecipitate the bacteriocin, 40–100 μl of antiserum is added to the total cell extract described above or to the cell-free culture medium. All samples are then incubated overnight at 4°C. To each sample is added 30–50 μl of protein A–Sepharose CL4B (Sigma #P-3391), and samples are incubated at room temperature on a rocker for 2 hours. Samples are centrifuged 30 seconds in a microcentrifuge, the supernatant is gently pipetted off, and the pellet is washed four times with 1 ml of 50 mM Tris-HCl, pH 8.0, 1 M NaCl, 1% Triton X-100, and 1 mM EDTA. Pellets are then washed twice with 1 ml of 10 mM Tris-HCl, pH 8.0, to lower the salt concentration, then dried under vacuum. Dried samples are resuspended in 30–50 μl Laemmli loading buffer, boiled 2 minutes to release the antibody and antigen from protein A, then centrifuged. A fraction of the samples (10–30 μl) is loaded on a SDS–PAGE gel. ^{35}S-Labeled immunoprecipitated proteins can then be visualized by autoradiography.

E. Determination of Kinetics of Export and Cellular Location of Bacteriocin

To determine the kinetics of export of the bacteriocin, ^{35}S labeling can be performed as a pulse chase, followed by immunoprecipitation of cell lysates and culture media. Specifically, the ^{35}S pulse described above is followed (after 50–60 seconds) by addition of cold methionine (or cysteine) to give a final concentration of 0.4 mM (pH 7). Following the addition of the chase, samples are removed at various intervals and are processed for immunoprecipitation. Furthermore, to examine the cellular export pathway, cells can be fractionated after pulse-chase labeling. Again using immunoprecipitation, the bacteriocin content can be compared in cytoplasmic, periplasmic, inner membrane, and outer membrane fractions. Thus the bacteriocin can be localized to specific cellular compartments in time course experiments and the cellular pathway of peptide export can be elucidated.

F. Identification of Export Signals within the Bacteriocin Structural Gene: Generation of PhoA Protein Fusions

The isolation of export-deficient mutants was discussed in Section III,C as a method for identifying export signals in the secreted protein. A complementary strategy for identifying export information is the use of protein fusions between N-terminal sequences of the target protein and C-terminal sequences of alkaline phosphatase (PhoA) (Manoil and Beckwith, 1985). Because the PhoA moiety in these constructs lacks its signal sequence, export of the hybrid protein can occur only if sufficient export information is contributed by the target gene. Furthermore, export of these hybrids can be detected readily because PhoA is enzymatically active only after crossing the inner membrane (Boyd *et al.*, 1987a), and because its activity is easily assayed in bacterial colonies using the chromogenic substrate 5-bromo-4-chloro-3-indoyl phosphate (XP). Thus, to identify export signals within a target protein, fusions containing different segments of this protein are generated and the PhoA activities of the chimeras are analyzed.

This strategy, it must be noted, has limitations for the study of proteins secreted by dedicated export mechanisms. For example, a target protein may encode C-terminal export signals as in the case of hemolysin (Gray *et al.*, 1986), or when fused to PhoA, the target protein may not be properly modified after translation. Steric impairment of translocation is also possible. Therefore the inability of a fusion product to be exported must be interpreted cautiously. Pugsley and Cole (1986) found, for example, that colicin N–PhoA fusions are not secreted into the medium, despite the fact that each is normally a secreted product. Furthermore, we have found that MccB17–PhoA fusions are not translocated, even when the hybrid protein contains the entire MccB17 peptide (R. Skvirsky and R. Kolter, unpublished observation). On the other hand, we have found that ColV–PhoA protein fusions are indeed translocated across the inner membrane (Gilson *et al.*, 1990). Thus in this instance, PhoA is translocated by a signal sequence-independent pathway that involves a dedicated export apparatus.

PhoA fusions can be generated *in vivo* using the transposon Tn*phoA* (Manoil and Beckwith, 1985). Tn*phoA* mutagenesis has been described in detail (Manoil and Beckwith, 1985, 1986). To isolate Tn*phoA* insertions in a bacteriocin structural gene, clones that are blue on XP media and deficient in bacteriocin activity are screened. Once transpositions into the structural gene for the target protein are obtained, export of the hybrid proteins can be assessed by determining PhoA activities in enzyme assays (Michaelis *et al.*, 1983). The specific sites of Tn*phoA* insertions are determined by DNA sequencing at the fusion junction, using a primer specific for the *phoA* portion of Tn*phoA*: 5'-AATATCGCCCTGAGCA-3'.

If PhoA fusions at specified sites in the target gene are required, then fusions can be constructed *in vitro*. This is accomplished, for example, using a pUC19-derived vector (pSWF1b) that has a polylinker inserted upstream of a truncated *phoA* gene. The *phoA* lacks its promoter and signal sequences. Appropriate restriction fragments of the target gene can be inserted into the polylinker region of this vector to produce protein fusions.

IV. Conclusion

The methodologies presented here have been described primarily as they apply to the bacteriocin secretion systems under study in our laboratory. However, many of the same approaches can be applied to the investigation of other bacteriocins, as well as to other signal sequence-independent export systems. Furthermore, it may be possible to exploit the dedicated export mechanisms of MccB17 and ColV to excrete the products of cloned genes. The finding that ColV–PhoA proteins fusions are translocated suggests that the ColV system in particular may be used effectively to export heterologous proteins. Thus the investigation of signal sequence-independent export systems may have practical application in the construction of exportable protein fusions; in addition, these pathways provide relatively simple models for investigating fundamental principles underlying protein secretion.

ACKNOWLEDGMENTS

We wish to acknowledge support from the Cystic Fibrosis Foundation (Grant Z138), the National Science Foundation (Grant DMB-8813612), and the National Institutes of Health (Grant AI25944). R. C. S. was a Science Scholar at the Mary Ingraham Bunting Institute, and R. K. was the recipient of an American Cancer Society Faculty Research Award.

REFERENCES

Blight, M. A., and Holland, I. B. (1990). Structure and function of haemolysin B, P-glycoprotein and other members of a novel family of membrane translocators. *Mol. Microbiol.* **4,** 873–880.

Boyd, D., Guan, C., Willard, S., Wright, W., Strauch, K., and Beckwith, J. (1987a). Enzymatic activity of alkaline phosphatase precursor depends on its cellular location. *In* "Phosphate Metabolism and Cellular Regulation in Microorganisms" (A. Torriani-Gorini, F. G. Rothman, S. Silver, A. Wright, and E. Yagil, eds.) pp. 89–93. Am. Soc. Microbiol. Washington, D. C.

Boyd, D., Manoil, C., and Beckwith, J. (1987b). Determinants of membrane protein topology. *Proc. Natl. Acad. Sci. U.S.A.* **84,** 8525–8529.

Clewell, D. B., and Helinski, D. R. (1969). Super-coiled circular DNA-protein complex in *Escherichia coli*: Purification and induced conversion to an open circular form. *Proc. Natl. Acad. Sci. U.S.A.* **62,** 1159–1166.

Connell, N., Han, Z., Moreno, F., and Kolter, R. (1987). An *E. coli* promoter induced by the cessation of growth. *Mol. Microbiol.* **1,** 195–201.

Davagnino, J., Herrero, M., Furlong, D., Moreno, F., and Kolter, R. (1986). The DNA replication inhibitor microcin B17 is a forty-three amino acid protein containing sixty percent glycine. *Proteins* **1,** 230–238.

Felmlee, T., Pellet, S., and Welch, R. A. (1985). Nucleotide sequence of an *Escherichia coli* chromosomal hemolysin. *J. Bacteriol.* **163,** 94–105.

Fowler, R. G., Degnen, G. E., and Cox, E. C. (1974). Mutational specificity of a conditional *Escherichia coli* mutator, mutD5. *Mol. Gen. Genet.* **103,** 179–191.

Garrido, M. C., Herrero, M., Kolter, R., and Moreno, F. (1988). The export of the DNA replication inhibitor microcin B17 provides immunity for the host cell. *EMBO J.* **7,** 1853–1862.

Gilson, L., Mahanty, H. K., and Kolter, R. (1987). Four plasmid genes are required for colicin V synthesis, export, and immunity. *J. Bacteriol.* **169,** 2466–2470.

Gilson, L., Mahanty, H. K., and Kolter, R. (1990). Genetic analysis of an MDR-like export system: The secretion of colicin V. *EMBO J.* **9,** 3875–3884.

Giulian, G. G., Moss, R. L., and Greaser, M. (1983). Improved methodology for analysis and quantitation of proteins on one-dimensional silver-stained slab gels. *Anal. Biochem.* **129,** 277–287.

Glaser, P., Sakamoto, H., Bellalou, J., Ullmann, A., and Danchin, A. (1988). Secretion of cyclolysin, the calmodulin-sensitive adenylate cyclase-haemolysin bifunctional protein of *Bordetella pertussis*. *EMBO J.* **7,** 3997–4004.

Gray, L., Mackman, N., Nicaud, J.-M., and Holland, I. B. (1986). The carboxy-terminal region of haemolysin 2001 is required for secretion of the toxin from *Escherichia coli*. *Mol. Gen. Genet.* **205,** 127–133.

Hardy, K. G. (1975). Colicinogeny and related phenomena. *Bacteriol. Rev.* **39,** 464–515.

Harlow, E., and Lane, D. (1988). "Antibodies: A Laboratory Manual." Cold Spring Harbor Lab., Cold Spring Harbor, New York.

Hartlein, M., Schiebl, S., Wagner, W., Rdest, U., Kreft, J., and Goebel, W. (1983). Transport of hemolysin by *Escherichia coli*. *J. Cell. Biochem.* **22,** 87–97.

Hernandez-Chico, C., San-Millan, J. L., Kolter, R., and Moreno, F. (1986). Growth phase and OmpR regulation of transcription of the Microcin B17 genes. *J. Bacteriol.* **167,** 1058–1065.

Herrero, M., and Moreno, F. (1986). Microcin B17 blocks DNA replication and induces the SOS system in *Escherichia coli*. *J. Gen. Microbiol.* **132,** 393–402.

Herrero, M., Kolter, R., and Moreno, F. (1986). Effects of microcin B17 on microcin B17-immune cells. *J. Gen. Microbiol.* **132,** 403–410.

Laemmli, U. K. (1970). Cleavage of structural proteins during the assembly of the head of bacteriophage T4. *Nature (London)* **227**, 680–685.

Letoffe, S., Delepelaire, P., and Wandersman, C. (1990). Protease secretion by *Erwinia chrysanthemi*: The specific secretion functions are analogous to those of *Escherichia coli* α-haemolysin. *EMBO J.* **9**, 1375–1382.

Malamy, M. H., and Horecker, B. L. (1964). Release of alkaline phosphatase from cells of *Escherichia coli* upon lysozyme spheroplast formation. *Biochemistry* **3**, 1889–1893.

Manoil, C., and Beckwith, J. (1985). TnphoA: A transposon probe for protein export signals. *Proc. Natl. Acad. Sci. U.S.A.* **82**, 8129–8133.

Manoil, C., and Beckwith, J. (1986). A genetic approach to analyzing membrane protein topology. *Science* **233**, 1403–1408.

Mayr-Harting, A., Hedges, A. J., and Berkeley, R. C. W. (1972). Methods for studying bacteriocins. *Methods Microbiol.* **7A**, 315–422.

Michaelis, S., Inouye, H., Oliver, D., and Beckwith, J. (1983). Mutations that alter the signal sequence of alkaline phosphatase in *Escherichia coli*. *J. Bacteriol.* **154**, 366–374.

Miller, J. H. (1972). "Experiments in Molecular Genetics." Cold Spring Harbor Lab., Cold Spring Harbor, New York.

Osborn, M. J., and Munson, R. (1974). Separation of the inner (cytoplasmic) and outer membranes of gram-negative bacteria. *In* "Methods in Enzymology" (S. Fleischer and L. Packer, eds.), Vol. 31, pp. 642–653 Academic Press, New York.

Osborn, M. J., Gander, J. E., Parisi, E., and Carson, J. (1972). Mechanism of assembly of the outer membrane of *Salmonella typhimurium*: Isolation and characterization of cytoplasmic and outer membrane. *J. Biol. Chem.* **247**, 3962–3972.

Pugsley, A. P., and Cole, S. T. (1986). β-Galactosidase and alkaline phosphatase do not become extracellular when fused to the amino-terminal part of colicin N. *J. Gen. Microbiol.* **132**, 2297–2307.

Pugsley, A. P., d'Enfert, C., Reyss, I., and Kornacker, M. G. (1990). Genetics of extracellular protein secretion by Gram-negative bacteria. *Annu. Rev. Genet.* **24**, 67–90.

Pugsley, A. P., and Reyss, I. (1990). Five genes at the 3′ end of the *Klebsiella pneumoniae* pulC operon are required for pullulanase secretion. *Mol. Microbiol.* **4**, 365–379.

Pugsley, A. P., and Schwartz, M. (1985). Export and secretion of proteins by bacteria. *FEMS Microbiol. Rev.* **32**, 3–38.

Randall, L. L., Hardy, S. J. S., and Thom, J. R. (1987). Export of proteins: A biochemical review. *Annu. Rev. Microbiol.* **41**, 507–541.

Ross, G. W., and O'Callahan, C. H. (1976). β-lactamase assays. *In* "Methods in Enzymology" (J. H. Hash, ed.), Vol. 43, pp. 69–85. Academic Press, New York.

San-Millan, J. L., Kolter, R., and Moreno, F. (1985). Plasmid genes involved in microcin B17 production. *J. Bacteriol.* **163**, 1016–1020.

Shaw, W. V. (1976). Chloramphenicol acetyltransferases from chloramphenicol-resistant bacteria. *In* "Methods in Enzymology" (J. H. Hash, ed.), Vol. 43, pp. 737–755. Academic Press, New York.

Thomas, J. O., and Kornberg, R. D. (1978). The study of histone-histone associations by chemical cross-linking. *In* "Methods in Cell Biology" (S. Stein, J. Stein, and L. J. Kleinsmith, eds.), Vol. 18, pp. 429–440. Academic Press, New York.

Wandersman, C., and Delepelaire, P. (1990). TolC, an *Escherichia coli* outer membrane protein required for hemolysin secretion. *Proc. Natl. Acad. Sci. U.S.A.* **87**, 4776–4780.

Yang, C. C., and Konisky, J. (1984). Colicin V-treated *Escherichia coli* does not generate membrane potential. *J. Bacteriol.* **158**, 757–759.

Yanisch-Perron, C., Vieira, J., and Messing, J. (1985). Improved M13 phage cloning vectors and host strains: Nucleotide sequence of the M13mp18 and pUC19 vectors. *Gene* **33**, 103–119.

Chapter 10

Transcription of Full-Length and Truncated mRNA Transcripts to Study Protein Translocation across the Endoplasmic Reticulum

REID GILMORE, PAULA COLLINS, JULIE JOHNSON, KENNAN KELLARIS, AND PETER RAPIEJKO

Department of Biochemistry and Molecular Biology
University of Massachusetts Medical School
Worcester, Massachusetts 01655

METHODS IN CELL BIOLOGY, VOL. 34

I. Introduction

The development of RNA transcription systems employing bacterio-phage RNA polymerases has had a tremendous impact in the fields of molecular and cellular biology during the past decade. These systems allow the user to transcribe RNA from any available DNA template that is cloned downstream from a bacteriophage RNA polymerase promoter. Radiolabeled or biotinylated RNA transcripts can be used as hybridization probes for Northern and Southern blots (Melton *et al.*, 1984), as substrates in RNA processing studies (Krieg and Melton, 1984a), and for *in situ* mapping (Henikoff *et al.*, 1986). Antisense RNA transcripts have been used to block translation of mRNA *in vivo* (Melton, 1985). This chapter will focus on the transcription of translatable mRNA for use in studies of protein translocation across the rough endoplasmic reticulum (RER).

Prior to the advent of the RNA transcription systems, investigators who studied protein translocation and organelle biosynthesis had to rely upon RNA isolated from tissues, tissue culture cells, or microorganisms as a source of mRNA for translation. Although some tissues can serve as an excellent source for the isolation of a major mRNA (e.g., bovine pituitary as a source for prolactin mRNA), the majority of mRNAs are minor constituents of these preparations. Consequently, investigators who stud-ied biosynthesis of these less abundant proteins were at a disadvantage because they had to immunoprecipitate the relevant *in vitro* translation product. Krieg and Melton demonstrated that DNA templates could be transcribed with SP6 RNA polymerase to synthesize RNA that was func-tionally active as mRNA both *in vitro* in the reticulocyte lysate translation system and *in vivo* after microinjection into *Xenopus* oocytes (Krieg and Melton, 1984b). Immunoprecipitation is not necessary after translation of a mRNA transcript because a single major translation product will be produced. This enhancement in synthesis of the relevant translation pro-duct results in a 5- to 500-fold decrease in the exposure time needed to visualize a polypeptide after resolution on a SDS–polyacrylamide gel. The *in vitro* transcription systems have also allowed the investigator to translate artificially constructed gene fusion products. The functions of RER signal and stop-transfer sequences and mitochondrial- and peroxisomal-targeting sequences have been examined in a systematic manner after combining these sequence elements with suitable passenger domains from other pro-teins (Hurt *et al.*, 1984; Mize *et al.*, 1986; Perara and Lingappa, 1985; Small *et al.*, 1988; Yost *et al.*, 1990).

Translation of mRNA transcripts may allow synthesis of sufficient pro-tein to monitor a functional activity. Synthesis of 0.5–2.5 fmol of a polypeptide per microliter of translation reaction is feasible. Clearly, a sensitive assay is needed to detect such a low quantity of protein. Three notable examples wherein functional studies have been possible are cited here. Partially assembled signal recognition particle (SRP) complexes lacking the 9 and 14-kDa subunits can be functionally reconstituted by *in vitro* translation of these latter polypeptides (Strub and Walter, 1990). *In vitro*-translated SRP receptor α-subunit can reconstitute the translocation activity of SRP receptor-deficient microsomal membranes (Andrews *et al.*, 1989). The DNA-binding region within the GCN4 protein was localized by transcription–translation of the full-sized protein and a series of internal deletion mutants (Hope and Struhl, 1986).

In vitro transcription systems have additionally provided a means of obtaining truncated mRNAs, which have been invaluable for the investigation of protein translocation across the endoplasmic reticulum. We use the term truncated mRNA to refer to a mRNA of a defined size that lacks a termination codon. Translation of a truncated mRNA results in the synthesis of a ribosome-bound nascent polypeptide. The translation phase and the nascent chain transport phase of a protein translocation reaction could be investigated separately by incubating preassembled ribosome–nascent chain complexes with microsomal membranes (Mueckler and Lodish, 1986a,b; Perara *et al.*, 1986). Important conclusions concerning the mechanism of protein translocation were drawn from these experiments. It was found that the continued synthesis of a polypeptide does not provide the energy for transport of the protein across the membrane. Second, ribonucleotide triphosphates were shown to be essential for translocation of proteins across the endoplasmic reticulum. Further research has demonstrated that both ATP hydrolysis and guanine ribonucleotides are involved in the translocation reaction (Chirico *et al.*, 1988; Connolly and Gilmore, 1989; Hansen *et al.*, 1986; Schlenstedt and Zimmermann, 1987; Zimmermann *et al.*, 1988). Truncated mRNAs have also been key reagents in the identification of proteins that mediate nascent chain transport. Partially translocated nascent chains are obtained when truncated mRNAs are translated in the presence of microsomal membranes (Connolly *et al.*, 1989). These translocation intermediates can be cross-linked to a 35- to 39-kDa integral membrane glycoprotein termed the signal sequence receptor (Krieg *et al.*, 1989; Wiedmann *et al.*, 1987).

Other applications for truncated mRNA transcripts have interesting potential. The authors do not know of experiments in which truncated mRNAs have been used as tools to investigate protein import into organelles other than the endoplasmic reticulum. Perhaps ribosome-bound peptidyl tRNAs might prove useful in these other systems as an alternate method for obtaining partially translocated precursor proteins. Antisera raised against prolactin can immunoprecipitate nascent chains containing the first 56 residues of the mature polypeptide, but not nascent chains containing only the first 20 residues of mature prolactin. Thus, a series of truncated mRNAs encoding successively larger fragments of a protein would provide a means for mapping the epitopes recognized by a set of monoclonal antibodies. A third potential application would be the identification of functional domains within a protein using nascent chains encoded by truncated mRNAs. Radiolabeled nascent chains that correspond to potential domains of a protein can be released from the ribosome by the addition of puromycin (100 μM) or EDTA (10 mM) and can thus be assayed for function.

II. *In Vitro* Transcription of mRNA

A. RNA Polymerases and Transcription Vectors

At the current time, three bacteriophage RNA polymerases are in common use for the production of mRNA transcripts: the SP6, T3, and T7 RNA polymerases. Each of these RNA polymerases recognizes a different bacteriophage-specific promoter that is distinct from eukaryotic and prokaryotic promoters (Butler and Chamberlin, 1982; McAllister *et al.*, 1981). Thus, these enzymes will only initiate transcription at the correct promoter sequence and transcribe any DNA sequence that is cloned downstream from the promoter. Purified SP6, T3, and T7 RNA polymerases are currently available from a number of commercial sources (e.g., Promega Biotech, Stratagene, International Biotechnologies, Inc., Boehringer Mannheim, and New England Biolabs). The three RNA polymerases are similar with respect to specificity of promoter recognition and synthesis rates.

The specific recognition of a unique promoter sequence by SP6 RNA polymerase led Melton and colleagues to construct two RNA transcription vectors (SP64 and SP65) that contain a SP6-specific promoter adjacent to a polylinker sequence (Melton *et al.*, 1984). These investigators demonstrated that the SP6 RNA polymerase could be used to synthesize significant quantities of mRNA for translation (Krieg and Melton, 1984b) or radiolabeled RNA for use as hybridization probes (Melton *et al.*, 1984). RNA transcription vectors are now available from several companies, including Promega Biotech, International Biotechnologies, Inc., Boehringer Mannheim, and Stratagene. Current versions of RNA transcription vectors typically contain promoters for two different bacteriophage RNA polymerases flanking a polylinker site. This arrangement allows the insertion of a cDNA sequence between the two bacteriophage RNA polymerase promoters. Positive-strand transcripts for translation or negative-strand transcripts for hybridization can then be transcribed from the DNA template by selection of the appropriate RNA polymerase. The reader is referred to promotional literature from the relevant companies as a source of further information concerning specific transcription vectors.

B. Transcript Features that Influence Translation Efficiency

Several factors have been identified that interfere with subsequent translation of mRNA transcripts. Transcripts that contain poly(G) tracts at the 5′ end of the mRNA translate poorly in the wheat germ translation system (Galili *et al.*, 1986; Holland and Drickamer, 1985). The polynucleotide tract at the 5′ end of the mRNA apparently prevents protein synthesis by forming a stable secondary structure that can interfere with initiation of translation (Galili *et al.*, 1986). The homopolymer tracts arise from the addition of dG and dC tails by terminal deoxyribonucleotidyl transferase during construction of cDNA libraries. The 5′ polynucleotide tract should be removed prior to insertion of the cDNA into the multiple cloning sequence of the transcription vector. Removal of the 5′ homopolymer tracts can be accomplished by restriction enzyme digestion if an appropriate site is located in the 5′ sequence upstream from the initiation codon. Alternatively, the homopolymer tract can be removed by digestion of the cDNA insert with *Bal*31 nuclease. Out-of-frame AUG codons that precede the normal initiation codon can also interfere with *in vitro* translation of mRNA transcripts, particularly when the codon is in a favorable context for initiation of translation (Kozak, 1987). The removal of upstream AUG codons prior to insertion of the cDNA sequence into the transcription vector will solve this problem.

C. Transcription of mRNA with Bacteriophage RNA Polymerases

The RNA transcription vector containing a cDNA cloned downstream from the bacteriophage RNA polymerase promoter should be linearized by digestion with a restriction enzyme prior to transcription to obtain "runoff" transcripts. Although linearization of the DNA template is not essential for the transcription reaction, transcripts from unlinearized templates will contain RNA complementary to the vector as well as to the cDNA insert. The linearization site will correspond to the 3′ end of the

RNA transcript, making the choice of which enzyme to use dependent upon whether the investigator wishes to transcribe complete or truncated mRNA transcripts. In order to obtain full-length mRNA transcripts, the vector should be linearized within the polylinker sequence distal to the RNA polymerase promoter with a restriction enzyme that lacks sites within the cDNA insert. It has been reported that aberrant transcription products can be produced by initiation at the ends of DNA fragments that contain 3′ overhangs (Schenborn and Mierendorf, 1985). Promoter-independent initiation could occur from either strand of the linearized DNA template, so both negative- and positive-strand mRNAs were produced (Schenborn and Mierendorf, 1985). Negative-strand RNAs will interfere with translation of the mRNA transcript due to the formation of duplex structures. For this reason, it is preferable to linearize the DNA template with a restriction enzyme that leaves either a blunt end or a 5′ overhang.

cDNAs linearized at restriction endonuclease cleavage sites within the protein-coding region provide templates for the synthesis of truncated mRNAs. The mRNA transcript will extend between the RNA polymerase start site and the first downstream cleavage site for the restriction enzyme. The investigator should ensure that the restriction enzyme does not cleave the transcription vector in the RNA polymerase promoter sequence. For example, the T7 RNA polymerase promoter contains a *Hin*fI site, consequently linearization with *Hin*fI cannot precede transcription with T7 RNA polymerase. Our laboratory has prepared truncated mRNA transcripts from DNA templates that have been linearized with enzymes that leave either 5′ or 3′ overhangs or enzymes that leave blunt ends. Translatable RNA transcripts have been obtained from templates with all three end types. However, the most active RNAs have been obtained after linearization with enzymes that yield either blunt ends or 5′ overhangs.

Typical RNA transcription reactions for the preparation of uncapped mRNA for translation contain 10–20 μg of linearized DNA template. We do not gel purify the DNA fragment containing the promoter and the cDNA insert prior to the transcription reaction. Instead, the linearized transcription vector is extracted first with phenol-CHCl$_3$ and then with CHCl$_3$. The aqueous phase containing the DNA is adjusted to 0.3 M NaOAc and precipitated with two volumes of ethanol, either for 12 hours at $-20°$C or 2 hours at $-80°$C. The ethanol-precipitated DNA is washed once with 70% ethanol to remove excess salt and is dried and resuspended in 10 mM Tris and 1 mM EDTA (TE buffer) at a concentration of 1 μg/μl. Removal of excess salt from the DNA template is important because the RNA polymerases are most active at low ionic strength (Melton *et al.*, 1984). A 200-μl transcription reaction will contain 40 μl of 5 × transcription buffer (200 mM Tris-Cl, pH 7.5, 50 mM NaCl, 30 mM MgCl$_2$, and 10 mM

spermidine), 10–20 μg of DNA template in 20 μl of TE, 2 μl of 1 M DTT, 4 μl of a neutralized NTP stock (25 mM each of ATP, GTP, UTP, and CTP), 8 μl of placental RNase inhibitor (RNasin, 40,000 U/ml), and 50–100 Units of RNA polymerase. The transcription reaction is adjusted to 200 μl with sterile H_2O. Ribonucleotide stocks are neutralized with Tris base. The transcription reactions are incubated at 37°C for 1 hour. Typical yields of RNA in a transcription reaction range from 2 to 5 μg of RNA/μg of DNA template. Although RNase inhibitor is included during the transcription reaction, the investigator should still take precautions to prevent the introduction of ribonucleases into the reagents used for RNA transcription and RNA isolation.

The uncapped mRNA transcripts described in the preceding paragraph will translate in both the wheat germ and reticulocyte lysate translation systems (Melton *et al.*, 1984). The translation efficiency of capped mRNA transcripts is clearly higher using these *in vitro* systems (Melton *et al.*, 1984). Uncapped mRNA transcripts are not suitable for microinjection into *Xenopus* oocytes due to degradation of the transcript (Melton *et al.*, 1984). The most commonly used method for capping mRNA transcripts is to reduce the final GTP concentration in the transcription reaction to 50–100 μM and to include 500 μM 5′-GpppG-3′ (Holland and Drickamer, 1985; Perara and Lingappa, 1985) or 5′-7meGpppG-3′ (Nielson and Shapiro, 1986). The cap analogs are incorporated into the 5′ end of the mRNA transcript with little (R. Gilmore, unpublished observation), or no (Nielson and Shapiro, 1986) inhibition of the transcription reaction. Unmethylated caps are methylated *in vitro* by inclusion of 8 μM S-adenosylmethionine in the cell-free translation reaction (Hansen *et al.*, 1986).

D. Isolation of mRNA Transcripts for Translation

Several different methods are used to prepare RNA transcripts for translation. Our laboratory extracts the transcription reaction with an equal volume of phenol: $CHCl_3$, followed by extraction with $CHCl_3$. The aqueous phase is adjusted to 0.3 M NaOAc and the nucleic acid is precipitated overnight at $-20°C$ by the addition of 2.3 volumes of ethanol. The precipitated nucleic acid is collected by centrifugation for 15 minutes in a microcentrifuge. The nucleic acid pellet is dried in a Speed Vac and is resuspended in 30 μl of sterile water. RNA is selectively precipitated by the addition of 10 μl of 12 M LiCl (8–12 hours at 4°C). After centrifugation, the supernatant is carefully removed and the pellet is dried. The RNA is dissolved in sterile water, adjusted to 0.4 M NH_4OAc, and precipitated with 2.3 volumes of ethanol. The RNA pellet is dried once again, dissolved in sterile water, and the concentration of RNA is determined by the absorbance at 260 nm (40 μg/ml of RNA has an OD_{260} of 1.0). The purified RNA transcripts are stored at $-80°C$ at a concentration of approximately 1 μg/μl. Transcripts that we have prepared and stored in this manner remain active as mRNA for translation for several years.

Alternatively, the DNA template can be removed by digestion with RNase-free DNase (Melton *et al.*, 1984). Following a 10-minute digestion, the transcription reaction can be extracted with phenol:CHCl$_3$, adjusted to 0.7 M NH$_4$OAc, and precipitated with 2.3 volumes of ethanol (Melton *et al.*, 1984). Unincorporated nucleotide triphosphates and deoxynucleotide monophosphates are removed by chromatography on a Sephadex G-100 column (Melton *et al.*, 1984). Although we have used the gel filtration method in our laboratory, the recovery of translatable mRNA was somewhat variable. A third "coupled transcription–translation" procedure does not require removal of the DNA template. Transcription products are added directly to the reticulocyte translation system at a final concentration of 20% (Perara and Lingappa, 1985). Although one might think that the cap analogs included during transcription would interfere with subsequent translation, the fivefold dilution apparently reduces their concentration to a level that is not inhibitory (Perara and Lingappa, 1985).

III.   Translation of mRNA Transcripts

A.   *In Vitro* Translation of mRNA Transcripts

The procedures for translation of mRNA transcripts are essentially identical to the methods that have been used to translate mRNA that has been isolated from tissues, cells, or microorganisms. Procedures for the preparation and use of cell-free translation systems from wheat germ (Erickson and Blobel, 1983), reticulocyte lysate (Jackson and Hunt, 1983; Pelham and Jackson, 1976), and yeast (Gasior *et al.*, 1979; Moldave and Gasior, 1983) have been described in detail by other investigators. The reader is referred to these previous articles for additional information concerning cell-free translation systems. A 25-μl wheat germ translation that we use to translate mRNA transcripts will contain 7.5 μl of staphylococcal nuclease-digested wheat germ S23, 5 μl of 5$\times$ energy mix, 5 μl of 5$\times$ buffer mix, 7μl of sterile water, and 0.5 μl of mRNA transcript (200 to 500 ng of uncapped mRNA). The staphylococcal nuclease-digested wheat germ S23 is prepared as described previously (Erickson and Blobel, 1983), and is in 40 mM HEPES/KOH (pH 7.5), 100 mM KOAc, 5 mM Mg(OAc)$_2$, and 4 mM DTT. The 5$\times$ energy mix corresponds to 13 mM ATP, 1.4 mM GTP, 78 mM creatine phosphate, 0.1 mM each of 19 amino acids (excluding methionine), 40 mM dithiothreitol, 20 mM KOH, 4.5 μCi/ μl of [^{35}S]methionine, and 0.48 mg/ml of creatine phosphokinase. The KOH is added to neutralize the ribonucleotides, and this reagent should be added to the energy mix prior to the addition of the creatine phospho-kinase. The 5× buffer mix corresponds to 0.135 M HEPES/KOH (pH 7.5), 350 mM KOAc, 5 mM Mg(OAc)$_2$, 2.6 mM spermidine, 4000 U/ml of placental RNase inhibitor (RNasin, Promega Biotech), 1 mg/ml calf liver tRNA, 0.5 μg/ml each of pepstatin, chymostatin, antipain and leupeptin, and 5 μg/ml of aprotinin. The final K$^+$ and Mg^{2+} concentrations using the above protocol are 100 and 2.5 mM, respec-tively. The translation efficiency of mRNA transcripts can often be im-proved by determining the optimal K$^+$ and Mg^{2+} ion concentration for a given mRNA. KOAc should be tested between 60 and 120 mM, whereas Mg(OAc)$_2$ should be tested between 1.5 and 2.5 mM to optimize transla-tion of a mRNA transcript. The *in vitro* synthesis of protein should be maximal when 100–500 ng of uncapped mRNA or 50–200 ng of capped mRNA transcript is added to a 25-μl translation reaction. mRNA trans-cripts should be added to the translation reaction as the last reagent. All reagents for *in vitro* translation should be prepared with sterile water, and the investigator should take precautions to avoid contamination of transla-tion reagents and mRNA transcripts with ribonucleases.

B. Translation of Truncated mRNAs

Truncated mRNA transcripts have been translated *in vitro* with the wheat germ (Connolly *et al.*, 1989; Connolly and Gilmore, 1986), reticulo-cyte lysate (Mueckler and Lodish, 1986a; Perara *et al.*, 1986), and yeast (Hansen and Walter, 1988) translation systems. The conditions that are used for translation of full-length mRNA transcripts are applicable to the translation of truncated mRNA transcripts with minor modifications. We typically add 200–400 ng of uncapped mRNA to a 25-μl *in vitro* translation reaction. The truncated mRNAs lack a termination codon, so the normal termination reaction of protein synthesis cannot occur. Because a single nascent chain will be produced per mRNA molecule, the transla-tion efficiency of the truncated mRNAs will be lower, provided that aber-rant termination reactions do not occur. As one might expect, the *in vitro*-synthesized polypeptides accumulate as ribosome-bound peptidyl-tRNA. Multiple ribosomes can initiate upon a mRNA in the *in vitro* systems, so we anticipate that polyribosomes will accumulate in the ab-sence of termination. The stability of ribosome–peptidyl-tRNA complexes assembled by translation of truncated mRNAs will be addressed in Sec-tion III,D.

When the translation products from a truncated mRNA are analyzed by SDS–polyacrylamide gel electrophoresis, we observe that the predominant translation product has a mobility consistent with the size of protein that should be encoded by the mRNA transcript. However, several additional translation products may also be produced. The presence of larger translation products, particularly the full-size translation product, is diagnostic of an incomplete restriction digestion prior to transcription. The mRNA that encodes the full-size translation product will translate more efficiently due to the presence of a termination codon. Frequently, a ladder of translation products will be obtained from translation of the truncated mRNA. The presence of multiple ribosomes upon the truncated mRNA can account for these additional polypeptides that migrate more rapidly than the primary translation product. The ribosome will protect between 30 and 45 nucleotides of a mRNA from nuclease digestion (Cancedda and Shatkin, 1979), and the maximum packing density of ribosomes upon a translating mRNA has been estimated to be one ribosome per 60 nucleotides (Lodish, 1971). Based upon these two estimates, we would expect to see a ladder of translation products differing in size by 15 to 25 amino acid residues per step. The presence of these additional polypeptides can interfere with translocation experiments. Fortunately, the accumulation of these ladders can be minimized by limiting the number of ribosomes that initiate upon a single mRNA. Because uncapped mRNAs initiate translation with lower efficiency, they provide a convenient means of reducing the accumulation of nascent chain ladders. Alternatively, 7-methylguanosine-5'-monophosphate can be added to a final concentration of 2 mM several minutes after the start of translation to inhibit multiple initiations of capped mRNAs. Synchronization of translation reactions with 7-methylguanosine-5'-monophosphate has been described in detail by previous investigators (Rothman and Lodish, 1977; Walter and Blobel, 1981). Initiation at internal methionine residues may also lead to the presence of additional, more rapidly migrating, polypeptides when translating both full-length and truncated mRNA transcripts. We do not know of a technique that can eliminate initiation at internal methionine residues.

C. Truncation of mRNAs with Oligodeoxyribonucleotides and RNase H

A restriction endonuclease site may not be located in the cDNA sequence at a location that is desirable for the investigator. An alternative method for producing truncated nascent chains of a defined length is to supplement the translation system with an oligodeoxyribonucleotide that is complementary to the mRNA (Gilmore and Blobel, 1985; Haeuptle *et al.*, 1986). This technique is a modification of the hybrid-arrest translation approach that has been used to confirm the identity of cDNA clones (Patterson *et al.*, 1977). When we first applied this method (Gilmore and Blobel, 1985), we presumed that the duplex formed between the oligonucleotide and the mRNA interfered with movement of the ribosome along the mRNA. Subsequent studies demonstrated that the RNA–DNA heteroduplex served as a substrate for an endogenous RNase H-like activity present in the wheat germ extract (Haeuptle *et al.*, 1986). Optimization of the oligonucleotide-mediated truncation procedure showed that a 10-fold molar excess of a 20mer oligonucleotide was sufficient to yield complete cleavage of a mRNA transcript in the wheat germ translation system (Haeuptle *et al.*, 1986). Less efficient cleavage of mRNAs was obtained in the reticulocyte lysate translation system (Haeuptle *et al.*, 1986).

D.   Stability of Ribosome–Peptidyl-tRNA Complexes

When translation products from truncated mRNAs were analyzed by sucrose density gradient centrifugation, we observed that the majority of the newly synthesized polypeptide cosedimented with either 80 S ribosomes or larger polyribosomes (Connolly and Gilmore, 1986). However, some of the product encoded by the truncated mRNA sedimented much less rapidly, indicating that it was no longer ribosome associated. The cetyltrimethylammonium bromide (CTABr) precipitation method described below was used to determine what proportion of the translation products remained bound to the ribosome as peptidyl-tRNA. We typically observed that less than 20% of the translation product encoded by the truncated mRNA was released from the ribosome during the first 15 minutes of translation. After a 1-hour translation, the amount of product that was no longer in a tRNA linkage frequently exceeded 50%. Other laboratories have reported even higher levels of peptidyl-tRNA release from ribosomes after 1-hour translations (Haeuptle *et al.*, 1986). Differences in the *in vitro* translation systems may account for the varying stability of peptidyl-tRNAs. For experiments where the quantity of the translation product is more important than the continued association with the ribosome, longer translation times are advantageous. If nascent chains that remain bound to the ribosome as peptidyl-tRNAs are desired, translation periods of 10 to 15 minutes are recommended.

Precipitation of the translation products with CTABr is a convenient method to determine what proportion of the newly synthesized polypeptides remains bound to the ribosome as peptidyl-tRNA (Gilmore and Blobel, 1985; Hobden and Cundliffe, 1978). Cell-free translation products (25 μl) are mixed with 250 μl of 2% (w/v) cetyltrimethylammonium bromide (Sigma Chemical Company) by vortexing in a 1.5-ml microcentrifuge tube. RNA and any protein covalently linked to RNA are precipitated by the addition of 250 μl of 0.5 M sodium acetate (pH 5.4) containing 50 μg of yeast tRNA as a carrier. The samples are incubated at 30°C for 10 minutes to allow the CTABr–RNA precipitate to aggregate. The precipitate is collected during a 5-minute centrifugation at room temperature. The supernatant can be transferred to a microcentrifuge tube and precipitated with 100 μl of 100% (w/v) trichloroacetic acid to recover proteins that are not linked to RNA. The CTABr precipitate and the TCA precipitate of the CTABr supernatant can be analyzed by SDS–polyacrylamide gel electrophoresis after the pellets are washed twice with 250 μl of ice-cold acetone:HCl (19:1) to remove CTABr. Failure to remove the CTABr from the samples with result in a grossly distorted migration of the proteins on the polyacrylamide gel.

IV.   Ribonucleotide-Dependent Translocation Systems

A.   Depletion of Ribonucleotides from Translation Products

Ribonucleotide-dependent steps in protein translocation across the endoplasmic reticulum could be investigated once experimental systems were developed that allowed separation of the translation phase and the protein transport phase of the translocation reaction. Three different systems have allowed investigators to test the roles of ribonucleotides. Translation of a truncated mRNA yields ribosome-bound peptidyl-tRNAs that serve as suitable substrates for transport across mammalian microsomal membranes (Mueckler and Lodish, 1986b; Perara *et al.*, 1986). Several extremely short polypeptides can be translocated across mammalian microsomal membranes following completion of synthesis (Schlenstedt and Zimmermann, 1987; Wiech *et al.*, 1987). Completed prepro-α-factor can be posttranslationally translocated across yeast microsomal membranes (Rothblatt and Meyer, 1986; Water and Blobel, 1986). Following translation of either a ribosome-bound peptidyl-tRNA or a completed precursor protein, further protein synthesis is prevented by the addition of cycloheximide (250 μM to 1 mM), emetine (1 mM), or puromycin (100 μM). Puromycin causes termination of peptidyl-tRNAs, so this inhibitor can only be used when investigating translocation of completed polypeptides.

Several different methods have been used to demonstrate that ribonucleotides are essential for the transport reaction. Hydrolysis of ribonucleotides by exogenous enzymes is a rapid and efficient method to deplete ATP and GTP pools (Mueckler and Lodish, 1986b; Rothblatt and Meyer,

1986; Schlenstedt and Zimmermann, 1987; Wiech *et al.*, 1987). The combination of glucose plus hexokinase yields ADP and GDP as hydrolysis products whereas AMP and GMP will be the products that remain after apyrase treatment. Glycerol kinase from *Escherichia coli* will selectively hydrolyze ATP (Thorner and Paulus, 1973). However, the nucleotide diphosphate kinase (NDP-kinase) activities present in the *in vitro* translation systems will synthesize ATP from ADP at the expense of GTP. Thus, hydrolysis of ATP by glucerol kinase leads to the eventual depletion of both ATP and GTP pools. Following enzymatic hydrolysis of NTP pools, microsomal membranes are added to the translation products to determine whether the transport reaction requires ATP and GTP. The primary disadvantage of this method is the continued presence of ribonucleotide diphosphates that will interfere with subsequent reconstitution experiments. For example, if ATP is added back to a translation reaction following hydrolysis of ribonucleotides by hexokinase, GTP will be produced from GDP by action of NDP-kinase.

The preferred method for depletion of ribonucleotides from translocation substrates is gel filtration chromatography (Connolly and Gilmore, 1986; Hansen *et al.*, 1986; Mueckler and Lodish, 1986b; Perara *et al.*, 1986; Waters and Blobel, 1986). In our laboratory, 1.0-ml Sephacryl S-200 columns are poured in disposable plastic syringes. The bed support for the column consists of a piece of 10-μm nylon mesh cloth that is held in place on the bottom of the syringe barrel with a short length of Tygon tubing. The gel filtration column is precoated with 100 μl of a 10-mg/ml solution of bovine serum albumin, and then equilibrated with 1.5 ml of 50 mM TEA, 150 mM KOAc, 2.5 mM Mg(OAc)$_2$, 1 mM DTT, and 0.002% Nikkol (buffer A). Translation products (100 μl) are applied to the column and allowed to enter the bed. The column is washed with an additional 300 μl of buffer A. The void volume fraction containing ribosome-bound nascent polypeptides is eluted with an additional 150–175 μl of buffer A. Ribonucleotides and translation products that have been released from the ribosome will elute in subsequent fractions. Gel filtration beads with a lower exclusion limit (Sephadex G-10 or G-25) are used to separate completed translation products (e.g., prepro-α-factor) from ribonucleotides (Hansen *et al.*, 1986; Waters and Blobel, 1986).

B. Reconstitution of Ribonucleotide-Dependent Translocation Reactions

Both the ATP-hydrolysis and GTP-dependent events in protein translocation have been reconstituted upon readdition of the appropriate nucleotide (Connolly and Gilmore, 1986; Hansen *et al.*, 1986; Mueckler and Lodish, 1986b; Perara *et al.*, 1986; Waters and Blobel, 1986). The ribo-nucleotide-depleted translation products are incubated with microsomal membranes after supplementation with exogenous ribonucleotide triphosphates or ribonucleotide triphosphate analogs. Secretory protein transport or membrane protein integration reactions are detected by one of the following assays: signal sequence cleavage, addition of high-mannose oligosaccharide, protection from externally added protease, or cosedimentation with membrane vesicles. Signal sequence cleavage and addition of high-mannose oligosaccharide are detected by changes in mobility of polypeptides on SDS–polyacrylamide gels (Hansen *et al.*, 1986; Mueckler and Lodish, 1986b; Perara *et al.*, 1986; Waters and Blobel, 1986). Protease protection assays (Hansen *et al.*, 1986; Perara *et al.*, 1986) and membrane cosedimentation assays (Connolly and Gilmore, 1986) have been described in detail in preceding publications. The parameters that need to be explored in ribonucleotide-dependent transport reactions are (1) the specificity of the ribonucleotide, (2) the concentration of ribonucleotide that is needed to elicit a response, and (3) the time course of the transport reaction.

The specificity for a given ribonucleotide triphosphate is examined by testing ribonucleotides and ribonucleotide analogs. GTP-dependent steps in protein translocation have been studied in our laboratory for several years (Connolly and Gilmore, 1986; Hoffman and Gilmore, 1988; Wilson *et al.*, 1988). The SRP receptor-mediated displacement of SRP from the signal sequence was shown to be GTP-dependent (Connolly and Gilmore, 1989). GTP, dGTP, ITP, and the nonhydrolyzable analogs GMPPNP and GMPPCP support nascent chain insertion into the membrane (Connolly and Gilmore, 1986; R. Gilmore, unpublished data). GDP, in contrast, is a competitive inhibitor of nascent chain insertion in reactions containing GMPPNP (Connolly and Gilmore, 1986). Initially, we were surprised to find that the nonhydrolyzable analog GTPγS cannot replace GTP (R. Gilmore, unpublished data). GTPγS is generally considered to be less resistant to hydrolysis than either GMPPNP or GMPPCP. Moreover, commercially available GTPγS preparations are grossly contaminated with GDP (10%). Thus, GTPγS may not be the best choice as a nonhydrolyzable GTP analog, particularly when working with relatively crude systems that include cytosol and membrane fractions. GMP, imidodiphosphate, GDP-mannose, ATP, and AMPPNP are neither inhibitory nor stimulatory with respect to nascent chain insertion (Connolly and Gilmore, 1986; R. Gilmore, unpublished data). Partial reconstitution of the ATP hydrolysis-dependent step in protein translocation is obtained upon addition of other ribonucleotide triphosphates (GTP, UTP, and CTP), but not by addition of the nonhydrolyzable ATP analogs, AMPPNP or AMPPCP (Hansen *et al.,* 1986; Mueckler and Lodish, 1986b; Water and Blobel, 1986). Interconversion of nucleotide triphosphates by NDP-kinase may be responsible for partial reconstitution of the ATP hydrolysis-dependent transport reaction.

The concentration of nucleotide needed to support a given reaction should fall within a reasonable range if the ribonucleotide effect is physiologically relevant. For example, the GTP-dependent translocation reaction requires approximately 3 μM GMPPNP (Connolly and Gilmore, 1986). GTP-binding proteins typically have affinity contants in this range for GMPPNP. Ribonucleotide specificity becomes more difficult to address as the concentration of nucleotide increases due to the presence of contaminating ribonucleotides. ATP preparations typically contain 1–2% GTP. When 1 mM ATP is added to a transport reaction, a significant concentration of GTP is also present.

The time course of transport may allow the investigator to distinguish between relevant and irrelevant transport pathways. For example, membrane insertion of the nascent hemagglutinin–neuraminidase (HN) protein of Newcastle disease virus could occur when GTP, GMPPNP, and ATP, but not AMPPNP, were added (Wilson *et al.,* 1988). A time course experiment demonstrated that the GMPPNP-dependent reaction occurred approximately 40-fold faster than did the ATP-dependent reaction (Wilson *et al.,* 1988). The ATP-dependent insertion reaction of HN nascent chains may be due to NDP-kinase-mediated synthesis of GTP from membrane-bound GDP. Posttranslational transport reactions that are either very slow or very inefficient relative to cotranslational transport reactions therefore deserve close scrutiny to ensure that actual transport events are being monitored. If the time course for a posttranslational reaction is significantly slower than the time course of the corresponding cotranslational reaction, one must question whether uncoupling the translation and transport phases of the translocation reaction is not creating a requirement for cytosolic factors and ribonucleotide hydrolysis rather than revealing requirements that also exist in the coupled system.

REFERENCES

Andrews, D. W., Lauffer, L., Walter, P., and Lingappa, V. R. (1989). *J. Cell Biol.* **108,** 797–810.
Butler, E. T., and Chamberlin, M. J. (1982). *J. Biol. Chem.* **257,** 5772–5778.
Cancedda, R., and Shatkin, A. J. (1979). *Eur. J. Biochem.* **91,** 41–50.
Chirico, W. J., Waters, M .G., and Blobel, G. (1988). *Nature (London)* **322,** 805–810.
Connolly, T., and Gilmore, R. (1986). *J. Cell Biol.* **103,** 2253–2261.
Connolly, T., and Gilmore, R. (1989). *Cell (Cambridge, Mass.)* **57,** 599–610.
Connolly, T., Collins, P., and Gilmore, R. (1989). *J. Cell Biol.* **108,** 299–307.

Erickson, A. H., and Blobel, G. (1983). *In* "Methods in Enzymology" (S. Fleischer and B. Fleischer, eds.), Vol. 96, pp. 38–50. Academic Press, New York.

Galili, G., Kawata, E. E., Cueller, R. E., Smith, L. D., and Larkins, B. A. (1986). *Nucleic Acids Res.* **14**, 1511–1524.

Gasior, E., Herrera, F., Sadnik, I., McLaughlin, C., and Moldave, K. (1979). *J. Biol. Chem.* **254**, 3965–3969.

Gilmore, R., and Blobel, G. (1985). *Cell (Cambridge, Mass.)* **42**, 497–505.

Haeuptle, M.-T., Frank, R., and Dobberstein, B. (1986). *Nucleic Acids Res.* **14**, 1427–1448.

Hansen, W., and Walter, P. (1988). *J. Cell Biol.* **106**, 1075–1081.

Hansen, W., Garcia, P. D., and Walter, P. (1986). *Cell (Cambridge, Mass.)* **45**, 397–406.

Henikoff, S., Keene, M. A., Fechtel, K., and Fristrom, J. W. (1986). *Cell (Cambridge, Mass.)* **44**, 33–42.

Hobden, A. H., and Cundliffe, E. (1978). *Biochem. J.* **170**, 57–61.

Hoffman, K., and Gilmore, R. (1988). *J. Biol. Chem.* **263**, 4381–4385.

Holland, E. C., and Drickamer, K. (1985). *J. Biol. Chem.* **261**, 1286–1292.

Hope, I. A., and Struhl, K. (1986). *Cell (Cambridge, Mass.)* **46**, 885–894.

Hurt, E. C., Pesold-Hurt, B., and Schatz, G. (1984). *EMBO J.* **3**, 3149–3156.

Jackson, R. J., and Hunt, T. (1983). *In* "Methods in Enzymology" (S. Fleischer and B. Fleischer, eds.), Vol. 96, pp. 50–74. Academic Press, New York.

Kozak, M. (1987). *Nucleic Acids Res.* **15**, 8125–8148.

Krieg, P. A., and Melton, D. A. (1984a) *Nature (London)* **308**, 203–206.

Krieg, P. A., and Melton, D. A. (1984b). *Nucleic Acids Res.* **12**, 7057–7070.

Krieg, U. C., Johnson, A. E., and Walter, P. (1989). *J. Cell Biol.* **109**, 2033–2043.

Lodish, H. F. (1971). *J. Biol. Chem.* **246**, 7131–7138.

McAllister, W. T., Morris, C., Rosenberg, A. H., and Studier, F. W. (1981). *J. Mol. Biol.* **153**, 527–544.

Melton, D. A. (1985). *Proc. Natl. Acad. Sci. U.S.A.* **82**, 144–148.

Melton, D. A., Krieg, P. A., Rebagliati, M. R., Maniatis, T., Zinn, K., and Green, M. R. (1984). *Nucleic Acids Res.* **12**, 7035–7056.

Mize, N. K., Andrews, D. W., and Lingappa, V. R. (1986). *Cell (Cambridge, Mass.)* **47**, 711–719.

Moldave, K., and Gasior, E. (1983). *In* "Methods in Enzymology" (R. Wu, L. Grossman, and K. Moldave, eds.), Vol. 101, pp. 644–650. Academic Press, New York.

Mueckler, M., and Lodish, H. F. (1986a). *Cell (Cambridge, Mass.)* **44**, 629–637.

Mueckler, M., and Lodish, H. F. (1986b). *Nature (London)* **322**, 549–552.

Nielson, D. A., and Shapiro, D. J. (1986). *Nucleic Acids Res.* **14**, 5936.

Patterson, B. M., Roberts, B. E., and Kuff, E. L. (1977). *Proc. Natl. Acad. Sci. U.S.A.* **74**, 4370–4374.

Pelham, H. R. B., and Jackson, R. J. (1976). *Eur. J. Biochem.* **67**, 247–256.

Perara, E., and Lingappa, V. R. (1985). *J. Cell Biol.* **101**, 2292–2301.

Perara, E., Rothman, R. E., and Lingappa, V. R. (1986). *Science* **232**, 348–352.

Rothblatt, J. A., and Meyer, D. I. (1986). *EMBO J.* **5**, 1031–1036.

Rothman, J. E., and Lodish, H. F. (1977). *Nature (London)* **269**, 775–780.

Schenborn, T., and Mierendorf, R. C. (1985). *Nucleic Acids Res.* **13**, 6223–6236.

Schlenstedt, G., and Zimmermann, R. (1987). *EMBO J.* **6**, 699–703.

Small, G. M., Szabo, L. J., and Lazzarow, P. B. (1988). *EMBO J.* **7**, 1167–1173.

Strub, K., and Walter, P. (1990). *Mol. Cell. Biol.* **10**, 777–784.

Thorner, J., and Paulus, H. (1973). *In* "The Enzymes" (P. D. Boyer, eds.), Vol. 8, pp. 487–508. Academic Press, New York.

Walter, P., and Blobel, G. (1981). *J. Cell Biol.* **91**, 557–561.

Waters, M. G., and Blobel, G. (1986). *J. Cell Biol.* **102,** 1543–1550.

Wiech, H., Sagstetter, M., Muller, G., and Zimmermann, R. (1987). *EMBO J.* **6,** 1011–1016.

Wiedmann, M., Kurzchalia, T. V., Hartmann, E., and Rapoport, T. A. (1987). *Nature (London)* **328,** 830–833.

Wilson, C., Connolly, T., Morrison, T., and Gilmore, R. (1988). *J. Cell Biol.* **107,** 69–77.

Yost, C. S., Lopez, C. D., Prusiner, S B., Myers, R. M., and Lingappa, V. R. (1990). *Nature (London)* **343,** 669–672.

Zimmermann, R., Sagstetter, M., Lewis, M., and Pelham, H. (1988). *EMBO J.* **7,** 2875–2880.

Chapter 11

Probing the Molecular Environment of Translocating Polypeptide Chains by Cross-Linking

DIRK GÖRLICH, TEYMURAS V. KURZCHALIA, MARTIN
WIEDMANN, AND TOM A. RAPOPORT

Central Institute of Molecular Biology
D-1115 Berlin-Buch, Germany

I. Introduction

Many proteins in a eukaryotic cell are synthesized in the cytosolic compartment but are then transported to certain cell organelles or secreted from the cell. Targeting is achieved by signal or targeting sequences that are contiguous amino acid sequences in the transported proteins, often located at the amino termini and cleaved off from precursor molecules (for

METHODS IN CELL BIOLOGY, VOL. 34

review, see Pugsley, 1989). It is now generally accepted that signal sequences are recognized by specific protein receptors either located in the cytoplasm or in the membrane of the target organelle (Blobel, 1980). The identification of these receptors, however, has been difficult. The best known example is the signal recognition particle (SRP), a cytoplasmic ribonucleoprotein complex that binds signal sequences of nascent polypeptides destined for translocation across the endoplasmic reticulum (ER) membrane, through its 54-kDa polypeptide component (SRP54) (Walter and Blobel, 1981; Kurzchalia *et al.*, 1986; Krieg *et al.*, 1986). Candidates for signal or targeting receptors have also been described for other systems (Söllner *et al.*, 1989, 1990; Adam *et al.*, 1989; Pain *et al.*, 1988; Wiedmann *et al.*, 1987b).

Intracellular protein transport often involves protein translocation across at least one membrane, a process that is poorly understood as yet. Two main hypotheses have been proposed: (1) the direct transport of polypeptides through the phospholipid bilayer without participation of membrane proteins (von Heijne and Blomberg, 1979; Engelman and Steitz, 1981) and (2) by way of a hydrophilic, protein-conducting channel formed by transmembrane proteins (Blobel and Dobberstein, 1975a; Rapoport, 1985). However, in spite of much speculation, the molecular environment of translocation polypeptide chains is not yet known.

The molecular dissection of the mechanism of protein transport relies on the use of *in vitro* systems. Following the pioneering work of Blobel and Dobberstein (1975b) by which a cell-free translation-translocation system was established for the ER membrane, similar systems have been worked out for the import of proteins into mitochondria (Maccecchini *et al.*, 1979), chloroplasts (Highfield and Ellis, 1978), and peroxisomes (Small *et al.*, 1987), and for the transport of proteins across the cytoplasmic membrane of *Escherichia coli.* (Müller and Blobel, 1984). These *in vitro* systems are used to identify signal or targeting receptors and to elucidate the mechanism of transport.

For the identification of signal sequence receptors and of candidates for constituents of a protein-conducting channel, cross-linking methods have proved to be useful, as they allow the identification of proteins that are located in close spatial proximity of the transported protein. Two different cross-linking approaches have been employed: (1) photocross-linking, with the probes incorporated into the transported polypeptide, and (2) cross-linking with bifunctional chemical reagents. In both cases, the transported protein is radioactively labeled. It is essential that all polypeptide chains are at the same stage of the transport process (transport intermediates), because the molecular environment of the transported polypeptide may change. Also, if protein transport occurs cotranslationally, as is generally the case for the ER membrane, all nascent chains must have the same length to produce distinct and interpretable cross-linked products.

In this article, we first discuss general features of the cross-linking approach, then outline the principles and problems of the two methods, and finally give detailed protocols for their application. The examples are taken from the work of our group on the protein transport across the ER membrane, but the protocols can in all likelihood be applied to other systems with appropriate modifications.

II. The Cross-Linking Methods—General Considerations

A. Translocation Intermediates

Several strategies can be used to produce intermediates (see also Gilmore *et al.*, Chapter 10).

1. TRAPPING THE INTERACTION OF A SIGNAL OR TARGETING SEQUENCE WITH A RECEPTOR

The recognition of a signal or targeting sequence by a receptor is an early event and further transport must be prevented.

If recognition occurs in the cytoplasm, as is the case for the binding of signal sequences by the SRP (Walter and Blobel, 1981), it is sufficient to omit ER membranes from the *in vitro* system and to produce nascent chains, the signal sequence of which has just emerged from the ribosome (Krieg *et al.*, 1989; Wiedmann *et al.*, 1989; S. High, D. Görlich, M. Wiedmann, T. A. Rapoport, and B. Dobberstein, unpublished results). Because about 40 amino acids are buried inside the ribosome (Malkin and Rich, 1967; Blobel and Sabatini, 1970) and because signal sequences contain about 20–30 residues (von Heijne, 1981), the nascent chain should be about 70 residues long. Such short N-terminal polypeptide fragments can be synthesized by translation of truncated mRNAs (Krieg *et al.*, 1989; Wiedmann *et al.*, 1989). A general property of truncated mRNA, which lacks a termination codon, is that the translating ribosome comes to halt when the 3' end of the mRNA is reached and that the nascent chain remains attached (Perara *et al.*, 1986; Mueckler and Lodish, 1986). Truncated mRNAs can be produced by *in vitro* Transcription of a plasmid cut with a restriction enzyme within the gene, or by addition of an oligonucleotide complementary to a certain region of the mRNA and cleavage of the mRNA by RNase H (Haeuptle *et al.*, 1986).

In the case of the SRP, polypeptide chain elongation is often arrested when the signal sequence has emerged from the ribosome and interacts with the particle (Walter and Blobel, 1981). Thus, cross-linking of the arrested fragment of the transported protein (70 amino acids; 70mer) with the SRP54 could be demonstrated even without the use of truncated mRNAs (Kurzchalia et al., 1986; Krieg et al., 1986).

If the interaction of a signal sequence with a membrane receptor is studied, further transport of the polpeptide chain can often be inhibited by removal of ATP or GTP and/or by lowering the temperature (see Hartl et al., 1989; Sanz and Meyer, 1989). Inhibitors have also been used. In the case of mitochondrial import, substances that dissipate the membrane potential have been employed (for review, see Hartl et al., 1989).

2. TRAPPING A POLYPEPTIDE DURING ITS TRANSFER THROUGH A MEMBRANE

In most cases studied so far, translocation of the N-terminus of a protein precedes that of the C-terminus. A translocation intermediate with a transmembrane orientation can therefore be produced if the C-terminus is prevented from being transferred across the membrane. The easiest way to prevent translocation of the C-terminus is to leave the nascent polypeptide attached to a ribosome, a condition that can be produced by translation of a truncated mRNA (Perara et al., 1986; Mueckler and Lodish, 1986; Connolly et al., 1989). Other procedures are based on the circumstance that folded protein domains or bulky molecules are not transported and will prevent translocation of the C-terminus of a protein when attached to it (Eilers and Schatz, 1986).

B.   Photocross-Linking

The photocross-linking method is based on the possibility of incorporating chemical groups into newly synthesized proteins by means of modified aminoacyl-tRNA (Johnson et al., 1976). Normally, owing to the specificity of the aminoacyl-tRNA synthetases, only the 20 naturally occurring amino acids are incorporated into polypeptides, and modifications in the side chains are not tolerated. However, if the modification is carried out *after* the amino acid is linked to its specific tRNA, the modified residue will be incorporated, because, beyond aminoacylation, all amino acids are accepted almost equally well by the translational machinery.

The principle of introducing photoreactive lysine derivatives into polypeptides is schematically outlined in Fig. 1a. The amino acid, bound

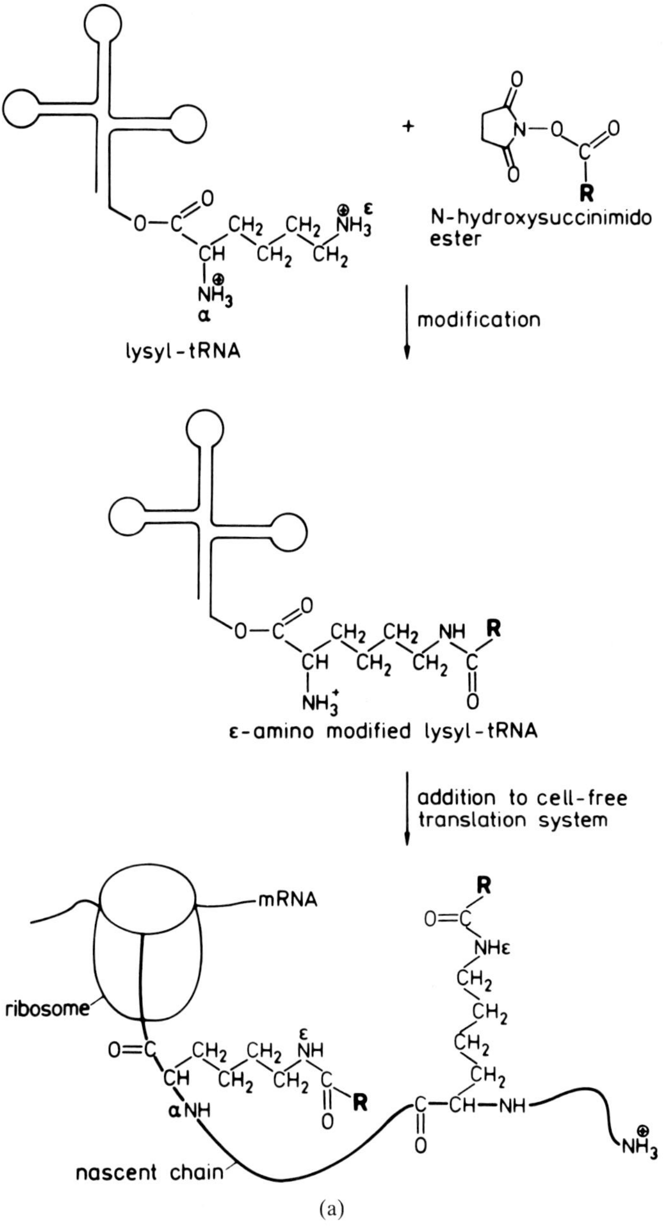

FIG. 1. Principle of the photocross-linking method (a) and photoreactive reagents that have been employed (b). (a) Lysyl-tRNA is modified in the ε-position of the amino acid with a N-hydroxysuccinimido ester. The three reagents that have been used are shown in (b). Modified lysyl-tRNA is added to a cell-free translation system and nascent chains carrying the photoreactive probes are synthesized. After irradiation, cross-linked products of the nascent chains and neighboring proteins are formed.

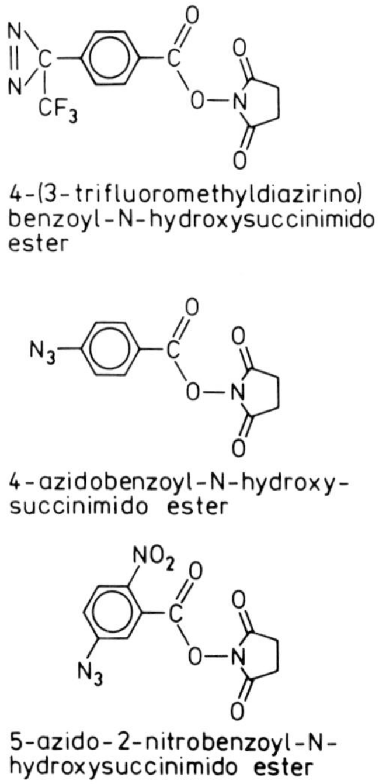

4-(3-trifluoromethyldiazirino)
benzoyl-N-hydroxysuccinimido
ester

4-azidobenzoyl-N-hydroxy-
succinimido ester

5-azido-2-nitrobenzoyl-N-
hydroxysuccinimido ester

(b)

FIG. 1. (*continued*)

to the specific tRNA, is modified at its ϵ-position with an *N*-hydroxy-succinimido ester of an acid carrying a photoreactive group. The modified lysyl-tRNA is then added to a cell-free translation system and lysine derivatives are incorporated at positions of the polypeptide where lysines normally occur. A labeled amino acid (preferably [^{35}S]methionine) is also incorporated and a translocation intermediate is produced as discussed before. Irradiation results in covalent cross-links between the nascent polypeptide chain and neighboring proteins (Krieg *et al.*, 1986; Kurzchalia *et al.*, 1986; Wiedmann *et al.*, 1987a).

Two photoreactive groups have been employed (Fig. 1b); upon irradiation these produce either a carbene or a nitrene. The diazirino reagent has the advantage that a very reactive carbene is produced with little if any selectivity for the reaction partner. Also, the diazirino compound is not sensitive to SH-reagents, in contrast to the azido compounds, and is less sensitive to dim light. On the other hand, the trifluoromethyldiazirinobenzoic acid is not yet commercially available and its synthesis is tedious. In practice, the carbene-producing reagent has often given cleaner results when compared with the azido reagents.

Among the two nitrene-producing azido compounds, the one containing the nitro group in the benzene ring should be preferred. Irradiation can be carried out at 320- to 350-nm wavelengths, rather than at 265–275 nm, with less UV damage to the system.

It should be noted that the method of "tRNA-mediated protein modification" can be used to introduce other chemical groups into newly synthesized proteins, such as biotin (Kurzchalia *et al.*, 1988) or fluorescent groups (T. V. Kurzchalia, unpublished results).

C. Cross-Linking with Bifunctional Reagents

The method is old but it has been applied only recently to protein translocation (Görlich *et al.*, 1990). A translocation intermediate is produced with a radioactively labeled polypeptide and cross-links to neighboring proteins are induced by bifunctional reagents.

A number of commercially available reagents can be employed, but carbodiimides and *m*-maleimidobenzoyl-sulfosuccinimide ester (sulfo-MBS) (see Fig. 2) have given the cleanest results. Carbodiimides will only produce amide bonds between neighboring carboxyl and amino groups without introducing spacers. Sulfo-MBS gives rise to cross-links between SH and amino groups. These constraints reduce the number of cross-linked products. Reagents such as dithiobis(succinimidyl propionate) (DSP) or disuccinimidyl suberate (DSS) which react with neighboring amino groups, may give a greater spectrum of products. Of course, different cross-linking reagents have to be tried to find the optimal one.

D. Comparison of the Two Cross-Linking Methods

The two methods have both merits and disadvantages. Photocrosslinking with the probes incorporated into the nascent chain allows one to study interactions of selected regions of the chain because the photoreactive lysine derivatives can be precisely positioned. This can be done either by selecting an appropriate protein (e.g., a secretory protein with lysine residues in the signal sequence) or by creating a lysine codon at the desired position by *in vitro* mutagenesis. The photoreaction does not strongly depend on suitable chemical groups of the interacting neighbors. The high reactivity of both nitrenes and carbenes ensures that only genuine neighbors are cross-linked but also generally results in low yields owing to the competing reaction with water. However, in some cases, wherein hydrophobic partners interact, the yield can be high (up to 10%) (Wiedmann

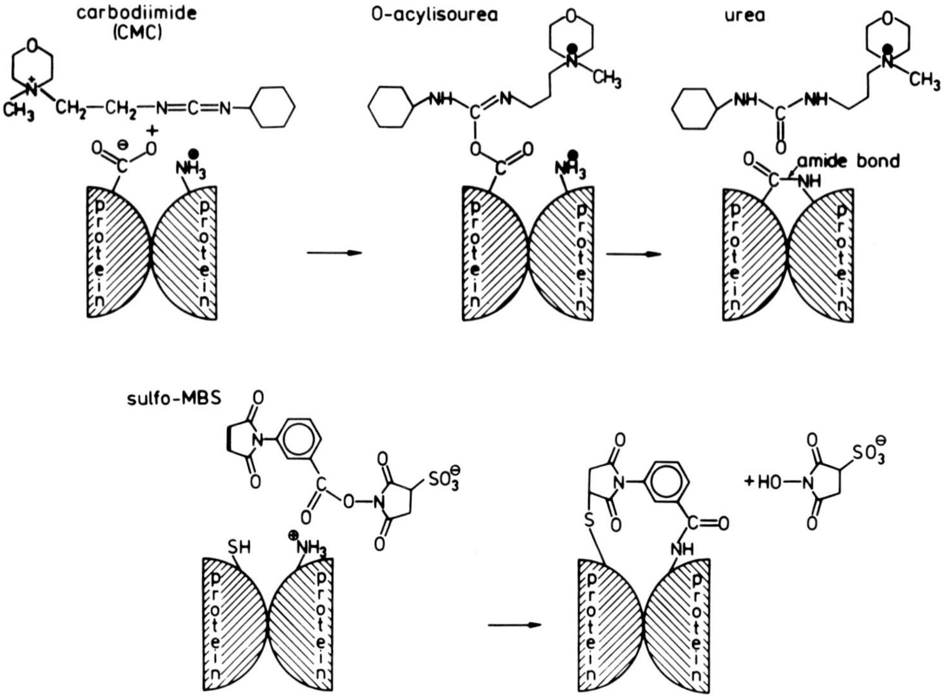

FIG. 2.   Principle of cross-linking by two bifunctional reagents. The reaction mechanisms
leading to cross-links between two neighboring proteins is shown for a carbodiimide (CMC)
and for sulfo-MBS. The first reagent leads to an amide bond between a carboxyl and an amino
group; the second leads to a cross-link between adjacent sulfhydryl and amino groups.

et al., 1987a). The photocross-linking method is probably the most gentle
one with little overall disturbance of the system; even translation can
resume after irradiation. On the other hand, it should be kept in mind that
the protein has undergone "mutations" by modifying the positively
charged lysines to uncharged hydrophobic derivatives.

   Cross-linking with bifunctional reagents, of course, depends on suitable
chemical groups of the neighboring proteins. Only rarely is it possible to
position precisely the probes in a protein. Also, the reagents must have
access to both partners, which may be a problem for membrane proteins or
for nascent chains still buried inside the ribosome. The yields of cross-
linked product can be high (5–10%), but it is possible that artifacts are
produced by the extensive chemical modifications introduced. It may be
necessary to remove interfering substances before treating the translocation intermediate with the reagent.

III. Cross-Linking Protocols

A. Photocross-Linking with Probes Incorporated into Nascent Polypeptide Chains

This method includes five main steps:

1. Charging of tRNA with lysine to produce lysyl-tRNA.
2. Chemical modification of the lysyl-tRNA in the ϵ-position of the amino acid using an N-hydroxysuccinimido ester of one of the reagents shown in Fig. 1b.
3. Purification of the modified lysyl-tRNA.
4. Incorporation of the modified lysines into cell-free synthesized polypeptide chains.
5. Irradiation and analysis of cross-linked products.

The following procedures refer to the use of 4-(3-trifluoromethyldiazirino)benzoic acid (TDBA). However, the same protocols can be used for the other two reagents. The N-hydroxysuccinimido esters of the azido compounds are commercially available (Pierce Chemical Co.).

1. CHARGING OF tRNA WITH LYSINE

The tRNA from bakers' yeast is charged with lysine using the tRNA synthetase of a crude rat liver extract.

   a. Preparation of a Crude Extract from Rat Liver.   All steps are done at 4°C.

PROCEDURE:

1. The liver of a male Wistar rat (7.6 g) is homogenized with a Potter homogenizer (15 strokes) in four volumes (30 ml) of a buffer containing 50 mM Tris-HCl, pH 7.5, 25 mM KCl, 5 mM MgCl$_2$, 1 mM dithiothreitol (DTT), 1 mM phenylmethanesulfonyl fluoride (PMSF), and 250 mM sucrose.
2. The homogenate is centrifuged for 20 minutes at 10,000 g.
3. The supernatant is decanted and centrifuged for 60 minutes at 50,000 rpm (R 65 rotor) in a Beckman ultracentrifuge.
4. The supernatant after the second centrifugation is collected, avoiding the fat floating at the top of the tube. It is applied to a 150-ml Sephadex G-25 (medium) (Pharmacia) column equilibrated in a buffer containing 50 mM Tris-HCl, pH 7.5, 1 mM EDTA, 1 mM DTT, and 10% (v/v) glycerol. The red-colored fractions of the void

volume are pooled (roughly the same volume as applied to the column). This crude liver extract is frozen in small aliquots and stored in liquid nitrogen. Repeated freezing and thawing diminishes the activity. A typical preparation of crude liver extract has optical densities at 280 and 260 nm of 10 and 8, respectively.

   b.   The Charging Reaction (Preparative Scale).

PROCEDURE:

1. Prepare the following mixture, which yields 10,000 μl:

3333 μl bakers' yeast tRNA (Boehringer Mannheim) at 12 mg/ml in H_2O
2000 μl 1 M Tris-HCl, pH 7.5
  250 μl 1 M KCl
  170 μl 1 M MgCl$_2$
   10 μl 1 M DTT
  500 μl 0.1 M ATP
  170 μl 0.6 M creatine phosphate
  120 μl 10 mg/ml creatine kinase
  500 μl [^{14}C]lysine 7.7 GBq/mmol (210 mCi/mmol), 8 MBq/ml
3000 μl crude liver extract

Note: Different batches of liver extract give different yields of charged tRNA. It is therefore necessary to determine in pilot experiments the amount of tRNA that can be aminoacylated to completion in a given reaction volume. It is useful to perform 10-μl test assays with increasing tRNA concentrations. The amount of charged [^{14}C]lysyl tRNA can be determined by the radioactivity precipitated with cold trichloroacetic acid (TCA). At low tRNA concentrations there is a linear relationship with the amount of lysyl-tRNA formed. In preparative work, the highest tRNA concentration that is still in the linear range should be employed.

2. The reaction mixture is incubated for 20 minutes at 37°C.
3. The reaction is stopped by placing the tube on ice and adding 1 ml of 2 M sodium acetate buffer, pH 5, and 0.5 ml 20% SDS.
4. The mixture is extracted twice with phenol (pH not adjusted) and twice with ether.
5. The tRNA is precipitated by addition of 25 ml of 96% ethanol, chilling to −20°C for 20 minutes, and centrifuging for 10 minutes at 10,000 g. The pellet is washed once in 70% ethanol.
6. The pellet is dissolved in 2 ml of 0.2 M sodium acetate buffer, pH 5.0. To determine roughly the efficiency of charging, a 10-μl aliquot is diluted with 1 ml of methanol and added to 5 ml of toluene scintillator; the radioactivity is determined in a liquid scintillation counter.

Typically, about 500,000 dpm/mg tRNA are obtained (20,000,000 dpm for a 10-ml charging assay).

7. The dissolved tRNA is applied to a 50-ml Sephadex G-25 column equilibrated with 0.2 M sodium acetate buffer, pH 5. Then 2-ml fractions are collected and both the optical density at 260 nm and the radioactivity are measured. The peak fractions of the void volume are pooled. Typically, 23,000 dpm per OD_{260} unit (350,000 dpm/mg tRNA) are obtained. The tRNA is precipitated at $-20°C$ by addition of 2.5 volumes of 96% ethanol.

8. After centrifugation for 20 minutes at 10,000 g, the tRNA is taken up in H_2O to 3,000,000 dpm/ml and is then ready for modification. The charged tRNA can be stored in 70% ethanol at $-20°C$ for several months without loss of activity.

2. CHEMICAL MODIFICATION OF THE LYSYL-tRNA

In this step, the cross-linking reagent is covalently attached to the ϵ-amino group of lysine. The reaction conditions ensure an optimal compromise between modification of the ϵ-amino group and hydrolysis of the aminoacyl-tRNA. Also, the α-amino group remains largely unaffected.

On a preparative scale, the modification is usually performed in a 5-ml volume. The amount of TDBA–lysyl-tRNA will be sufficient for about 250–500 cross-linking assays (50-μl translation mixtures). All the following steps are done in dim light to prevent the decay of the reagent. A crucial point is the purity of the dimethyl sulfoxide (DMSO); impurities diminish the yield significantly.

PROCEDURE:

1. 10 μmol of 4-(3-trifluoromethyldiazirino)benzoyl-N-hydroxy-succinimido ester (3.3 mg) are dissolved in 3 ml of DMSO and the solution is put on ice, upon which the DMSO solidifies.

2. Charged tRNA (1.5 ml) is added (4,500,000 dpm). The DMSO thaws slowly after addition of the aqueous solution consuming the heat, which is released when DMSO and water are mixed.

3. Start the reaction by addition of 0.5 ml of 0.5 M 3-(cyclohexyl-amino)propane sulfonic acid (CAPS) buffer, pH 10.5.

4. Stop the reaction after 10 minutes by addition of 5 ml of 2 M sodium acetate buffer, pH 5.0, and 25 ml of cold 96% ethanol. The mixture is kept at $-20°C$ for 10 minutes and centrifuged at 10,000 g for 20 minutes. The pellet is washed once with 70% ethanol.

Note: The modified tRNA can be used at this stage, but usually the efficiency of cross-linking will be higher if the tRNA is further purified by chromatography on BD cellulose. This step removes much of the unrelated tRNA species, which would inhibit translation at high concentrations, as well as unmodified and α-modified lysyl-tRNA, which also interfere. On the other hand, the purification step is time consuming and there is some loss of material. If the modified lysyl-tRNA is used without further purification, it should be dissolved in water at a concentration of 10,000 dpm/μl and stored at $-80°C$ protected from light.

3. PURIFICATION OF THE MODIFIED tRNA BY CHROMATOGRAPHY ON BD CELLULOSE

PROCEDURE:

1. The tRNA from the step before is dissolved in 2 ml of 0.2 M sodium acetate buffer, pH 5.0, and is applied to a 10-ml column of BD cellulose (benzoylated DEAE-cellulose; SERVACEL BD; Serva Fine Biochemicals Inc.) equilibrated with 50 mM sodium acetate buffer, pH 5.0, 0.4 M NaCl, and 10 mM MgCl$_2$. Fractions (4 ml) are collected and the radioactivity in the aliquots is determined. The column is washed with two volumes of equilibration buffer. All radioactivity should remain bound.
2. The unmodified lysyl-tRNA is eluted with four column volumes of 50 mM sodium acetate buffer, pH 5.0, 1.0 M NaCl, and 10 mM MgCl$_2$.
3. The modified tRNA is eluted with 50 mM sodium acetate buffer, pH 5.0, 1.0 M NaCl, 10 mM MgCl$_2$, and 30% (v/v) ethanol. A typical elution profile is shown in Fig. 3. Normally, about 70% of the lysyl-tRNA is modified.
4. The peak fractions of modified lysyl-tRNA are pooled and 2.5 volumes of 96% ethanol are added. The mixture is kept at $-20°C$ overnight and the tRNA is sedimented at 10,000 g for 30 minutes.
5. The pellet is dissolved in 1 ml of 0.2 M sodium acetate buffer, pH 5.0, reprecipitated with ethanol, and washed once with 70% ethanol. The tRNA is taken up in H$_2$O to 10,000 dpm/μl. The tRNA is now ready for use and should be stored at $-80°C$ protected from light.

4. TEST OF THE TDBA–LYSYL-tRNA BY CROSS-LINKING OF THE SIGNAL SEQUENCE OF PREPROLACTIN TO THE SRP54

The following protocol is designed to test the prepared TDBA–lysyl-tRNA and to determine the optimal tRNA concentration in the wheat germ translation system. It is necessary to find a compromise between the

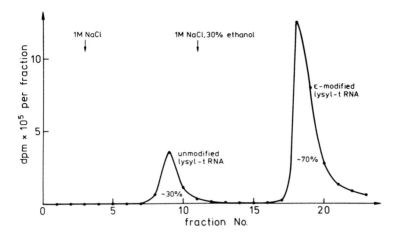

FIG. 3. Purification of modified lysyl-tRNA on BD cellulose. [^{14}C]Lysyl-tRNA is modified as described in the text. The sample is applied to a BD cellulose column and eluted as indicated in the figure and detailed in the text. About 70% of the radioactivity elutes as ε-modified lysyl-tRNA and can be used in photocross-linking experiments.

incorporated number of photoreactive groups on the one hand, and inhibition of translation by excess of tRNA on the other. The optimal concentration of TDBA–lysyl-tRNA may vary from batch to batch.

The test, which is routinely used in our laboratory, is based on the observation that the signal sequence of a short fragment of a secretory protein [in the example given below, a fragment of 86 amino acids (86mer) of preprolactin] can be cross-linked to SRP54 (Krieg *et al.*, 1989; Wiedmann *et al.*, 1989). The yield of the cross-linked product is sufficiently high to detect it among the total TCA-precipitated products without further enrichment.

a. Translation.

PROCEDURE:

1. Prepare the following "Premix":

7.5 μl amino acid mixture, 1 mM each, lacking methionine and lysine
3   μl 0.1 M ATP
1   μl 1 M DTT
1   μl 1 M HEPES/KOH, pH 7.6
2   μl 0.02 M GTP
4   μl 0.6 M creatine phosphate
3   μl 10 mg/ml creatine kinase

2. Prepare the following mix:

20   μl wheat germ extract (Erickson and Blobel, 1983)
4.4 μl "premix"
16.7 μl H_2O
1.6 μl 1.25 M potassium acetate
0.5 μl 5 μM SRP from dog pancreas (Walter and Blobel, 1983b)
4   μl [^{35}S]methionine 555 MBq/ml (15 mCi/ml)
0.8 μl TDBA–lysyl-tRNA, 10,000 dpm/μl
2   μl transcript coding for the 86mer of preprolactin, phenol extracted, ethanol precipitated, and dissolved in H_2O (plasmid pGEMBPL cut with PvuII and transcribed with T7 RNA polymerase) (Krieg *et al.*, 1989; Wiedmann *et al.*, 1989)

3. The translation mixture is incubated for 15 minutes at 26°C.

Note: The volume of wheat germ extract and the concentrations of magnesium and potassium should be optimized to ensure efficient translation. The values given above may be taken as starting points and each of the three parameters should be varied, keeping the others constant. The magnesium concentration, contributed by the wheat germ extract, is usually at the upper end of the optimum. The quantities of SRP and of the transcript should also be optimized.

If the azido compounds are used (Fig. 1b), the wheat germ extract should be prepared with 2 mM 2-mercaptoethanol (Kurzchalia *et al.*, 1986) or with 1 mM reduced glutathione (Krieg *et al.*, 1986) instead of with DTT, and DTT should be omitted from the premix.

The modified lysyl-tRNAs can also be employed in other *in vitro* translation systems (e.g., prepared from *Saccharomyces cerevisiae* or from *Candida maltosa*) (A. Müsch, M. Wiedmann, and T. A. Rapoport, unpublished results).

b.   Irradiation and Analysis.

PROCEDURE:

1. After translation, the sample is put on ice and divided into two equal aliquots. One will serve as the negative irradiation control, the other is irradiated on ice for 5 minutes with a 100- or 150- high-pressure mercury arc lamp with a 300- to 400-nm bandpass filter (about 10^5 erg/cm^2; distance from the lamp to the sample about 30 cm). Irradiation is conveniently carried out in caps of Eppendorf tubes placed on a cooled steel block.

Note: The duration of irradiation may have to be optimized (tests necessary). For the azido compounds it has been observed that irradiation at the temperature of liquid nitrogen increases the yield of cross-linked product (Krieg *et al.*, 1989).

2. The irradiated and nonirradiated samples are precipitated with TCA and submitted to SDS–polyacrylamide gel electrophoresis (10–15% linear acrylamide gradient gel) (Laemmli, 1970) and fluorography. In the irradiated sample, there should be visible a 62-kDa band representing the adduct of the 86mer (10 kDa) and the SRP54. This band should contain about 10% of the radioactivity present in the 10-kDa band.

Note: The quality of the modified lysyl-tRNA can also be checked by the incorporation of modified lysines into cell-free synthesized proteins, but this, of course, represents a less stringent control than the cross-linking test. The incorporation of modified lysines can be tested by using [^{3}H]lysine instead of [^{14}C]lysine to produce modified lysyl-tRNA with a high specific radioactivity of the amino acid (Kurzchalia *et al.*, 1986). Alternatively, two-dimensional O'Farrell gel electrophoresis can be used to test for the expected shift of the isoelectric point as a function of the number of incorporated lysine derivatives (T. V. Kurzchalia and R. Benndorf, unpublished results).

5. APPLICATION OF PHOTOCROSS-LINKING TO THE IDENTIFICATION OF MEMBRANE PROTEINS IN PROXIMITY TO TRANSLOCATING POLYPEPTIDE CHAINS

The following protocol has been used to identify proteins in close proximity to nascent polypeptide chains during their transfer through the membrane. The experiment is similar to the one described before, except that prior to irradiation, microsomal membranes are added to the 86mer of preprolactin to produce a translocation intermediate. If the total TCA-precipitable products are analyzed, it is found that the yield of 62-kDa cross-linked product between the 86mer and the SRP54 is greatly diminished (Wiedmann *et al.*, 1987b), indicating that the signal sequence leaves the SRP when it interacts with membranes. The cross-linked products with membrane proteins are hardly visible among the total products and therefore have to be enriched for. The main cross-linked product will be a 46-kDa adduct of the 86mer and an integral ER glycoprotein of M_r 34,000, called the signal sequence receptor (SSR) (Wiedmann *et al.*, 1987b).

a.   Irradiation of a Translocation Intermediate.

PROCEDURE:

1. Prepare a 50-μl translation mixture as described in Section II,4,a.
2. Incubate at 26°C for 10 minutes.
3. Add 8 equivalents (for definition, see Walter *et al.*, 1981) of salt- and EDTA-washed canine microsomes (EK-RM) (prepared as described by Walter and Blobel, 1983a).
4. Incubate at 26°C for 5 minutes more.
5. Irradiate as described in Section II,4,b. A negative irradiation control should be performed.

b.   Enrichment of Cross-Linked Products. The following procedures have been used to enrich for cross-linked products containing SSR either singly or in sequence. The protocols can be adopted for other cases with appropriate modifications.

1.   Carbonate extraction. This method enriches for products containing integral membrane proteins (Fujiki *et al.*, 1982). The sample is adjusted to a final concentration of 0.1 M Na_2CO_3 (pH 11.5) in a volume of up to 150 μl (prepare the carbonate buffer freshly). After incubation for 10 minutes on ice, the membranes are sedimented at 30 psi for 10 minutes in an Airfuge (Beckman). The pellet is resuspended thoroughly in 0.1 M Na_2CO_3 and the membranes are again sedimented as before.

2.   Binding to concanavalin A(Con A) Sepharose. This method enriches for products containing glycoproteins, which bind to Con A (Wiedmann *et al.*, 1987b; Görlich *et al.*, 1990). The sample is diluted to 1 ml with Con A buffer [50 mM HEPES/KOH, pH 7.5, 500 mM potassium acetate, 5 mM $MgCl_2$, 1 mM $MnCl_2$, 1 mM $CaCl_2$, and 1% Nonidet P-40 (NP-40)]. Con A Sepharose (25 μl) (Pharmacia) is added (thoroughly prewashed with Con A binding buffer). The mixture is gently shaken at 4°C overnight. After brief centrifugation (about 5 seconds at 3000 rpm in an Eppendorf centrifuge), the supernatant is discarded and the resin is washed five times with 1 ml of Con A buffer and finally with 1 ml of water. Elution of the glycoproteins is carried out with 50 μl of SDS–sample buffer (Laemmli, 1970) by incubation for 5 minutes at 95°C.

3.   Immunoprecipitation. The identity of a cross-linked product is best determined by immunoprecipitation with antibodies directed against both constituents of the product (i.e., against the translocating polypeptide and against the suspected membrane protein) (Wiedmann *et al.*, 1987b; Görlich *et al.*, 1990). The sample (less than 100 μl) is adjusted to 1% SDS final concentration and heated to 90°C for 5 minutes. It is then diluted to 1 ml to give final concentrations of 10 mM Tris-HCl, pH 7.5, 150 mM NaCl, 1 mM EDTA, 1% NP-40, and 0.1% SDS (IP-buffer). Antiserum

(3 μl) is added and the mixture is incubated for 1 hour at room temperature and overnight at 4°C. Protein A–Sepharose (20 μl) (Pharmacia) is added and then incubation is continued at 4°C for 30 minutes with gentle shaking. The resin is washed five times with IP-buffer and once with water. Elution is carried out with SDS–sample buffer.

> Note: The precise conditions of immunoprecipitation may have to be optimized for each antiserum. The salt concentration in the IP-buffer can sometimes be increased to 500 mM to lower the background; SDS denaturation of the sample may not be applicable for all antibodies, and the amount of antiserum can often be decreased; purified immunoglobulins sometimes have to be used. Affinity-purified antibodies should be preincubated with protein A–Sepharose and the resin washed before addition of the sample (affinity purification of antibodies often results in loss of their binding to protein A). DTT should be avoided in the sample in any case.

B. Cross-Linking with Bifunctional Reagents

Two protocols are given below which have been used successfully to probe the environment of polypeptide chains traversing the ER membrane with the bifunctional crosslinking reagent CMC (Görlich *et al.*, 1990). Again, crosslinking of the 86mer of preprolactin is taken as an example.

1. CROSS-LINKING BY CMC OF TRANSLOCATING PREPROLACTIN

PROCEDURE:

1. Prepare a 50-μl translation mixture as described in Section II,4,a, except that the amino acid mixture lacks only methionine and that TDBA–lysyl-tRNA is omitted.
2. Incubate at 26°C for 10 minutes.
3. Add 8 equivalents EK-RM.
4. Incubate at 26°C for 5 minutes more.
5. The translation mixture is layered on top of a 100-μl cushion (C-buffer: 50 mM HEPES/KOH, pH 7.5, 500 mM KCl, 500 mM sucrose, and 5 mM MgCl$_2$) and is centrifuged at 4°C in an Airfuge for 10 minutes at 25 psi. The supernatant is carefully removed and the pellet (containing the membrane-targeted nascent chains) is resuspended in 100 μl of MP-buffer (50 mM HEPES/KOH, pH 7.5,

140 mM KCl, 3 mM MgCl$_2$, and 250 mM sucrose). Keep 50 μl as a negative cross-linking control.

Note: This step removes several substances present in the translation mixture that bear free carboxyl or amino groups and would quench the carbodiimide reaction, e.g., potassium acetate and amino acids. The radioactive methionine would also interfere by causing an enormous background due to its random attachment to proteins catalyzed by the carbodiimide.

6. To the other 50 μl, add 200 μl CMC, 2mg/ml in MP-buffer (final concentration 3.8 mM); the reaction is allowed to proceed for 30 minutes on ice.

7. The reaction is quenched by addition of 15 μl 1 M glycine, pH 7.5, for 30 minutes.

8. Analysis can be carried out after TCA precipitation by SDS–gel electrophoresis and fluorography. A major crosslinked band of 46 kDa should be visible.

Note: If the above mentioned quenching conditions are used, the proteins are extensively modified: carboxyl groups are converted to amides, and amino groups are converted to their acyl derivatives. The modified proteins may be poorly recognized by antibodies raised against the authentic proteins. The following protocol may be used if immunoprecipitation is intended after cross-linking with CMC. The example refers to immunoprecipitation of the 46-kDa cross-linked product with prolactin antibodies.

2. IMMUNOPRECIPITATION AFTER CMC TREATMENT

PROCEDURE:

1. The procedure given previously is followed up to step 6.

2. The excess of carbodiimide is removed by centrifugation of the membranes through a sucrose cushion as in step 5, except that the cushion volume is reduced to 50 μl and the sample is distributed in two Airfuge tubes.

3. The pellets are resuspended in a total volume of 20 μ MP-buffer and are left for 30 minutes longer on ice.

4. Quenching of the reaction is carried by addition of 25 μl of 100 mM Na$_2$HPO$_4$ and 2% SDS (pH about 9), and heating for 5 minutes at 90°C.

5. The sample is diluted to 1 ml to give final concentrations of 10 mM Tris-HCl, 150 mM NaCl, 1 mM EDTA, 1% NP40, and 0.1% SDS (IP-buffer).

6. Antiprolactin serum (3 μl) is added and the sample is incubated for 1 hour at room temperature and overnight at 4°C.
7. All other steps are carried out as described in Section II,5,b,3.

3. CROSS-LINKING BY SULFO-MBS OF TRANSLOCATING β-LACTAMASE

The following protocol demonstrates the use of the bifunctional reagent sulfo-MBS (see Fig. 2) to identify proteins in the proximity of translocating chains of β-lactamase. The fragment employed (141 amino acids, 141mer) is long enough to produce signal peptide cleavage by the signal peptidase located at the lumenal side of the ER membrane resulting in a 118mer. The experiment also shows the application of the reticulocyte lysate system (Görlich et al., 1990).

PROCEDURE:

1. Prepare the following translation mixture:

35 μl nuclease-treated reticulocyte lysate (Promega) (Pelham and Jackson, 1976)
1 μl amino acid mixture minus methionine
4 μl methionine 555 MBq/ml (15 mCi/ml)
4 μl H$_2$O
2 μl transcript coding for the 141mer of β-lactamase (plasmid pLAC81M1 cut with AvaII and transcribed with SP6 RNA polymerase (Görlich et al., 1990)
4 μl EK-RM (1 equivalent/μl)
2. Incubate at 30°C for 15 minutes.
3. Place the tube on ice and adjust to final concentrations of 500 mM KCl and 5 mM MgCl$_2$ in a volume of 100 μl.
4. The mixture is layered on two 100-μl cushions of C-buffer and the membranes are pelleted for 10 minutes at 25 psi in an Airfuge.
5. The membrane are resuspended in 200 μl of MP-buffer. One-half is kept as a negative cross-linking control; to the other half add 5 μl of 1 mg/ml sulfo-MBS in MP-buffer (final concentration 0.12 mM) and the reaction is allowed to proceed for 10 minutes on ice.
6. The reaction is quenched by addition of 10 μl 500 mM glycine/HCl, pH 7.5 and 50 mM 2-mercaptoethanol for 30 minutes.
7. Analysis is carried out after TCA precipitation by SDS-gel electrophoresis and fluorography. Two main cross-linked products should be visible: a cross-link of about 46 kDa of the 118mer of β-lactamase (141mer lacking the signal sequence) and a 35-kDa membrane glycoprotein, and a product of about 75 kDa of the 118mer and a soluble 53-kDa glycoprotein (Görlich et al., 1990).

IV. Glossary

CMC   1-cyclohexyl-3-(2-morpholinoethyl)carbodiimide–metho-4-toluene sulfonate
DMS   dimethyl suberimidate
DSP   dithiobis(succinimidyl propionate)
DTT   dithiothreitol
EK-RM   microsomal membranes washed with 0.5 M salt and EDTA
ER   endoplasmic reticulum
NP-40   Nonidet P-40
SRP   signal recognition particle
SSR   signal sequence receptor
sulfo-MBS m-maleimidobenzoyl-sulfosuccinimide ester
TDBA   4-(3-trifluoromethyldiazirino)benzoic acid

BUFFERS

C buffer   50 mM HEPES/KOH, pH 7.5, 500 mM KCl, 5 mM MgCl$_2$, 500 mM sucrose
Con A buffer   50 mM HEPES/KOH, pH 7.5, 500 mM potassium acetate, 5 mM MgCl$_2$, 1 mM MnCl$_2$, 1 mM CaCl$_2$, 1% NP-40
IP buffer   10 mM Tris-HCl, pH 7.5, 150 mM NaCl, 1 mM EDTA, 1% NP-40, 0.1% SDS
MP buffer   50 mM HEPES/KOH, pH 7.5, 140 mM KCl, 3 mM MgCl$_2$, 250 mM sucrose

ACKNOWLEDGMENTS

We acknowledge the important contribution of Drs. A. S. Girshovich and E. S. Bochkareva in establishing the photocross-linking method (Kurzchalia *et al.*, 1986). We thank Dr. S. M. Rapoport for critical reading of the manuscript.

REFERENCES

Adam, S. A., Lobl, T. J., Mitchell, M. A., and Gerace, L. (1989). Identification of specific binding proteins for a nuclear location sequence. *Nature (London)* **337**, 276–279.
Blobel, G. (1980). Intracellular protein topogenesis. *Proc. Natl. Acad. Sci. U.S.A.* **77**, 1496–1500.
Blobel, G., and Dobberstein, B. (1975a). Transfer of proteins across membranes. I. Presence of proteolytically processed and unprocessed nascent immunoglobulin light chains on membrane-bound ribosomes of murine myeloma. *J. Cell Biol.* **67**, 835–851.
Blobel, G., and Dobberstein, B. (1975b). Transfer of proteins across membranes. II. Reconstitution of functional rough microsomes from heterologous components. *J. Cell Biol.* **67**, 852–862.
Blobel, G., and Sabatini, D. D. (1970). Controlled proteolysis of nascent polypeptides in rat

liver cell fractions. I. Location of the polypeptides within ribosomes. *J. Cell Biol.* **45,** 130–145.

Connolly, T., Collins, P., and Gilmore, R. (1989). Access of proteinase K to partially translocated nascent polypeptides in intact and detergent-solubilized membranes. *J. Cell Biol.* **108,** 299–307.

Eilers, M., and Schatz, G. (1986). Binding of a specific ligand inhibits import of a purified precursor protein into mitochondria. *Nature (London)* **322,** 228–232.

Engelman, D. M., and Steitz, T. A. (1981). The spontaneous insertion of proteins into and across membranes: The helical hairpin hypothesis. *Cell (Cambridge, Mass.)* **23,** 411–422.

Erickson, A. H., and Blobel, G. (1983). Cell-free translation of messenger RNA in a wheat germ system. *In* "Methods in Enzymology" (S. Fleischer and B. Fleischer, eds.), Vol. 96, pp. 38–50. Academic Press, New York.

Fujiki, Y., Hubbard, A. D., Fowler, S., and Lazarow, P. B. (1982). Isolation of intracellular membranes by means of sodium carbonate treatment: Application to endoplasmic reticulm. *J. Cell Biol.* **93,** 97–102.

Görlich, D., Prehn, S., Hartmann, E., Herz, J., Otto, A., Kraft, R., Wiedmann, M., Knespel, S., Dobberstein, B., and Rapoport, T. A. (1990). The signal sequence receptor has a second subunit and is part of a translocation complex in the endoplasmic reticulum as probed by bifunctional reagents. *J. Cell Biol.* **111,** 2283–2294.

Haeuptle, M.-T., Frank, R., and Dobberstein, B. (1986). Translation arrest by oligodeoxynucleotides complementary to mRNA coding sequences yields polypeptides of predetermined length. *Nucleic Acids Res.* **14,** 1427–1448.

Hartl, F.-U., Pfanner, N., Nicholson, D. W., and Neupert, W. (1989). Mitochondrial protein import. *Biochim. Biophys. Acta* **988,** 1–45.

Highfield, P. E., and Ellis, R. J. (1978). Synthesis and transport of the small subunit of chloroplast ribulose bisphosphate carboxylase. *Nature (London)* **271,** 420–424.

Johnson, A. E., Woodward, W. R., Herbert, E., and Menninger, J. R. (1976). N^ϵ-Acetyllysine transfer ribonucleic acid: A biologically active analog of aminoacyl transfer ribonucleic acids. *Biochemistry* **15,** 569–575.

Krieg, U. C., Walter, P., and Johnson, A. E. (1986). Photocrosslinking of the signal sequence of nascent preprolactin to the 54 kD polypeptide of the signal recognition particle. *Proc. Natl. Acad. Sci. U. S. A.* **83,** 8604–8608.

Krieg, U. C., Johnson, A. E., and Walter, P. (1989. Protein translocation across the endoplasmic reticulum membrane: Identification by photocross-linking of a 39-kD integral membrane glycoprotein as part of a putative translocation tunnel. *J. Cell Biol.* **109,** 2033–2043.

Kurzchalia, T. V., Wiedmann, M., Girshovich, A. S. Bochkareva, E. S., Bielka, H., and Rapoport, T. A. (1986). The signal sequence of nascent preprolactin interacts with the 54 kD polypeptide of the signal recognition particle. *Nature (London)* **320,** 634–636.

Kurzchalia, T. V., Wiedmann, M., Breter, H., Zimmermann, W., Bauschke, E., and Rapoport, T. A. (1988). tRNA-mediated labelling of proteins with biotin. A nonradioactive method for the detection of cell-free translation products. *Eur. J. Biochem.* **172,** 663–668.

Laemmli, U. K. (1970). Cleavage of structural proteins during the assembly of the head of bacteriophage T4. *Nature (London)* **227,** 680–685.

Maccecchini, M.-L., Rudin, Y., Blobel, G., and Schatz, G. (1979). Import of proteins into mitochondria: Precursor forms of the extramitochondrially made F_1-ATPase subunits in yeast. *Proc. Natl. Acad. Sci. U. S. A.* **76,** 343–347.

Malkin, L. I., and Rich, A. (1967). Partial resistance of nascent polypeptide chains to proteolytic digestion due to ribosomal shielding. *J. Mol. Biol.* **26,** 329–346.

Mueckler, M., and Lodish, H. F. (1986). Post-translational insertion of a fragment of the glucose transporter into microsomes requires phosphoanhydride bond cleavage. *Nature (London)* **322,** 549–552.

Müller, M., and Blobel, G. (1984). *In vitro* translocation of bacterial proteins across the plasma membrane of *Escherichia coli. Proc. Natl. Acad. Sci. U. S. A.* **81,** 7421–7425.

Pain, D., Kanwar, Y., and Blobel, G. (1988). Identification of a receptor for protein import into chloroplasts and its localization to envelope contact sites. *Nature (London)* **331,** 232–237.

Pelham, H. R. B., and Jackson, R. J. (1976). An efficient mRNA-dependent translation system from reticulocyte lysates. *Eur. J. Biochem.* **67,** 247–256.

Perara, E., Rothman, R. E., and Lingappa, V. R. (1986). Uncoupling translocation from translation: Implications for transport of proteins across membranes. *Science* **232,** 348–352.

Pugsley, A. (1989). "Protein Targeting." Academic Press, San Diego, California.

Rapoport, T. A. (1985). Extensions of the signal hypothesis–sequential insertion model versus amphipathic tunnel hypothesis. *FEBS Lett.* **187,** 1–10.

Sanz, P., and Meyer, D. I. (1989). Secretion in yeast: Preprotein binding to a membrane receptor and ATP-dependent translocation are sequential and separable events *in vitro. J. Cell Biol.* **108,** 2101–2106.

Small, G. H., Imanaka, T., Shio, H., and Lazarow, P. B. (1987). Efficient association of *in vitro* translation products with purified, stable *Candida tropicalis* peroxisomes. *Mol. Cell. Biol.* **7,** 1848–1855.

Söllner, T., Griffiths, G., Pfaller, R., Pfanner, N., and Neupert, W. (1989). MOM19, an import receptor for mitochondrial precursor proteins. *Cell (Cambridge, Mass.)* **59,** 1061–1070.

Söllner, T., Pfaller, R., Griffiths, G., Pfanner, N., and Neupert, W. (1990). A mitochondrial import receptor for the ADP/ATP carrier. *Cell (Cambridge, Mass.)* **62,** 107–115.

von Heijne, G. (1981). On the hydrophobic nature of signal sequences. *Eur. J. Biochem.* **116,** 419–422.

von Heijne, G., and Blomberg, C. (1979). Trans-membrane translocation of proteins. The direct transfer model. *Eur. J. Biochem.* **97,** 175–181.

Walter, P., and Blobel, G. (1981). Translocation of proteins across the endoplasmic reticulum. III. Signal recognition protein (SRP) causes signal sequence-dependent and site-specific arrest of chain elongation that is released by microsomal membranes. *J. Cell Biol.* **91,** 557–561.

Walter, P., and Blobel, G. (1983a). Preparation of microsomal membranes for cotranslational protein translocation. *In* "Methods in Enzymology" (S. Fleischer and B. Fleischer, eds.), Vol. 96, pp. 557–561. Academic Press, New York.

Walter, P., and Blobel, G. (1983b). Signal recognition particle: A ribonucleoprotein required for cotranslational translocation of protein, isolation and properties. *In* "Methods in Enzymology" (S. Fleischer and B. Fleischer, eds.), Vol. 96, pp. 682–691. Academic Press, New York.

Walter, P., Ibrahimi, I., and Blobel, G. (1981). Translocation of proteins across the endoplasmic reticulum. I. Signal recognition protein (SRP) binds to *in vitro* assembled polysomes synthesizing secretory protein. *J. Cell Biol.* **91,** 545–550.

Wiedmann, M., Kurzchalia, T. V., Bielka, H., and Rapoport, T. A. (1987a). Direct probing of the interaction between the singal sequence of nascent preprolaction and the signal recognition particle by specific cross-linking. *J. Cell Biol.* **104,** 201–208.

Wiedmann, M., Kurzchalia, T. V., Hartmann, E., and Rapoport, T. A. (1987b). A signal sequence receptor in the endoplasmic reticulum membrane. *Nature (London)* **328,** 830–833.

Wiedmann, M., Görlich, D., Hartmann, E., Kurzchalia, T. V., and Rapport, T. A. (1989). Photocrosslinking demonstrates proximity of a 34 kDa membrane protein to different portions of preprolaction during translocation through the endoplasmic reticulum. *FEBS Lett.* **257,** 263–268.

Chapter 12

Reconstitution of Secretory Protein Translocation from Detergent-Solubilized Rough Microsomes

CHRISTOPHER NICCHITTA, GIOVANNI MIGLIACCIO,
AND GÜNTER BLOBEL

The Laboratory of Cell Biology
Howard Hughes Medical Institute
Rockefeller University
New York, New York 10021

METHODS IN CELL BIOLOGY, VOL. 34

I. Introduction

In higher eukaryotes, the sorting of secretory proteins shares a common early event, the targeting of the nascent precursor to the rough endoplasmic reticulum (RER). This initial sorting process requires the interaction of the signal recognition particle (SRP) with the signal sequence of the nascent chain (Walter and Blobel, 1981). Following recognition of the secretory precursor, the SRP/nascent chain/ribosome complex is targeted to the RER by virtue of the interaction of the SRP with its RER-localized cognate receptor, the SRP receptor (docking protein) (Gilmore *et al.,* 1982; Meyer *et al.,* 1982). This targeting reaction results in the release of SRP and the subsequent association of the nascent chain with the components of the RER that mediate translocation (Gilmore *et al.,* 1982; Meyer *et al.,* 1982; Gilmore and Blobel, 1983).

The mechanism of translocation is unkown and there has been considerable debate over the past decade as to whether translocation requires the physical interaction of the nascent precursor with protein components of the RER, or whether translocation occurs by direct transfer through the lipid bilayer (von Heijne and Blomberg, 1979; Blobel, 1980; Engelman and Steitz, 1981). Although this central debate remains unresolved, recent evidence indicates that translocation across the rough endoplasmic reticulum is indeed mediated by resident integral membrane proteins. For example, analyses of the interaction of partially translocated secretory precursors with the RER membrane, based on the capacity for extraction of the nascent chain from the membrane by chaotropic agents, suggest that the polypeptide chain may traverse the bilayer in an aqueous, presumably proteinaceous, environment (Gilmore and Blobel, 1985). More direct biochemical, as well as genetic, evidence for an integral membrane protein requirement for protein translocation has also been provided. Studies with chemical alkylating agents have indicated the existence of protein components, other than the signal recognition particle receptor, whose activity is required for translocation (Hortsch *et al.,* 1986; Nicchitta and Blobel, 1989). The sulfhydryl-directed alkylating agent, *N*-ethylmaleimide, blocks translocation at a point subsequent to insertion of the nascent chain but prior to cleavage, suggesting that nascent chain translocation, *per se,* requires the activity of integral membrane proteins (Nicchitta and Blobel, 1989). In the yeast *Saccharomyces cerevisiae*, three genes, *SEC61*, *SEC62*, and *SEC63*, have been identified whose gene products are likely to be required for translocation (Deshaies and Schekman, 1989; Rothblatt *et al.,* 1989; Sadler *et al.,* 1989). These genes were identified by selection of mutant cells that were unable to translocate a signal sequence containing fusion construct. All the gene products, Sec61p, Sec62p, and Sec63p, contain potential transmembrane domains and may function as components of the translocation apparatus. To date, specific biochemical activities have not been ascribed to the Sec 61, 62, and 63 gene products and it is equally uncertain how chemical alkylation of RER membranes disrupts protein translocation.

A biochemical analysis of the mechanism of translocation, which implies knowledge of the identity and interactions of the components that mediate this process, requires experimental systems in which the phospholipid and protein components of the RER can be reversibly dissociated and reconstituted to yield functional membranes. In addition to providing insight into the mechanism of translocation, such studies may also allow an analysis of the structure/function relationships of the components that comprise and define the RER.

In this report we describe procedures for the reconstitution of secretory protein translocation from detergent-solubilized canine rough microsomes and present results of a series of studies on the biochemical activities of the reconstituted membranes.

II.   Preparation of Membranes

It is often observed, and perhaps not surprisingly, that the reconstitution of various biological transport functions from detergent-solubilized membranes yields vesicle preparations that function at activity levels significantly below that of native membranes. There are many possible explanations for such phenomena, including degradation/denaturation of the transport activity and inappropriate topological assembly during reconstitution. Given that a decrease in activity is likely to occur upon the reconstitution of any given membrane transport function, it is worthwhile to use native membrane preparations of the highest activity possible as a starting point for such studies. Following complete solubilization, the membrane components will, in all likelihood, comprise a common population of mixed micelles. Subsequent reconstitution should then lead to recovery of these components in ratios stoichiometrically similar to those present in the detergent-soluble fraction. On the basis of these considerations, it would appear that sufficient attention should be given to the purity of the starting membrane fraction. If, for example, the activity in question is dependent upon the presence of more than one species of protein in a given vesicle, the probability of recovering all necessary proteins in a single vesicle will vary as a function of the purity of the starting membrane fraction.

For our current studies on the reconstitution of secretory protein trans-location, we use a rough microsome fraction prepared from canine pan-creas according to the procedure of Walter and Blobel (1983a). To limit proteolysis during preparation of the membranes, all buffers are sup-plemented with 1 mM EGTA, 1 mM EDTA, 10 U/ml aprotinin, and 0.5 mM phenylmethylsulfonyl flouride (PMSF). PMSF is added to buffers immediately prior to use from a 0.2 M stock in ethanol. All procedures, including processing of the freshly excised pancreas, resuspension, and washing of the membranes, etc. are performed at 4°C. The rough micro-some preparation, at a concentration of 50 A$_{280}$ U/ml (determined in 1% SDS), is first washed in a low-salt buffer consisting of 50 mM triethanola-mine (TEA), pH 7.5, 1.5 mM Mg(OAc)$_2$, 10 U/ml aprotinin, and 0.5 mM PMSF. To remove bound ribosomes, the low-salt washed mem-branes are then subsequently washed by 1:1 dilution of the membrane suspension in a buffer consisting of 1 M KOAc, 50 mM TEA, pH 7.5, 10 mM EDTA, 10 U/ml aprotinin, and 0.5 mM PMSF. Membranes, di-luted in the high-salt/EDTA-supplemented buffer, are maintained on ice for 20 minutes and collected by centrifugation through a cushion of 0.5 M sucrose, 0.5 M KOAc, 50 mM TEA, and 10 U/ml aprotinin, for 1.5 hours at 140,000 g_{av} in the Beckman Ti50.2 rotor (4°C). In general, membranes that yield a firm pellet after the high-salt/EDTA wash appear to yield vesicles of somewhat higher translocation activity in both the native and reconstituted state. The high-salt/EDTA washed microsomal vesicles are resuspended to the original starting volume in 0.25 M sucrose, 50 mM TEA, and 1 mM DTT and are stored at -80°C in 1.0-ml aliquots. No decrease in activity is observed upon storage at -80°C for up to 1 year.

The morphology of a characteristic preparation of rough and high-salt/EDTA washed microsomes is shown in Fig. 1, panels A and B, respectively. Both preparations consist of closed vesicles of 0.15–0.35 μm in diameter, with the only apparent morphological difference being the presence or absence of ribosomes. By morphological criteria, the mem-brane fraction consists almost entirely of rough microsomes and no further purification is necessary to yield a membrane fraction suitable for recon-stitution studies.

Prior to further study, the membrane preparation is assayed for trans-location and glycosylation activity in *in vitro* cotranslational translocation assays using either a wheat germ translation system (Erickson and Blobel, 1983) or reticulocyte lysate system (Jackson and Hunt, 1983). Either translation system can be fairly easily prepared by the published protocols, which also include extensive documentation of the procedures for *in vitro* translation. Alternatively, both the wheat germ and reticulocyte lysate translation systems can be purchased from commercial sources (e.g.,

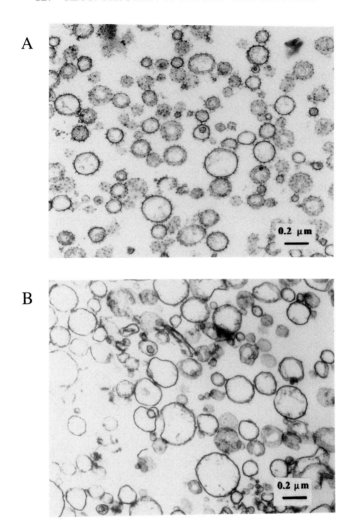

FIG. 1. Electron micrographs of canine pancreas rough microsomes and high-salt/EDTA washed rough microsomes. 50-µl aliquots of rough microsomes (panel A) or high-salt/EDTA washed rough microsomes (panel B) were fixed in 0.1 M cacodylate and 2.0% glutaraldehyde for 15 minutes at 4°C. The fixed membrane suspensions were collected by centrifugation for 15 minutes at 30 psi in the Beckman A-100/30 rotor and the pellets were gently washed. Samples were osmicated, block stained with uranyl acetate, and embedded in Epon/Araldite for subsequent processing.

Promega or Novagen, Madison, WI). For the majority of our studies on the reconstitution of translocation, we analyze the translocation activity of the reconstituted vesicles in a reticulocyte lysate translation system. Cell-free translations are performed in a final volume of 20 μl containing 8 μl of nuclease-treated reticulocyte lysate, 40 μCi of [^{35}S]methionine, 2 U of RNasin, and 100–300 ng of mRNA coding for bovine preprolactin. The ionic conditions of the translation/translocation assay are adjusted to 120 mM KOAc, 2.0 mM Mg(OAc)$_2$, and 1 mM DTT, and reactions are performed for 45 minutes at 25°C. Reconstituted vesicles are normally present at a concentration of 6–9 nmol of lipid phosphorous/20 μl reaction volume. The translation products are collected by fractionation of the reaction with ammonium sulfate, at 66% saturation, followed by analysis of the pellet fraction on 12.5% SDS–PAGE gels. Treatment of the reticulocyte lysate translation reactions with ammonium sulfate, as described above, separates the translation products from the endogenous globin and eases preparation of the samples for SDS–PAGE.

A typical high-salt/EDTA washed microsome preparation (RMek) will exhibit close to 90% processing of preprolactin and/or glycosylation of the yeast secretory precursor prepro-α-factor at a concentration of 1 μl of the membrane stock/20 μl reaction volume.

III. Reconstitution of Translocation-Competent Vesicles

A. Solubilization

RMek preparations, as described above, are rapidly thawed and treated with staphylococcal nuclease to digest bound RNA. For nuclease digestion, the membrane suspension is supplemented with CaCl$_2$ to a concentration of 1.0 mM. Staphylococcal nuclease is then added to a final concentration of 20 U/ml from a stock (5–10,000 U/ml) prepared in water. The nuclease digestion is performed for 10 minutes at room temperature and the reaction is quenched by addition of EGTA to a final concentration of 2.5 mM. The membrane suspension is then chilled on ice and collected by centrifugation for 20 minutes at 120,000 g_{av} in the Beckman TLA 100.2 rotor at 4°C. The pelleted membranes are then resuspended, on ice, in a buffer consisting of 0.4 M sucrose, 0.45 M KCl, 10 mM Tris-Cl, pH 8.0, at 4°C, and 0.5 mM MgCl$_2$. To ensure uniform resuspension, the vesicle preparation is gently homogenized, by hand, in a 1.0 ml Teflon glass homogenizer with 5–10 slow passes of the pestle. Solubilization is performed by addition of sodium cholate, on ice, to a final concentration of 0.8%. The sodium cholate is prepared as a 10% (w/v) stock in water. It is important that the sodium cholate be of the highest purity possible. Commercial preparations of cholic acid can be purified by repeated recrystallization from 70% ethanol and neutralization with NaOH to pH 7.4. Alternatively, a purified grade of sodium cholate can be obtained from commercial sources such as Calbiochem (La Jolla, CA). The membrane suspension becomes transparent immediately upon addition of the detergent. The membrane/detergent mixture is mixed by repeated inversion and is maintained on ice for a period of 15–30 minutes.

B. Centrifugation

Centrifugation of the detergent-treated membranes is performed to separate material sedimenting at > 50 S from the soluble fraction. Typically, 100-μl aliquots of the cholate-treated membrane fraction are centrifuged for 20–30 minutes at 30 psi (4°C) in the A-100/30 rotor of the Beckman Airfuge (~185,000 g). These centrifugation conditions are somewhat arbitrary; under centrifugation conditions sufficient to yield sedimentation of particles of ~20 S, there was no significant decrease in the capacity of the detergent extract to yield translocation-competent membranes upon reconstitution. The supernatant fraction from the centrifugation step is referred to as the cholate extract (CE). Electron micrographs of the supernatant fraction obtained from centrifugation under the described conditions reveal amorphous regions of electron-dense material devoid of any lamellar structure (Nicchitta and Blobel, 1990). In addition, membrane protein components present in the CE, such as the SRP receptor and the signal peptidase complex, migrate in sucrose gradients with S values similar to those determined for the purified components, indicating that solubilization of the membranes has indeed occurred (Nicchitta and Blobel, 1990). By the criteria of sedimentation and morphological analysis, the supernatant fraction obtained from sodium cholate-solubilized RMek appears to represent a mixed micelle fraction, the necessary starting point for reconstitution.

C. Reconstitution of Solubilized Extracts

The reconstitution of translocation-competent vesicles from the CE can be achieved either by dialysis or by treatment of the CE with a hydrophobic resin, such as SM-2 BioBeads (BioRad, Richmond, CA). Dialysis is performed in 8000 molecular-weight cut-off dialysis tubing (Spectra-Por 1, 1-cm flat width, 0.32 ml/cm, Spectrum Medical Industries, Los Angeles, CA) that is pretreated by boiling for 2–5 minutes in 0.2 M sodium bicarbonate and 5 mM EDTA. In a typical reconstitution experiment, a 100-μl aliquot of the CE is placed in a 5-cm length of treated dialysis tubing and the tubing is sealed, with the minimal possible enclosed air volume, with

Spectra-Por closures (Spectrum Medical Industries, Los Angeles, CA). Dialysis is performed against 800–1000 volumes of 0.25 M sucrose, 0.4 M KCl, 10 mM PIPES, pH 7.25, 0.5 mM MgCl$_2$, and 1.5 mM DTT (dialysis buffer), and is typically done at room temperature (20–23°C) for a period of 8–12 hours. Reconstituted vesicles are recovered by centrifugation in the Beckman A-100/30 rotor for 10 minutes at 25 psi (140,000 g_{av}). The membrane pellet is resuspended with a micropipet in 50 μl of 0.25 M sucrose and 50 mM TEA, pH 7.5, transferred to a 0.5-ml sample tube, and gently homogenized with a Teflon pestle (Kontes, Vineland, NJ). Aliquots of the reconstituted membranes (rRM) are removed for assays of protein and phospholipid recovery, and the remaining vesicles are frozen in liquid nitrogen.

The reassembly of translocation-competent vesicles is relatively rapid, as depicted in Fig. 2, panels A and B. Panel A shows the translocation activity of vesicles recovered from dialysis for a period of 4 hours (lanes a–d), 6 hours (lanes e–h), 8 hours (lanes i–l), 10 hours (lanes m–p), or 23 hours (lanes q–t). Quantitation of the preprolactin processing activity, vesicle yield (micromoles of phospholipid/milliliter), and detergent concentration (cpm of [^{3}H]cholate) are depicted in panel B. Translocation is assayed as the membrane-dependent appearance of the processed, mature prolactin as well as protection of the mature prolactin from digestion by exogenous protease. Translocation-competent vesicles were recovered after as little as 4 hours of dialysis (Fig. 2, lanes a–d). The membrane fraction obtained after 4 hours of dialysis processed preprolactin at approximately 80% efficiency (lane b), and 25% of the mature prolactin was protected from digestion by exogenous protease in the absence (lane c), but not the presence (lane d), of detergent. The recovery of preprolactin processing occurs coincident with the decrease in cholate concentration of the CE. There is a slight, but insignificant, increase in processing activity in vesicles collected after dialysis for 10 hours (panel B). After extended dialysis, there was a decrease in the translocation activity of the rRM (lanes q–t) and in the recovery of phospholipid (panel B). Vesicles recovered after 23 hours of dialysis displayed approximately 50% of the processing activity of vesicles reconstituted for periods of 10 hours or less (compare lanes q–t with lanes a–p). Translocation into the lumen of the reconstituted vesicles was assessed by protease treatment of the individual reactions. Under all conditions, approximately 30% of the processed precursor was protected from protease protection in the absence (lanes c,g,k,o, and s) but not the presence (lanes d,h,l,p, and t) of added detergent. On the basis of these data, we commonly dialyze samples for 8 hours, a time point that provides for efficient recovery of translocation activity, as well as thorough removal of the detergent.

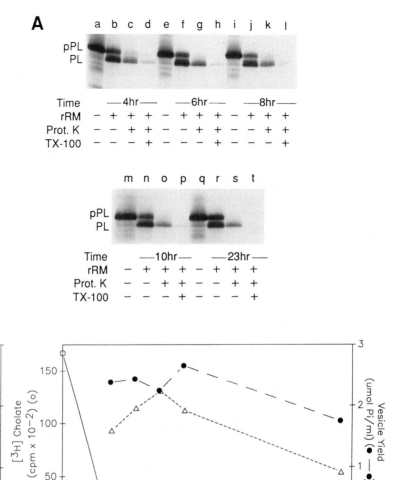

FIG. 2. Time course of recovery of translocation activity during detergent dialysis. The 0.8% cholate extract of RMek was prepared as described in the text, supplemented with [³H]cholate (300,000 cpm) and separated into five 100-μl aliquots for subsequent reconstitution by dialysis. At the indicated time points, aliquots were removed from dialysis, collected by centrifugation, resuspended in 0.25 M sucrose and 50 mM TEA, and frozen for subsequent analysis. Translation/translocation assays (panel A) were performed as described in the legend to Fig. 3. [³H]Cholate was determined by liquid scintillation analysis of aliquots of the vesicle suspension (panel B). Quantitation of the dried gels was performed by direct radiometric analysis of the dried gel with an AMBIS Radioanalytic Imaging System (Automated Microbiology Systems, Inc., San Diego, CA).

Translocation-competent vesicles can also be recovered from the CE by treatment of the CE with SM-2 BioBeads. For reconstitution by this procedure, SM-2 BioBeads are sequentially washed in methanol, water, and dialysis buffer. Approximately 400 µl of buffer-equilibrated BioBeads is packed into a 0.8 × 4-cm polypropylene column (BioRad PolyPrep) and briefly centrifuged to remove excess buffer. Cholate extract (200 µl) is then added to the packed resin, the column is capped to prevent elution of the CE, and the closed column is incubated at 4°C for 2 hours, to allow binding of the detergent to the hydrophobic resin. Reconstituted vesicles are then recovered by centrifugation of the column for 2 minutes at 2000 g. The turbid eluate is diluted with 1.5 volumes of cold water and the reconstituted vesicles are collected by centrifugation in a refrigerated microfuge for 1 hour at 18,500 g (4°C).

The composition of the solubilization buffer is somewhat arbitrary and we have observed that significant variations in the composition of the solubilization buffer can be made without deleterious effects on the recovery of translocation-competent vesicles. For instance, polyols, such as sucrose or glycerol, are commonly added to protein isolation buffers to stabilize protein activity and are thought to act by reducing the activity of water, particularly relevant in the case of hydrophobic membrane proteins (Ambudkar and Maloney, 1986). Although we have not attempted to reconstitute translocation-competent vesicles from detergent extracts prepared in buffers lacking polyols, we have observed that there is considerable latitude in the type, and concentration, of added polyols that can be used during reconstitution. As shown in Fig. 3, vesicles reconstituted from detergent extracts prepared in the presence of either 0.4 M sucrose (lanes 1–4), 1.15 M (8% w/v) glycerol (lanes 5–8), or 0.4 M myo-inositol (lanes 9–12) display relatively efficient processing of the precursor, preprolactin (compare lanes 2, 6, and 10). The mature prolactin is also partially protected from digestion by exogenous protease in the absence (lanes 3, 7, and 11), but not the presence of detergent (lanes 4, 8, and 12), indicating translocation of the cleaved precursor into the vesicle lumen.

There are a variety of experimental variables that affect the reconstitution process. We have previously reported that membranes reconstituted from detergent extracts prepared at cholate concentrations in excess of 0.8% exhibit very poor protection of mature prolactin from digestion by exogenous protease (Nicchitta and Blobel, 1990). When solubilization is performed at detergent concentrations below ~0.6%, the efficiency of solubilization drops and the reconstituted vesicles, although of high translocation activity, are recovered in relatively low yield (Nicchitta and Blobel, 1990). We have also observed that the recovery of translocation-competent vesicles by the dialysis procedure is sensitive to the salt concentration of the dialysis buffer. At KCl concentrations below 0.3 M or above 0.8 M, the translocation activity of the rRM is significantly lower than that

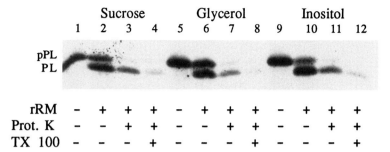

<table>
<tr><td></td><td colspan="3">Sucrose</td><td></td><td colspan="3">Glycerol</td><td></td><td colspan="3">Inositol</td></tr>
<tr><td></td><td>1</td><td>2</td><td>3</td><td>4</td><td>5</td><td>6</td><td>7</td><td>8</td><td>9</td><td>10</td><td>11</td><td>12</td></tr>
</table>

	1	2	3	4	5	6	7	8	9	10	11	12
rRM	−	+	+	+	−	+	+	+	−	+	+	+
Prot. K	−	−	+	+	−	−	+	+	−	−	+	+
TX 100	−	−	−	+	−	−	−	+	−	−	−	+

FIG. 3. Effects of variations in the polyol composition during solubilization on the activity of reconstituted vesicles. High-salt/EDTA washed rough microsomes (RMek) were solubilized in buffers containing 0.45 M KCl, 10 mM Tris-Cl, pH 8.0, 0.5 mM MgCl$_2$, and either 0.4 M sucrose, 8% glycerol, or 0.4 M inositol by addition of sodium cholate to a final concentration of 0.8%. A high-speed supernatant was prepared from each solubilization reaction and dialyzed against 1000 volumes of 0.25 M sucrose, 0.4 M KCl, 20 mM PIPES, pH 7.2, 0.5 mM MgCl$_2$, and 1.5 mM DTT for 10 hours at room temperature. Vesicles were collected by centrifugation and assayed for translocation activity in a [^{35}S]methionine-supplemented reticulocyte lysate translation system. Where indicated, vesicle were present at a concentration of 5–8 nmol lipid phosphorous/20 μl of reaction volume. Protease protection was performed for 30 minutes at 4°C in the presence of 50 μg/ml proteinase K. Reactions were quenched by addition of PMSF to 4 mM and the samples were fractionated by addition of ammonium sulfate to 66% saturation. The pellet fraction from this step was analyzed by SDS–PAGE on a 12.5% gel and by subsequent authoradiography of the dried gel.

observed in rRM reconstituted at 0.4 M. In addition, the yield of vesicles decreases as a function of increasing salt concentration (Nicchitta and Blobel, 1990). With these considerations in mind, the optimum conditions for reconstitution by the dialysis procedure represent a compromise between the recovery of translocation activity and the yield of vesicles. In general, the most reasonable compromise between these two variables is obtained by dialysis of a CE prepared at 0.8% sodium cholate for 8 hours in a dialysis buffer containing 0.4 M KCl.

IV. Structural Characterization of Reconstituted Membranes

A. Morphology

An electron micrograph of a representative preparation of reconstituted membranes is shown in Fig. 4. On average, the rRM have a broader size distribution and are usually recovered in association with amorphous, electron-dense regions of unknown composition.

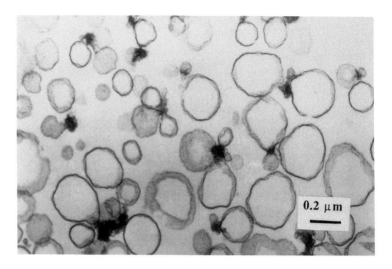

Fɪɢ. 4.   Electron micrograph of vesicles reconstituted by detergent dialysis. A 50μl aliquot of vesicles reconstituted by detergent dialysis was processed for electron microscopy as detailed in the legend to Fig. 1.

B.   Protein/Lipid Composition

Vesicles reconstituted from cholate extracts of RMek contain a complement of integral membrane proteins of approximately equal relative enrichment to native RMek. An immunoblot analysis of three integral membrane proteins, the α-subunit of the SRP receptor, the α-subunit of the signal sequence receptor, and the 22/23-kDA glycoprotein component of signal peptidase in RMek, the cholate extract, and reconstituted RM (rRM) is shown in Fig. 5. The recovery of these components is variable. When expressed with respect to the lipid mass, SRα and the signal peptidase complex are recovered at slightly higher enrichment than in the CE, whereas SSR is recovered at levels somewhat lower than that observed in the CE (data not shown). We have also assayed for the recovery of the β-subunit of the SRP receptor, the β-subunit of SSR and the 18-, 21-, and 25-kDa subunits of the signal peptidase complex during reconstitution. These proteins are all recovered at levels similar to their respective α-subunits (data not shown).

During reconstitution, lumenal proteins of the RER, specifically BiP and abundant glycoproteins of 94, 66, and 49 kDa are not recovered (Nicchitta and Blobel, 1990). As the reconstituted vesicles are active in *in vitro* assays

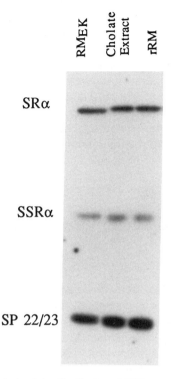

Fig. 5. Immunoblot analysis of the distribution of SRP receptor, signal sequence receptor, and the 22/23-kDa component of the signal peptidase complex in RMek, a cholate
extract, and reconsituted vesicles. Aliquots of high-salt/EDTA washed membranes (RMek)
an 0.8% cholate extract and reconstituted membranes (rRM), equivalent to 20 nmol of lipid
phosphorous, were fractionated by SDS–PAGE on a 12.5% gel and transferred to nitrocellulose for subsequent immunoblot analysis. Strips of the nitrocellulose membrane were incubated with polyclonal antisera directed against either the α-subunit of the SRP receptor
(SRα), the α-subunit of the signal sequence receptors (SSRα), or the 22/23-kDa glycoprotein
component of signal peptidase complex (SP 22/23). Detection of bound antibodies was
accomplished with ^{125}I-labeled protein A and autoradiography of the dried blot.

of translocation, it appears that these lumenal proteins are not absolutely
necessary for translocation of secretory precursors. Similar conclusions
have been presented in recent studies on the reconstitution of protein
translocation from β-octyl glucoside-treated rough microsomes (Yu et al.,
1989, Zimmerman and Walter, 1990). It must be noted, however, that a
temperature-sensitive mutation in the KAR2 gene in yeast, which encodes
the yeast homologue of BiP, causes a block in translocation (Vogel et al.,
1990). It appears therefore, that BiP may be necessary for translocation
in yeast.

The recovery of total protein (assayed by the BCA procedure; Pierce Chemical Co, IL) and phospholipid [assayed as organic phosphorous by the procedure of Ames and Dubin (1960)] is approximately 40–60% of the input levels. The protein:lipid ratio of the reconstituted membranes is approximately equal to that of native membranes and varies from 1 mg protein:1 μmol phospholipid to 1 mg protein:1.3 μmol phospholipid. The phospholipid composition of the reconstituted vesicles, 75% phosphatidylcholine, 15% phosphatidylethanolamine, <10% phosphatidylinositol, and <5% phosphatidylserine, is similar to that of RMek (Nicchitta and Blobel, 1990).

V. Analysis of Activity: Reconstituted Membranes

A. Signal Recognition Particle Requirement

In RMek, translocation and cleavage of preprolactin is a strictly cotranslational process that is dependent upon targeting of the precursor to the membrane by the signal recognition particle (Walter and Blobel, 1981). Cotranslational, SRP-independent cleavage of preprolactin to prolactin can also occur if the microsomes are first solubilized to permit access of the lumenally disposed signal peptidase to the nascent chain (Gilmore *et al.*, 1982). An SRP requirement for the processing event can thus provide strong evidence that the reconstitution procedures allow for the appropriate topological orientation of signal peptidase and that the precursor has been translocated. The SRP dependence for translocation of preprolactin into RMek and rRM reconstituted by either the dialysis or the detergent extraction (SM-2 BioBeads) procedure is shown in Fig. 6. A paired comparison of native membranes (RMek) (lanes 9–12) and vesicles reconstituted by detergent dialysis (lanes 13–16) is shown in the lower panel. In the absence of SRP (lanes 9 and 13), less than 10% of the precursor is processed. The processing observed in the absence of SRP is also observed in the absence of membranes, suggesting that it may represent a signal peptide cleaving activity present in the wheat germ extract, as previously observed by Prehn *et al.* (1987). Addition of SRP to the translation reaction markedly stimulates the degree of processing in both RMek (lanes 10) and reconstituted membranes (lane 14). The mature prolactin is protected from digestion by exogenous protease in the absence (lane 11 and 15), but not the presence (lanes 12 and 16), of detergent. As is evident from comparison of lanes 11 and 15, reconstituted membranes do not protect the translocated substrate at the efficiency oberved in RMek. In the experiment depicted in Fig. 6, 100% of the mature prolactin was protected from protease digestion in RMek, whereas 30% was protected in the rRM prepared by dialysis. At present it is unclear whether such differences reflect a defect in membrane structure or perhaps the presence of cleaved but incompletely translocated chains. A similar paired experiment comparing the activity of native membranes (RMek) (lanes 1–4) and

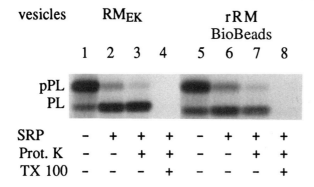

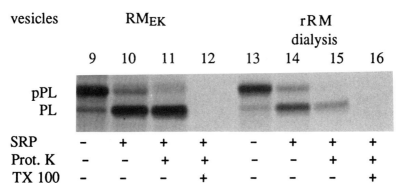

FIG. 6. Signal recognition particle (SRP) requirement for translocation into high-salt/EDTA washed microsomes and reconstituted vesicles. Vesicles reconstituted by either dialysis (lanes 13–16) or detergent extraction (lanes 5–8) were assayed for translocation activity in the presence or absence of 20 nM SRP (prepared as described by Walter and Blobel, 1983b) in a wheat germ translation system supplemented with [^{35}S]methionine. Following incubation for 1 hour at 25°C, reactions were quenched by addition of trichloroacetic acid to a final concentration of 10%, and samples were analyzed by SDS–PAGE on 12.5% gels followed by autoradiography of the dried gel. For comparison, the activity of native membranes is indicated in lanes 1–4 and 9–12. Protease digestion conditions are as indicated in the legend to Fig. 3.

vesicles reconstituted by the SM-2 BioBeads procedure (lanes 5–8) is shown in the upper panel of Fig. 6. As observed with membranes prepared by dialysis, membranes prepared by detergent extraction display cotranslational, SRP-dependent processing and translocation of preprolactin. In this experiment, approximately 50% of the mature prolactin is protected from digestion by exogenous protease. Such a high degree of protection in membranes reconstituted by this procedure is not, however, commonly observed. Protease protection levels vary on average from 20 to 40% with 25–30% protection representing the mean level of protection.

B.   Sedimentation

Sedimentation of the processed precursor with the vesicle fraction, following extraction of nonspecifically bound nascent chain/polysome complexes with EDTA, serves as an additional criterion to assay the translocation activity of reconstituted membranes. Although the assay does not provide insight into the whether the nascent chain has been completely translocated (cleaved partially translocated precursors being indistinguishable from cleaved, fully translocated precursors), it does provide indication of a productive association of the nascent chain with the translocation apparatus of the vesicles. In the experiment depicted in Fig. 7,

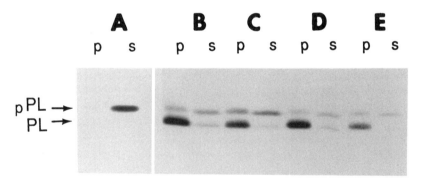

FIG. 7.   Sedimentation analysis of preprolactin translation products in the presence or absence of control or reconstituted membranes. Preprolactin was translated in a reticulocyte translation system, as described in the legend to Fig. 3, in the absence of membranes, or in the presence of RMek or reconstituted membranes (5–8 nmol of lipid phosphorous). After translation, reactions were chilled on ice and supplemented with either EDTA to 20 mM, to disrupt the ribosome/membrane junction and remove nonspecifically bound translation products, or an equivalent volume of water. After a 10-minute incubation at 4°C, reactions were overlayed onto 0.5 M sucrose cushions of equivalent ionic composition and centrifuged in the Beckman Airfuge to separate the membrane and supernatant fractions. The pellet and supernatant fractions were analyzed on 12.5% SDS–PAGE gels followed by autoradiography of the dried gel. A, minus membrane control; B, plus RMek; C, plus rRM; D, RMek plus 20 mM EDTA; E, rRM plus 20 mM EDTA.

preprolactin was translated in the presence or absence of RMek and rRM. Following synthesis, the translation reactions were adjusted to 20 mM EDTA or an equivalent volume of water was added. Under these conditions,the ribosomes disassemble and the ribosome/membrane junction is disrupted (Sabatini *et al.,* 1966; Connolly and Gilmore, 1986). The translation reactions were then overlayed onto a 50-μl cushion of 0.5 M sucrose and 50 mM TEA, pH 7.5, with or without 20 mM EDTA, and were centrifuged for 8 minutes at 20 psi in the Beckman Airfuge A-100/30 rotor (approximately 120,00 g_{av}). The supernatant and the cushion were removed and the translation products were collected from this fraction by precipitation with ammonium sulfate at 66% saturation. The pellet arising from the ammonium sulfate fractionation is washed with 10% trichloroacetic acid prior to analysis. The membrane pellet is solubilized by addition of 15 μl of 0.5 M Tris, 5% SDS to the Airfuge tube, heating at 50°C for 10 minutes, and removal of the solubilized membrane fraction with a micropipet. When the translation is performed in the absence of membranes (Fig. 7A), the precursor remains in the supernatant (S) fraction. In the presence of RMek (Fig. 7B), processing of preprolactin is observed and the mature prolactin is found predominately in the pellet (P) fraction. In the experiment shown in Fig. 7, rRM (C) were indistinguishable from RMek, with the majority of the mature prolactin sedimenting with the membrane fraction. The association of either processed or unprocessed preprolactin with RMek (D) and rRM (E) is not reversed in the presence of EDTA, indicating that the binding is likely to represent a productive interaction with the translocation apparatus of the vesicles.

C. Glycosylation

Glycosylation represents an additional, lumenally oriented, enzymatic reaction that occurs upon translocation of suitable precursors and can be considered as further evidence for translocation of the nascent chain. In contrast to native membranes, which are commonly capable of nearly quantitative glycosylation of prepro-α-factor, vesicles reconstituted by the cholate dialysis procedure described above are deficient in this reaction sequence (data not shown). Preincubation of the membrane fraction with the nucleotide sugar precursors of the high-mannose oligosaccharide core did not correct this defect (G. Migliaccio and G. Blobel, unpublished observations). At present, it is unclear why the glycosylation reaction is not recovered during reconstitution. Possible explanations include denaturation of the oligosaccharyl transferase, the loss of a lumenal protein that may be essential, such as the recently described glycosylation site binding protein (Geetha-Habib *et al.,* 1988), poor recovery of the lipid precursors of the glycosylation pathway, and/or improper topological assembly of the proteins necessary for this process. It should be noted that functional microsomal vesicles recovered by dilution of partially solubilized membranes were also found to be deficient in cotranslational glycosylation (Zimmerman and Walter, 1990).

VI. Detergent Requirements for Reconstitution

The reconstitution studies described above were done entirely with a bile salt, sodium cholate, a detergent that has proved useful in reconstitution studies. Sodium cholate, because of its charged nature, is, however, somewhat limited in its suitability for various chromatographic procedures. Furthermore, the vesicles reconstituted from sodium cholate extracts of RMek have certain functional shortcomings, including reduced protease protection of the cleaved precursor and loss of glycosylation activity. For these reasons we have investigated the detergent specificity of the reconstitution process using the zwitterionic cholate analog CHAPS and the nonionic detergents Triton X-100 and Nikkol.

Detergent extracts of RMek were prepared, as described in Section III,A, at CHAPS concentrations ranging from 0.4 to 1.2% (w/v) and the detergent extracts were reconstituted by the dialysis procedure described in Section III,C. Unilammelar vesicles, morphologically similar to those obtained from sodium cholate extracts of RMek, were obtained under all conditions (data not shown). The translocation activity of the various vesicle fractions was determined in an SRP-supplemented wheat germ translation system. As shown in Fig. 8, all vesicle preparations displayed SRP-dependent processing of preprolactin (Fig. 8, lanes 2, 6, 10, 14, 18, and 22). Protease protection of the mature prolactin was only observed, however, in the vesicle fractions obtained from reconstitution of the 0.6 and 0.4% CHAPS extracts (lanes 15 and 19). At present, it is unclear why vesicles reconstituted from detergent extracts prepared at CHAPS concentrations of 0.8–1.2% do not protect mature prolactin from protease digestion.

Reconstitution of the detergent extracts prepared from Triton X-100 and Nikkol-solubilized RMek was performed by the detergent extraction procedure described in Section III,C. Although vesicles were obtained from both the Triton X-100 and Nikkol extracts, these vesicles were inactive in translocation assays performed in an SRP-supplemented wheat germ system (data not shown). Perhaps more significantly, the sum of preprolactin and prolactin synthesized in the presence of SRP and the Triton X-100 or Nikkol-derived vesicles was markedly lower than that observed in either

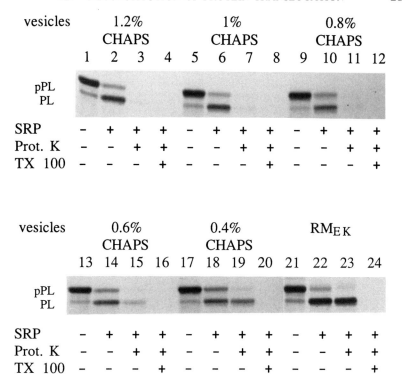

FIG. 8. Effects of CHAPS concentration during solubilization on the translocation activity of reconstituted membranes. Aliquots of RMek were solubilized as described in the legend to Fig. 3 at the indicated CHAPS concentration. After high-speed centrifugation, vesicles were reconstituted by dialysis and assayed for translocation in an SRP-supplemented wheat germ translation system.

RMek- or cholate-derived membranes. These data suggest that the interaction of the SRP/ribosome/nascent chain complex with the SRP receptor and associated components of the Triton or Nikkol membranes is compromised and that release of elongation arrest does not occur. There are at least two possible interpretations of these results: the proteins associated with the release of elongation arrest have lost activity or have been reconstituted in a topologically inappropriate orientation. Protocols for the purification of the SRP receptor in an active conformation, i.e., capable of elongation arrest release, require solubilization of rough microsomes with 1% Nikkol and 0.25 M KOAc. It is therefore unlikely that this protein has become denatured during the relatively brief reconstitution process. At this juncture, it appears more probable that the nonionic detergents Triton X-100 and Nikkol do not provide for the topologically appropriate assembly of the integral membrane proteins relevant to translocation.

VII. Discussion/Conclusions

By the procedures described above, the process of protein translocation can be reconstituted from detergent-solubilized pancreatic rough microsomes. The reconstituted vesicles reproduce many, but not all, of the biochemical activities associated with rough microsomes. Most notably lacking in the reconstituted membranes is the capacity for glycosylation. The reconstituted membranes are also less efficient than native membranes in affording protection from exogenous proteases to the translocated precursor. The biochemical basis for these defects is not understood. Such knowledge is likely, however, to provide insight into the structural and biochemical aspects of RER function and is certainly worthy of further investigation.

A noteworthy aspect of the reconstitution procedures described above is the rather strict detergent requirement. To date we have had success only with the bile salt (sodium cholate) and its zwitterionic relative (CHAPS). Reconstitution of membrane proteins solubilized with Triton X-100, Nikkol, or β-octyl glucoside (data not shown) have yet to yield translocation-competent membranes, although efficient recovery of the integral membrane protein components has been observed. Negative data are rarely compelling, however, and the lack of recovery of translocation-competent vesicles with these detergents may simply be due to an inappropriate choice of conditions. Alternatively, the success obtained with cholate may reflect its unique structural characteristics. Sodium cholate is a rigid, steroidlike molecule containing two distinct structural asymmetries. The hydroxyl and methyl groups present in the molecule are distributed on opposite, planar faces, and on one end of the molecule is a short, aliphatic chain terminating in a carboxyl group. Perhaps it is this rigid, planar structure that allows solubilization of the protein components of the RER in strict association with an annulus of asymmetrically oriented, bound lipid, and it is this phenomenon that allows the proper topological reassembly of the solubilized components into translocation-competent membranes. The success in reconstituting translocation-competent membranes from sodium cholate-solubilized membranes may also arise, at least in part, through the relatively moderate critical micelle concentration (CMC) of sodium cholate. It is well accepted that reconstitution of membrane proteins from detergent extracts occurs upon removal of detergent and the subsequent aggregation and assembly of the detergent-depleted micelles. It would be expected, therefore, that the recovery of a given transport function during reconstitution would be sensitive to the local changes in the detergent, lipid, and protein concentrations that accompany the gradual removal of the detergent. As the rate of detergent removal is in part a function of the CMC, it is reasonable to consider this parameter in future studies of the reconstitution of the translocation process.

An alternative procedure for the reconstitution of protein translocation from rough microsomes solubilized with β-octylglucoside has recently been reported (Yu *et al.*, 1989). By this method, membranes are recovered that are active in signal peptide processing and glycosylation of nascent precursors. Centrifugation of the detergent-treated membranes prior to reconstitution, however, leads to a loss of translocation activity in the subsequently reconstituted supernatant fraction. Yu *et al.* (1989) suggest that this loss of activity may represent the presence of large (>50 S) macromolecular complexes whose structure is not disrupted by the detergent treatment and whose intact structure is required for the reconstitution of translocation activity. It does not appear, however, that such a macromolecular complex need necessarily exist for reconstitution to occur. In the cholate extract of RMek, the protein components migrate with S values similar to the purified species, indicating that the components of the translocation apparatus can be reversibly disassembled (Nicchitta and Blobel, 1990).

The results described herein on the reconstitution of translocation represent a fairly crude state of knowledge of the parameters that determine the success of such experiments. At the present time we lack clear insight into both the mechanism of translocation and the identity of the proteins that mediate this process. It is also uncertain what role or function the lipid components of the vesicle may serve in this process. Although these questions no doubt provide significant experimental challenges, it is our expectation that these questions will be amenable to further study. Recent studies on the fractionation of the phospholipid and protein components of a detergent extract of rough microsomes as well as the reconstitution of protein subfractions provide an encouraging beginning (Nicchitta *et al.*, 1991).

REFERENCES

Ambudkar, S. V., and Maloney, P. C. (1986). Bacterial anion exchange: Use of osmolytes during solubilization and reconstitution of phosphate linked antiport from *Streptococcus lactis. J. Biol. Chem.* **261,** 10079–10086.

Ames, B. N., and Dubin, D. T. (1960). The role of polyamines in the neutralization of bacteriophage deoxyribonucleic acid. *J. Biol. Chem.* **235,** 769–775.

Blobel, G. (1980). Intracellular protein topogenesis. *Proc. Natl. Acad. Sci. U.S.A.* **77,** 1496–1500.

Connolly, T., and Gilmore, R. (1986). Formation of a functional ribosome-membrane junction during translocation requires the participation of a GTP binding protein. *J. Cell Biol.* **103,** 2253–2260.

Deshaies, R. J., and Schekman, R. W. (1989). SEC62 encodes a putative membrane protein required for protein translocation into the yeast endoplasmic reticulum. J. Cell Biol. 109, 2653–2664.

Engelman, D. M., and Steitz, T. A. (1981). The spontaneous insertion of proteins into and across membranes: The helical hairpin hypothesis. Cell (Cambridge, Mass.) 23, 411–422.

Erickson, A. H., and Blobel, G. (1983). Cell free translation of messenger RNA in a wheat germ system. In "Methods in Enzymology" (S. Fleischer and B. Fleischer, eds.) Vol. 96, pp. 38–50. Academic Press, New York.

Geetha-Habib, M., Noiva, R., Kaplan, H., and Lennarz, W. J. (1988). Glycosylation site binding protein, a component of oligosaccharyl transferase, is highly similar to three other 57 kD luminal proteins of the ER. Cell (Cambridge, Mass.) 54, 1053–1060.

Gilmore, R., and Blobel, G. (1983). Transient involvement of signal recognition particle and its receptor in the microsomal membrane prior to protein translocation. Cell (Cambridge, Mass.) 35, 677–685.

Gilmore, R., and Blobel, G. (1985). Translocation of secretory proteins across the microsomal membrane occurs through an environment accesible to aqueous perturbants. Cell (Cambridge, Mass.) 42, 497–505.

Gilmore, R., Blobel, G., and Walter, P. (1982). Protein translocation across the endoplasmic reticulum. I. Detection in the microsomal membrane of a receptor for the signal recognition particle. J. Cell. Biol. 95, 463–469.

Hortsch, M., Avossa, D., and Meyer, D. I. (1986). Characterization of secretory protein translocation: Ribosome–membrane interaction in endoplasmic reticulum. J. Cell Biol. 103, 241–253.

Jackson, R. J., and Hunt, T. (1983). Preparation and use of nuclease treated reticulocyte lysates for the translation of eukaryotic messenger RNA. In "Methods in Enzymology" (S. Fleischer and B. Fleischer, eds.), Vol. 96, pp. 50–75. Academic Press, New York.

Meyer, D. I., Krause, E., and Dobberstein, B. (1982). Secretory protein translocation across membranes: the role of the "docking protein". Nature (London) 297, 647–650.

Nicchitta, C. V., and Blobel, G. (1989). Nascent secretory chain binding and translocation are distinct processes: Differentiation by chemical alkylation. J. Cell Biol. 108, 789–795.

Nicchitta, C. V., and Blobel, G. (1990). Assembly of translocation-competent proteoliposomes from detergent-solubilized rough microsomes. Cell (Cambridge, Mass.) 60, 259–269.

Nicchitta, C. V., Migliaccio, G., and Blobel, G. (1991). Biochemical fractionation and assembly of the membrane components which mediate nascent chain targeting and translocation. Cell (Cambridge, Mass.) 65, 587–598.

Prehn, S., Wiedmann, M., Rapoport, T. A., and Zwieb, C. (1987). Protein translocation across wheat germ microsomal membranes requires an SRP like component. EMBO J. 6, 2093–2097.

Rothblatt, J. A., Deshaies, R. J., Sanders, S. L., Daum, G., and Schekman, R. (1989). Multiple genes are required for proper insertion of secretory proteins into the endoplasmic reticulum in yeast. J. Cell Biol. 109, 2641–2652.

Sabatini, D. D., Tashiro, Y., and Palade, G. E. (1966). On the attachment of ribosomes to microsomal membranes. J. Mol. Biol. 19, 503–515.

Sadler, I. A., Chiang, T., Kurihara, J., Rothblatt, J., Way, J., and Silver, P. (1989). A yeast gene essential for protein assembly into the endoplasmic reticulum and the nucleus has homology to DnaJ, and E. coli heat shock protein. J. Cell Biol. 109, 2665–2675.

Vogel, J. P., Misra, L. M., and Rose, M. D. (1990). Loss of BiP/GRP78 function blocks translocation of secretory proteins in yeast. J. Cell Biol. 110, 1885–1895.

von Heijne, G., and Blomberg, C. (1979). Trans-membrane translocation of proteins. (The direct transfer model.) Eur. J. Biochem. 97, 175–181.

Walter, P., and Blobel, G. (1981). Translocation of proteins across the endoplasmic reticulum III. Signal recognition protein (SRP) causes signal sequence dependent and site specific arrest of chain elongation that is released by microsomal membranes. *J. Cell Biol.* **91**, 557–561.

Walter, P., and Blobel, G. (1983a). Preparation of microsomal membranes for cotranslational protein translocation. *In* "Methods in Enzymology" (S. Fleischer and B. Fleischer, eds.), Vol. 96, pp. 84–93. Academic Press, New York.

Walter, P., and Blobel, G. (1983b). Signal recognition particle: A ribonucleoprotein required for cotranslational processing of proteins: Isolation and properties. *In* "Methods in Enzymology" (S. Fleischer and B. Fleischer, eds.), Vol. 96, pp. 682–691.

Yu, Y., Zhang, Y., Sabatini, D., and Kreibich, G. (1989). Reconstitution of translocation-competent membrane vesicles from detergent solubilized dog pancreas rough microsomes. *Proc. Natl. Acad. Sci. U.S.A* **86**, 9931–9935.

Zimmerman, D. L., and Walter, P. (1990). Reconstitution of protein translocation from partially solubilized microsomal vesicles. *J. Biol. Chem.* **265**, 4048–4053.

Chapter 13

Analysis of Protein Topology in the Endoplasmic Reticulum

HANS PETER WESSELS, JAMES P. BELTZER,
AND MARTIN SPIESS

Department of Biochemistry,
Biocenter, University of Basel,
CH-4056 Basel, Switzerland

I. Introduction

The rough endoplasmic reticulum (RER) is one of the very few compartments competent for insertion of proteins (Verner and Schatz, 1988). Polypeptides destined to the ER, the different Golgi compartments, the

METHODS IN CELL BIOLOGY, VOL. 34

plasma membrane, the exterior of the cell, endosomes, and lysosomes are initially inserted into the RER before they are sorted to their final location. Insertion into the ER is initiated by a signal sequence that is either at the amino terminus or at an internal location in the polypeptide. The signal directs the nascent protein–ribosome complex to the RER, mediated by the signal recognition particle (SRP) and the SRP receptor (Walter *et al.,* 1984; Wickner and Lodish, 1985). In the ER membrane it is recognized by the signal sequence receptor (Wiedmann *et al.,* 1987). In the subsequent insertion process, the topology of the protein with respect to the membrane is determined.

In addition to targeting, the signal sequence plays a role in topogenesis by initiating transfer across the membrane of either the amino-terminal or carboxy-terminal portion of the polypeptide. In the latter case, the signal may be cleaved on the lumenal side of the membrane by a specific signal peptidase, thus generating a new amino terminus in the ER lumen. A second topogenic element is the stop-transfer sequence, a hydrophobic segment in the polypeptide chain that blocks its further translocation across the membrane. In the final protein, stop-transfer sequences and uncleaved signals are embedded in the lipid bilayer as transmembrane segments.

The combination of cleaved and uncleaved signals with or without a following stop-transfer sequence can account for secretory proteins, single-spanning membrane proteins with either orientation in the membrane, and proteins that span the membrane twice with both ends in the cytoplasm. More complex, multispanning proteins appear to achieve their topology by a succession of alternating insertion signals and stop-transfer sequences (Wessels and Spiess, 1988; Lipp *et al.,* 1989). The main characteristic of both topogenic elements is a stretch of apolar amino acids, typically 7–15 in cleaved amino-terminal signals and around 20 in uncleaved signals and stop-transfer sequences.

In this article we describe some of the tools that have been used to determine the topology of membrane proteins and to study the properties of the topogenic sequences involved. In recent years a wealth of sequences of membrane proteins has been obtained due to the relative ease of cDNA cloning. To deduce the topology of a protein from its sequence, however, is a difficult task that is often only based on hydropathy plots (Kyte and Doolittle, 1982). The methods described here may help to determine the topology of relatively simple membrane proteins. In addition, they might be used for mutant proteins (fusion or deletion constructs) to answer more specific questions such as whether a certain sequence functions as a signal, whether it is cleaved and where, or whether a particular segment is translocated across the membrane.

II.   Insertion of Wild-Type and Mutant Proteins
into the ER

A.   *In Vitro* Expression and Insertion
into Microsomes

In vitro systems have been used to great advantage both for the characterization of topogenic sequences in secretory and membrane proteins as well as for the identification of components of the insertion machinery. cDNAs cloned into a suitable plasmid vector can be transcribed and translated *in vitro* in only a few hours. If no membranes are added to the translation reaction, the proteins are made without any of the modifications occuring in the ER (e.g., signal cleavage, glycosylation) and can serve as a control. The involvement of the SRP and the SRP receptor in the insertion process of a given protein can be easily tested only in *in vitro* systems. In addition, modified amino acyl-tRNAs can be supplemented to the *in vitro* translation for incorporation into proteins (e.g., to cross-link nascent proteins to components of the translocation machinery) (Wiedmann *et al.,* 1987; Krieg *et al.,* 1989).

The most frequently used translation systems are the rabbit reticulocyte lysate (RL) and the wheat germ (WG) extract (described below). Systems based on other mammalian cells (Garoff *et al.,* 1978) or yeast (Rothblatt and Meyer, 1986) have also been described. Target membranes used with WG and RL are usually microsomes prepared from dog pancreas. The plant and the rabbit translation systems have different characteristics. WG, in contrast to RL, does not contain SRP activity. As a result, WG is needed for experiments to test the involvement of SRP in the insertion process or to assay the interaction of SRP with nascent polypeptides. Binding of added SRP to signal sequence and ribosome was shown to be stronger in WG, leading to "translation arrest" at lower SRP concentrations than in RL (Meyer, 1985; Wolin and Walter, 1989). In addition, the rate of translation differs considerably in the two systems. We found translation to proceed at approximately 30 amino acids per minute in WG (Wessels and Spiess, 1988), which is roughly 10 times slower than in RL and *in vivo*. This low rate of synthesis can be exploited in time course experiments. In addition, recent studies suggest that the stop-transfer activity of certain hydrophobic sequences depends on the translation system used (Spiess *et al.,* 1989; Lopez *et al.,* 1990). The more homologous combination of rabbit RL with canine membranes appears to reflect more closely the *in vivo* situation. Yet, results obtained with any *in vitro* system should be dealt with critically and, if possible, be confirmed *in vivo*.

1. *In Vitro* Transcription Using SP6 RNA Polymerase

Solutions: $5 \times$ transcription buffer (200 mM Tris-HCl, pH 7.5, 30 mM MgCl$_2$, 10 mM spermidine, 100 mM NaCl); rNTPs (2.5 mM each of ATP, CTP, GTP, UTP); DTT (0.5 M); RNasin (RNase inhibitor); cap [5'-(7-methyl)-guanosine-5'-guanosine-triphosphate (7mGpppG, RNA cap analog), 25 A_{260} U/ml]; SP6 RNA polymerase.

The cDNA is subcloned into a plasmid containing a specific promoter for the RNA polymerase of bacteriophage SP6 (plasmids of the pSP and pGEM series; Promega). Plasmid DNA (20 μg, purified on CsCl gradient) is linearized by restriction enzyme digestion at a site downstream of the coding sequence, ethanol precipitated, and redissolved in H$_2$O.

Transcription reaction:

H$_2$O to a final volume of 50 μl
  2 μg of DNA
  10 μl $5 \times$ transcription buffer
  10 μl rNTPs
   1 μl DTT
  50 units RNasin
16.7 μl cap, if required
  20 units SP6 RNA polymerase

After 1 hour at 40°C, 50 μl H$_2$O is added and the nucleic acids are precipitated by the addition of 10 μl of 3 M Na-acetate (pH 5.2) and 200 μl ethanol at −70°C for ≥20 minutes. After centrifugation for 10 minutes, the pellet is dried and redissolved in 25μl H$_2$O. Typically a yield of 3–4 μg (corresponding to approximately six to eight transcripts of a ~1500 bp cDNA) is obtained. If required, the yield can be determined by incorporation of radioactive nucleotides.

We found that capping the mRNA greatly enhances the translation efficiency in RL (approximately 10-fold), but does not have a significant effect in the WG system. To cap the mRNA, the RNA cap analog 7mGpppG is simply added to the transcription reaction. It is then incorporated by the RNA polymerase in place of the initial GTP (Konarska *et al.*, 1984).

Other transcription systems based on RNA polymerases and promoters of other phages (e.g., T7) or *Escherichia coli* (Stueber *et al.*, 1984) are also commonly used for *in vitro* translation.

2. Preparation of Microsomes

The preparation of dog pancreas microsomes is described in detail by Walter and Blobel (1983a). Salt-washed, SRP-free microsomes and SRP are prepared according to Walter and Blobel (1983b), whereby the additional purification of the microsomes on a sucrose gradient can be omitted for most purposes. Preparation of salt-washed microsomes and purification of SRP may be advisable when the crude microsomes exhibit a strong inhibitory effect on translation. The separate components added to the translation inhibit much less. To digest endogenous mRNA contaminating the membranes, the microsome preparation is incubated with 2 mM CaCl$_2$ and 0.6 U/μl staphylococcal nuclease for 10 minutes at room temperature. The digestion is stopped by adding 4 mM EGTA. The membranes are stored at $-70°C$ in 50 μl aliquots. Dog pancreas microsomes are also commercially available.

3. *In Vitro* TRANSLATION AND MEMBRANE INSERTION

Preparation of WG and its use for translation are described by Anderson *et al.* (1983). RL is prepared and used as described by Pelham and Jackson (1976). Endogenous mRNA in WG and RL is digested using staphylococcal nuclease as described above for microsomes. Both systems are also available commercially. Translation reactions are typically performed with [^{35}S]methionine in a volume of 12.5 μl for 90 minutes at 30°C. It is advisable to titrate the optimal concentrations of K-acetate, mRNA, and microsomes. We usually add 0.25 equivalents of microsomes (as defined by Walter and Blobel, 1983a,b) and (in combination with salt-washed microsomes) 0.25 μl SRP (as eluted from ω-aminopentyl agarose) per reaction. *In vitro* transcribed mRNA (1 μl) (corresponding to approximately 100 ng; capped or uncapped) is added to WG reactions and 0.2 μl of capped mRNA is added to RL reactions. After translation, the samples are taken up directly in sample buffer and analyzed by SDS–gel electrophoresis and fluorography, because essentially a single product is synthesized. Gels should be extensively fixed to allow hydrolysis and diffusion of [^{35}S]methionyl-tRNA, which migrates in the middle of the gel. Alternatively, the translation products may also be immunoprecipitated (for example, using domain-specific antibodies as described below). The entire WG reaction and one-quarter of a RL reaction will yield a sufficient signal after fluorography for 1–2 days.

The WG and RL systems have different properties (compare lanes 1 and 3 with lanes 8 and 9 in Fig. 1). Aberrant initiation at internal ATG sequences can be observed in RL translations (asterisk in Fig. 1). The WG system, however, is more prone to premature termination resulting in a smear below the full-size band, and is less efficient in protein insertion into microsomes. In addition, glycosylation intermediates can often be observed, which sometimes may help to determine the number of oligosaccharides that are attached to the polypeptides.

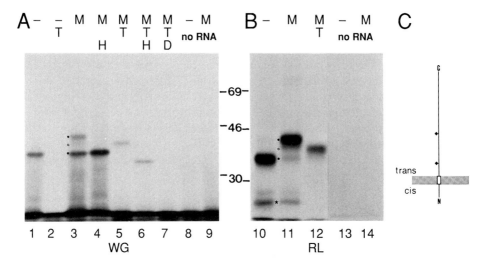

FIG. 1. *In vitro* translation and membrane insertion of the ASGP receptor H1. *In vitro*-transcribed mRNA was translated either in the WG (panel A) or the RL system (panel B) in the absence (−) or presence (M) of microsomes. Some samples were treated with trypsin (T; without or with detergent, D) or digested with endo H (H), as indicated. Fluorographs of the translation products after gel electrophoresis are shown. The position of marker proteins and their molecular masses (kDa) are indicated. The asterisk denotes a translation product generated by initiation at an internal ATG codon. A schematic representation of the deduced membrane topology of the ASGP receptor H1 is shown in panel C. The cytoplasmic and the lumenal sides of the membrane are specified by cis and trans, respectively. Glycosylation sites are indicated by diamonds. Reproduced in modified form from Wessels and Spiess (1988).

B. *In Vivo* Expression

The use of cultured cells to study insertion of proteins into the ER offers a number of advantages. Translation and insertion is very efficient, and artifacts in folding, formation of disulfide bonds, processing, and modification are unlikely to occur unless a protein is heavily overexpressed. There are well-established protocols for the transfection and expression of cloned cDNAs in cultured cells lines (Sambrook *et al.*, 1989; Roth, 1989; Cullen, 1987). The production of stably expressing cell lines, which have integrated the transfected DNA into the genome, allows continued experimentation under identical conditions but requires time-consuming selection, isolation, and screening of clonal cell lines with suitable levels of expression. Transient expression systems are therefore especially useful when many different proteins or a series of mutants are to be analyzed.

The choice of the cell line and of the expression vector depends on many different considerations discussed in detail by Sambrook *et al.* (1989) and

Cullen (1987). We have successfully expressed heterologous proteins in stably transfected NIH3T3 fibroblasts using the retroviral shuttle vector pLJ (Korman *et al.*, 1987) and in the transiently producing COS cell system (Cullen, 1987) using the vector plasmid pECE (Ellis *et al.*, 1986).

To study the membrane insertion and the topology of a particular protein in the ER of cultured cells, newly synthesized proteins are labeled with [^{35}S]methionine. With labeling times of less than 30 minutes, essentially all the radioactive secretory or membrane proteins are still localized in the ER and the oligosaccharide moieties of glycoproteins are still of the high-mannose type. If the cells are homogenized, right-side-out microsomes are formed. These can be used for further analysis (e.g., protease digestion; see below) even without purification, because after immunoprecipitation, gel electrophoresis, and fluorography, only the labeled ER form of the protein of interest will be detected. In addition, if labeling of the cells is followed by a chase period in the presence of unlabeled methionine, proteins can be followed in their intracellular transport. Proteins reaching the plasma membrane can also be accessed from the extracytoplasmic side on the surface of intact cells.

1. METABOLIC LABELING AND IMMUNOPRECIPITATION

For metabolic labeling, transfected cells grown in 35-mm wells are rinsed twice with PBS, starved in methionine-free medium for 30 minutes, and incubated with 50–300 μCi/ml [^{35}S]methionine in starvation medium for 10–30 minutes at 37°C. The cells are then washed twice with ice-cold PBS. If the cells are to be immunoprecipitated, 0.5 ml lysis buffer (1% Triton X-100, 0.5% deoxycholate, and 2 mM PMSF in PBS) are added to each well at 4°C for 20 minutes on a rocker. The cells are scraped into an Eppendorf tube, vortexed, kept on ice for 1 hour, and centrifuged for 15 minutes. The supernatant is mixed with 500 μl of "immunomix" (lysis buffer supplemented with 0.5% SDS and 1 mg/ml BSA), and with 2.5 μl of antiserum for 2 hours to overnight. A 1:1 suspension (40 μl) of protein A–Sepharose is added and incubated for 30 minutes to 1 hour with rocking. The Sepharose is quickly pelleted (20 seconds) and washed three times with "immunowash" (1% SDS, 1% Triton X-100, 0.5% deoxycholate, 1 mg/ml BSA, and 1 mM PMSF in PBS) and twice with PBS.

2. PREPARATION OF MEMBRANES

Cells grown in 35-mm wells are washed twice with cold PBS, scraped into 1 ml of PBS containing 2 mM PMSF, and broken in a 1-ml Dounce homogenizer (15 strokes). The membranes are pelleted in an Eppendorf centrifuge for 10 minutes and resuspended in 200 μl PBS.

III.   Probing the Transmembrane Topology

In this section we describe different methods to gain information on the topology of a membrane protein. Each addresses a specific question. The combination of several methods can help to deduce the transmembrane structure.

A.   Membrane Integration

The first question to be asked about a protein translated either *in vitro* or *in vivo* is whether it was inserted at all into the ER, and, if so, whether it was integrated into the membrane or translocated into the lumen of the ER as a soluble protein. Reisolation of microsomes will tell whether the protein is associated in any way with the membranes. Saponin extraction releases soluble proteins (cytosolic or lumenal) from transfected cells. The most stringent test to discriminate integral membrane proteins from soluble or only peripherally associated ones is alkaline extraction.

1.   REISOLATION OF MICROSOMES

This procedure is adapted from Gilmore and Blobel (1985). After *in vitro* translation in the presence of microsomes, the reaction is diluted to 50 μl with 50 mM triethanolamine, pH 7.5, 150 mM KOAc, 2.5 mM Mg(OAc)$_2$, and 1 mM DTT. It is loaded onto a 100-μl cushion of 0.5 M sucrose in the same solution in a nitrocellulose tube (precoated with 1% BSA) for the A-100/30 rotor and centrifuged for 5 minutes at 4°C in a Beckman Airfuge at 25 psi. The entire supernatant, including the cushion, is removed and precipitated with an equal volume of 20% TCA. The pellet fraction and the TCA pellet of the supernatant are taken up in SDS–gel sample buffer and analyzed by gel electrophoresis and fluorography. To increase the stringency of this procedure, higher salt concentrations can be used in the cushion (Gilmore and Blobel, 1985).

2.   SAPONIN EXTRACTION OF TRANSFECTED CELLS

Cells are washed with PBS at 4°C and incubated with 500 μl of 0.1% saponin per 35-mm well. This treatment permeabilizes cellular membranes and allows soluble proteins to diffuse out, but the cells are not entirely disintegrated and membrane proteins are retained. The saponin extract is centrifuged for 10 minutes in an Eppendorf centrifuge. The supernatant is mixed with 500 μl of "immunomix" for immunoprecipitation. The extracted cells are lysed and immunoprecipitated as described above.

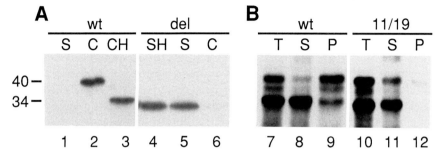

FIG. 2.   Membrane integration of the ASGP receptor H1. (A) The wild-type (wt) ASGP receptor H1 and a deletion (del) mutant protein lacking the hydrophobic segment were expressed in COS-7 cells. The cells were labeled for 30 minutes with [^{35}S]methionine and extracted with 0.1% saponin. The extract (S) and the rest of the cells (C) were immunoprecipitated and analyzed by gel electrophoresis and fluorography. Some immunoprecipitates were, in addition, digested with endo H (H). (B) The wild-type (wt) ASGP receptor H1 and a mutant protein with a shortened hydrophobic segment [11/19; described in detail by Spiess and Handschin (1987)] were translated in the WG system in the presence of microsomes and subjected to alkaline extraction. Aliquots of the original total sample (T), and the supernatant (S) and the membrane pellet (P), after centrifugation, were analyzed by gel electrophoresis and fluorography. The positions of the 40-kDa glycoprotein and the 34-kDa polypeptide of H1 are indicated.

Figure 2A shows the result of saponin extraction of COS-7 cells expressing either the wild-type asialoglycoprotein (ASGP) receptor H1 (lanes 1–3), a single-spanning membrane protein, or a deletion mutant lacking the transmembrane segment (lanes 4–6). The deletion mutant was completely released into the saponin (S) extract and the wild-type protein was entirely retained in the cells (C). Glycosylation of the wild-type receptor protein, resulting in a reduced electrophoretic mobility and sensitivity to deglycosylation by endo-β-N-acetylglucosaminidase H (endo H; H), confirms its insertion into the ER membrane (see below).

3.   ALKALINE EXTRACTION

Under strongly alkaline conditions (pH > 11), biological membranes have been shown to be disrupted to open sheets and to be stripped of peripheral proteins (Fujiki *et al.*, 1982). This method proved to be a useful empirical procedure to distinguish between integral membrane proteins and soluble or peripheral proteins.

In vitro translation reactions or membrane preparations of transfected cells are diluted to 50 μl with 100 mM HEPES, pH 11.5, and the pH is adjusted to 11–11.5 with 0.1 N NaOH (spotting 1-μl aliquots on pH

paper). After 10 minutes on ice, the samples are loaded onto a 100-μl
cushion of 0.2 M sucrose, 60 mM HEPES, pH 11, 150 mM KOAc, 2.5 mM
Mg(OAc)$_2$, and 1 mM DTT in a nitrocellulose tube (precoated with 1%
BSA) for the A-100/30 rotor and centrifuged for 10 minutes at 4°C in a
Beckman Airfuge at 30 psi. The entire supernatant, including the cushion,
is removed and precipitated with an equal volume of 20% TCA. The pellet
fraction and the TCA pellet of the supernatant are taken up in SDS–gel
sample buffer and analyzed by gel electrophoresis and fluorography. [This
procedure was adapted from Gilmore and Blobel (1985).]

Alkaline extraction of the ASGP receptor H1 translated in WG in the
presence of microsomes is shown in Fig. 2B (lanes 7–9). A large fraction of
the total protein synthesized (T) had an apparent molecular weight of
34,000, corresponding to the unmodified polypeptide. In addition, two
forms (37 and 40 kDa) were produced, corresponding to the once- and
twice-glycosylated protein. Upon alkaline extraction, nearly all the glycosylated species were recovered in the membrane pellet, whereas the
unmodified protein was almost entirely left behind in the supernatant. This
indicates that only the glycosylated forms were integrated into the ER
membrane.

However, for a mutant version of H1, whose transmembrane domain
was shortened by two residues (pSA11/19) (Spiess and Handschin, 1987),
the glycosylated protein was also largely extracted from the membrane
(lanes 10–12). Because glycosylation and partial protease resistance of this
form clearly showed that it was integrated in the bilayer with a transmembrane topology (Spiess and Handschin, 1987), this result indicates that
integral membrane proteins with hydrophobic domains too short to com-
fortably span the bilayer may be extracted by alkali treatment. Alkaline
extraction alone may thus sometimes yield ambiguous results (particularly
for artificial protein constructs), requiring confirmation by an independent
approach.

B. Glycosylation

The most obvious modification occurring in the ER is N-glycosylation of
asparagine residues in the sequence context Asn-X-Ser/Thr (where X is
any amino acid except Pro). Many but not all occurrences of this sequence
in translocated protein domains are modified. The oligosaccharide is assembled as a dolichyl pyrophosphate precursor on the lumenal side of the
ER membrane and transfered in one enzymatic step to the polypeptide
during or very shortly after translocation into the ER. Addition of each
oligosaccharide results in an increase of the apparent molecular weight
upon SDS–gel electrophoresis of approximately 3 kDa (compare lanes 1

and 3 and lanes 10 and 11 in Fig. 1) Using *in vitro* systems, it is often possible (particularly when using the WG system) to discern glycosylation intermediates that allow the number of oligosaccharide moieties transfered to be counted off the fluorographs (Wessels and Spiess, 1988).

In addition, N-glycosylation can easily be demonstrated by enzymatic digestion with endo-β-N-acetylglucosaminidase H, which specifically removes all but the first sugar residue of the high-mannose-type oligosaccharides added in the ER (Fig. 1, lanes 4 and 6). *In vitro* translation reactions are diluted with H_2O to 40 μl and mixed with 40 μl of 100 mM Na-citrate, pH 6, and 2% SDS; they are boiled for 2 minutes and then incubated for 2–3 hours at 37°C with 1–3 mU endo H. After addition of SDS–gel sample buffer and boiling for 2 minutes, the samples are analyzed by gel electrophoresis and fluorography. To deglycosylate immunoprecipitates of *in vitro* translations or of membrane preparations of transfected cells, the protein A–Sepharose pellet is directly taken up in 80 μl of 50 mM Na-citrate, pH 6, and 1% SDS; it is boiled for 2 minutes and briefly centrifuged. The supernatant is removed and incubated with endo H as above. Using a series of different enzyme concentrations, partially deglycosylated products can be generated (cf. Spiess *et al.*, 1989). Upon electrophoresis and fluorography, they form a ladder that allows determination of the number of oligosaccharides attached to the protein.

C. Protease Protection

Because microsomes are closed vesicles with a defined cytoplasmic-side-out orientation, proteins and polypeptide domains exposed on the outside can be selectively digested by proteases (Figs. 1 and 3). Samples are chilled on ice and incubated with 0.1 volume of 0.5 mg/ml TPCK-treated trypsin for 45 minutes on ice. Digestion is stopped by adding 0.1 volume of 2 mg/ml soybean trypsin inhibitor. Alternatively, the same amount of proteinase K, a very unspecific protease, can be used (20-minute incubation time) and is stopped by TCA precipitation or addition of one volume of PBS containing 5 mM PMSF (freshly added). To demonstrate that the protein is inherently sensitive to the protease used, a control digestion should be performed in the presence of 0.5% of a noionic detergent, such as Nonidet P-40 or Triton X-100, to disrupt the membrane.

D. Domain-Specific Antibodies

Upon protease digestion of the cytoplasmic surface of microsomes, membrane proteins are fragmented and only the polypeptide segments protected inside the vesicle and buried within the membrane are recovered.

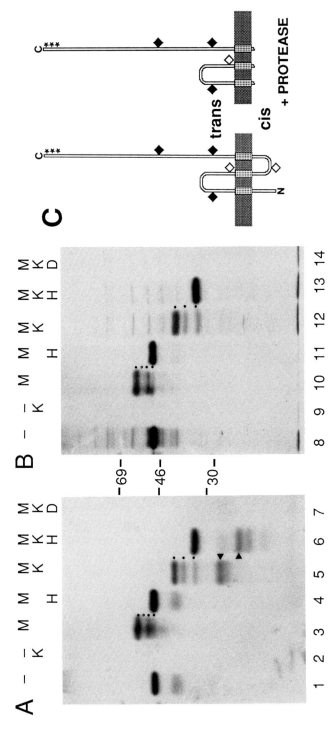

FIG. 3. Analysis of membrane topology using domain-specific immunoprecipitation. An artificial, potentially threefold membrane-spanning protein composed of sequences of the ASGP receptor H1 was translated in the RL system in the absence(−) or presence (M) of microsomes. Some samples were treated with proteinase K (K; without or with detergent, D) or digested with endo H (H), as indicated. The samples were immunoprecipitated either with a polyspecific antiserum (panel A) or with a domain-specific antiserum raised against a carboxy-terminal peptide (indicated by asterisks in panel C). Fluorographs of the translation products after gel electrophoresis are shown. The position of marker proteins and their molecular masses (kDa) are indicated. A schematic representation of the deduced membrane topology is shown in panel C. Potential glycosylation sites are indicated by diamonds; the ones that are used are filled. Reproduced in modified form from Beltzer et al. (1989).

To identify which portions of the original protein are resistant, domain-specific antibodies, such as monoclonal antibodies with known epitopes or antisera raised against synthetic oligopeptides, can be very useful. In Fig. 3, the topology of a potentially three-fold membrane-spanning protein is analyzed. Half of each sample was immunoprecipitated with a polyspecific antiserum (panel A) and the other half with an antiserum raised against a synthetic undecapeptide corresponding to the carboxy-terminal sequence of the protein (panel B). The twice-glycosylated fragment of approximately 38 kDa (indicated by dots in lanes 5, 6, 12, and 13) is identified as the carboxy-terminal portion of the protein. The once-glycosylated fragment (arrowheads in lanes 5 and 6) thus must correspond to the translocated loop segment from the first transmembrane domain to the second. Another example for the use of domain-specific antibodies is the study by Rothman *et al.* (1988).

E. Signal Cleavage

Amino-terminal signal sequences for targeting proteins to the ER are in most cases cleaved off by signal peptidase in the ER lumen. Depending on the molecular weight of the protein and the extent of glycosylation, the change in electrophoretic mobility upon signal cleavage may or may not be discernible. To distinguish between a cleaved signal and an amino-terminal signal-anchor sequence, and to identify the exact site of processing, the amino terminus of *in vitro* translation products has to be sequenced. The protein is translated with and without microsomes in the presence of suitable radioactive amino acids, which are chosen for their diagnostic value (e.g., there might be a typical sequence of leucine residues following the presumed cleavage site) and for their availability at high specific radioactivity. We have successfully used [^{3}H]leucine (140 Ci/mmol) and [^{3}H]glutamic acid (50 Ci/mmol) for this purpose (Schmid and Spiess, 1988). The labeled protein is subjected to automated Edman degradation, and the resulting radioactivity profile allows alignment of the labeled residues with the protein sequence.

RL translation with the appropriate labeled amino acid and with or without microsomes is upscaled 10- to 20-fold. The products are separated by gel electrophoresis, and the regions containing the proteins of interest are cut out, minced, and eluted overnight with 25 mM ammonium acetate and 0.1% SDS. Alternatively, the proteins can be electroeluted (Hunkapiller *et al.*, 1983). The eluate is lyophilized, taken up in a small volume of H$_2$O, and desalted on a small Sephadex G-25 column. If a single product is synthesized, the protein is immunoprecipitated after translation and the protein A–Sepharose pellet is boiled for 2 minutes with a minimal volume of 1% SDS and 2% β-mercaptoethanol. After centrifugation the pellet is washed with a small volume of the same solution. To remove most of the detergent, the sample is precipitated with five volumes of cold acetone during 5 hours at −20°C. The dried pellet is redissolved by boiling in 50 μl 0.2% SDS, and this material is subjected to automated Edman degradation. The radioactivity contained in the phenylhydantoin derivatives collected after each cycle is determined by scintillation counting.

IV. Discussion

Combining the different methods described above, the topology of simple proteins in the ER membrane can be deduced, as is illustrated for proteins that span the membrane once and threefold in Figs. 1 and 3, respectively. Analysis of the cDNA sequence of the asialoglycoprotein receptor subunit H1 (Spiess *et al.*, 1985) suggested the lack of an amino-terminal signal sequence. There was a single apolar segment of 20 residues further inside the polypeptide that could potentially span the membrane, and two potential sites for N-glycosylation toward the carboxy terminus. It could be shown that the protein is an integral membrane protein because it was neither extractable with 0.1% saponin from expressing cells, nor was it released from the microsomal membrane upon high-pH treatment (Fig. 2). Upon *in vitro* translation in the presence of microsomes, it was twice glycosylated (as judged from the reduced electrophoretic mobility and from endo H sensitivity; Fig. 1), indicating that at least the carboxy-terminal domain was translocated. Consistent with this, the glycosylated protein was also largely resistant to exogenous protease (except when the microsomal membrane was disrupted with detergent). Upon protease treatment, only a segment of approximately 3 kDa, corresponding to the size of the amino-terminal domain, was digested. These results taken together strongly indicate that the protein spans the membrane with the amino terminus in the cytoplasm and the carboxy terminus in the ER lumen (Fig. 1C). Characterization of a series of deletion and fusion proteins revealed that the apolar transmembrane segment functions as an internal signal necessary and sufficient for targeting to the ER and membrane insertion (Spiess and Lodish, 1986).

Figure 3 illustrates the analysis of an artificial protein composed of ASGP receptor sequences. It contains three hydrophobic, potentially membrane-spanning segments and a total of five glycosylation sites. By the same analysis combined with immunoprecipitation using a domain-specific antibody, the structure of this protein (as is shown in Fig. 3C) could be derived.

It is obvious that the topology of complex multispanning membrane proteins is very difficult to determine. Natural proteins, which based on their hydropathy plots are predicted to span the bilayer 7–12 times or more, often contain very short hydrophilic loops between transmembrane segments that are resistant to proteolysis. However, the methods described here may be of use in the functional analysis of subsegments of such proteins, e.g., tested in the context of a more amenable fusion protein. In addition, specific questions may also be answered by introducing specific tags into the protein, such as additional glycosylation sites or epitopes for antibodies. Even though the available methods are best suited to study simple proteins, the understanding gained on the characteristics of topogenic sequences could extend to more complex proteins.

ACKNOWLEDGMENT

This work was supported by Grant 31-26571.89 from the Swiss National Science Foundation.

REFERENCES

Anderson, C. W., Strous, J. W., and Dudock, B. S. (1983). In "Methods in Enzymology" (R. Wu, L. Grossman, and K. Moldave eds.) Vol. **101**, pp. 635–645. Academic Press, New York.

Beltzer, J. P., Wessels, H. P., and Spiess, M. (1989). *FEBS Lett.* **253**, 93–98.

Cullen, B. R. (1987). In "Methods in Enzymology" (S. L. Berger and A. R. Kimmel, eds), Vol. **152**, pp. 684–704. Academic Press, Orlando, Florida.

Ellis, L., Clauser, E., Morgan, D. O., Edery, M., Roth, R. A., and Rutter, W. J. (1986). *Cell (Cambridge, Mass.)* **45**, 721–732.

Fujiki, Y., Hubbard, A. L., Fowler, S., and Lazarow, P. B. (1982). *J. Cell Biol.* **93**, 97–102.

Garoff, H., Simons, K., and Dobberstein, B. (1978). *J. Mol. Biol.* **124**, 587–600.

Gilmore, R., and Blobel, G. (1985). *Cell (Cambridge, Mass.)* **42**, 497–505.

Hunkapiller, M. W., Lujan, E., Ostrander, F., and Hood, L. E. (1983). In "Methods in Enzymology" (C. H. W. Hirs and S. N. Timasheff eds.), Vol. **91**, pp. 227–236. Academic Press, New York.

Konarska, M. M., Padgett, R. A., and Sharp, P. A. (1984). *Cell (Cambridge, Mass.)* **38**, 731–736.

Korman, A. J., Frantz, J. D., Strominger, J. L., and Mulligan, R. C. (1987). *Proc. Natl. Acad. Sci. U.S.A.* **84**, 2150–2154.

Krieg, U. C., Johnson, A. E., and Walter, P. (1989). *J. Cell Biol.* **109**, 2033–2043.

Kyte, J., and Doolittle, R. F. (1982). *J. Mol. Biol.* **157**, 105–132.

Lipp, J., Flint, N., Haeuptle, M. T., and Dobberstein, B. (1989). *J. Cell Biol.* **109**, 2013–2022.

Lopez, C. D., Yost, C. S., Prusiner, S. B., Myers, R. M., and Lingappa, V. R. (1990). *Science* **248**, 226–229.

Meyer, D. I. (1985) *EMBO J* **4**, 2031–2033.

Pelham, H. R. B., and Jackson, R. J. (1976). *Eur. J. Biochem.* **67**, 247–256.

Roth, M. (1989) *In* "Expression Vectors for Epithelial Cells" (K. S. Matlin and J. D. Valentich, eds.), pp. 269–302. Alan R. Liss, New York.

Rothblatt, J. A., and Meyer, D. I. (1986). *EMBO J.* **5**, 1031–1036.

Rothman, R. E., Andrews, D. W., Calayag, M. C., and Lingappa, V. R. (1988). *J. Biol. Chem.* **263**, 10470–10480.

Sambrook, J., Fritsch, E. F., and Maniatis, T. (1989). "Molecular Cloning: A Laboratory Manual, 2nd ed., Vol. 3. Cold Spring Harbor Lab., Cold Spring Harbor, New York.

Schmid, S. R., and Spiess, M. (1988). *J. Biol. Chem.* **263**, 16886–16891.

Spiess, M., and Handschin, C. (1987). *EMBO J.* **6**, 2683–2691.

Spiess, M., and Lodish, H. F. (1986). *Cell (Cambridge, Mass.)* **44**, 177–185.

Spiess, M., Schwartz, A. L., and Lodish, H. F. (1985). *J. Biol. Chem.* **260**, 1979–1982.

Spiess, M., Handschin, C., and Baker, K. P. (1989). *J. Biol. Chem.* **264**, 19117–19124.

Stueber, D., Ibrahimi, I., Cutler, D., Dobberstein, B., and Bujard, H. (1984). *EMBO J.* **3**, 3143–3148.

Verner, K., and Schatz, G. (1988). *Science* **241**, 1307–1313.

Walter, P., and Blobel, G. (1983a). *In* "Methods in Enzymology" (S. Fleischer and B. Fleischer, eds.) Vol. **96**, pp. 84–93. Academic Press, New York.

Walter, P., and Blobel, G. (1983b). *In* "Methods in Enzymology" (S. Fleischer and B. Fleischer, eds.) Vol. **96**, pp. 682–691. Academic Press, New York.

Walter, P., Gilmore, R., and Blobel, G. (1984). *Cell (Cambridge, Mass.)* **38**, 5–8.

Wessels, H. P., and Spiess, M. (1988). *Cell (Cambridge, Mass.)* **55**, 61–70.

Wickner, W. T., and Lodish, H. F. (1985). *Science* **230**, 400–407.

Wiedmann, M., Kurzchalia, T. V., Hartmann, E., and Rapoport, T. A. (1987). *Nature (London)* **328**, 830–833.

Wolin, S. L., and Walter, P. (1989). *J. Cell Biol.* **109**, 2617–2622.

Chapter 14

Protein Import into Peroxisomes in Vitro

PAUL B. LAZAROW, ROLF THIERINGER,[1]
AND GERALD COHEN[2]

Department of Cell Biology and Anatomy
Mount Sinai School of Medicine
New York, New York 10029

TSUNEO IMANAKA

Department of Microbiology and Molecular Pathology
Teikyo University
Sagamiko, Kanagawa 199-01, Japan

GILLIAN SMALL

Department of Anatomy and Cell Biology
University of Florida at Gainesville
Gainesville, Florida 32610

[1] Present address: Department of Biochemistry, Merck, Sharpe & Dohme Pharmaceuticals, Rathway, New Jersey 07065.
[2] Present address: Sandoz Forschungsinstitut, G.m.b.H., A-1235 Vienna, Austria.

METHODS IN CELL BIOLOGY, VOL. 34

I. Introduction

The biogenesis of peroxisomes is a subject ready for study by *in vitro* methods. The main features of the assembly of this organelle are established (Lazarow and Fujiki, 1985) and progress has been made in elucidating the mechanisms (for recent reviews, see Borst, 1989; Lazarow, 1989; Lazarow and Moser, 1989; Small and Lewin, 1990; Osumi and Fujiki, 1990). However, many of the details are not yet known. The transport of newly synthesized proteins into peroxisomes has been reconstituted efficiently *in vitro*, permitting studies of the energy requirements for translocation (Imanaka *et al.,* 1987) and of the topogenic information (Small and Lazarow, 1987; Small *et al.,* 1988; Miyazawa *et al.,* 1989). A major advantage of the *in vitro* system is that it will permit detailed biochemical dissection of the import machinery.

Peroxisomes form by the elaboration and division of preexisting peroxisomes, in much the same fashion as new mitochondria and chloroplasts form. All peroxisomal proteins studied are encoded by nuclear genes, synthesized on free polyribosomes in the cell cytosol, and imported posttranslationally into preexisting peroxisomes. This applies to the integral membrane proteins and to the proteins that are concentrated in soluble form in the matrix space of the organelle. Phospholipids are synthesized in the endoplasmic reticulum and are transported to peroxisomes, perhaps by phospholipid carrier proteins. Thus, peroxisomes enlarge, and in most cells then divide to form new peroxisomes. In certain specialized cell types, the peroxisomes remain interconnected in an elaborate, contorted compartment called the peroxisome reticulum.

Most peroxisomal proteins are synthesized at their mature sizes. The topogenic information is integral to the mature proteins; it is not removed proteolytically after import. Many peroxisomal proteins contain a carboxy-terminal tripeptide, Ser-Lys-Leu (or a closely related tripeptide), that targets them to peroxisomes, or at least plays a major role in this targeting (Gould *et al.*, 1987, 1989, 1990; Miyazawa *et al.*, 1989). Other peroxisomal proteins lack this carboxy-terminal sequence: redundant targeting information has been found in the middle and at or near the amino terminus of a yeast acyl-CoA oxidase (Small and Lazarow, 1987; Small *et al.*, 1988). *In vitro* import studies and *in vivo* immunofluorescence localization studies have contributed to what we have learned about peroxisomal protein topogenesis. One of the major advantages of the *in vitro* import approach is the ability to quantitate the relative import efficiency of different protein fragments and of hybrids in which targeting information is attached to a passenger protein. Thus, one can ask not only whether a topogenic sequence is functional, but also how efficient it is.

Complementation analysis of fibroblasts from patients with genetic diseases caused by defects in peroxisomal assembly have identified six complementation groups thus far (Brul *et al.*, 1988; Roscher *et al.*, 1989). This suggests that at least six gene products are required for the biogenesis of the organelle. In one of these diseases, Zellweger syndrome, peroxisome membranes are assembled whereas the content proteins are not, resulting in empty membrane ghosts of peroxisomes (Santos *et al.*, 1988a,b). This means that some machinery is required for the assembly of matrix proteins that is unnecessary for membrane protein assembly. *In vitro* import studies have contributed to the analysis of this machinery by demonstrating two steps in the translocation of acyl-CoA oxidase into rat liver peroxisomes: an energy-independent binding step, and a translocation step that requires ATP hydrolysis (Imanaka *et al.*, 1987). Further understanding of this process will be facilitated by the ability to analyze each of the steps independently *in vitro*.

The principle of the *in vitro* reconstitution of peroxisomal protein assembly is simple: newly synthesized peroxisomal proteins (labeled with [^{35}S]methionine) are mixed with purified peroxisomes and the proteins enter the organelle in a time- and temperature-dependent fashion. Import is analyzed by digesting the nonimported proteins with a protease. Those proteins that have been imported into the peroxisomes are protected from proteolysis by the membrane of the organelle. This *in vitro* reconstitution of import requires a good supply of stable, purified peroxisomes, sufficient synthesis of radiolabeled peroxisomal proteins, appropriate incubation conditions for import, the selection of protease(s) that can digest nonimported proteins without destroying the organelle membrane, and careful controls. Each of these issues will be discussed below.

II.   Preparation of Purified Peroxisomes

Purification of peroxisomes consists of the following steps:

1. Induce peroxisome proliferation, if necessary. The greater the abundance of peroxisomes in a cell, the easier it is to isolate them in sufficient yield and reasonable purity.
2. Gently homogenize the cells.
3. Use differential centrifugation to prepare an organelle fraction enriched in peroxisomes. This step selects by organelle size.
4. Purify the peroxisomes by equilibrium density centrifugation.
5. Concentrate the peroxisomes in a medium suitable for import studies.

A.   Sources of Peroxisomes

Peroxisomes are fragile organelles, and therefore it is critical to treat them gently during purification. Moreover, the manner in which they are induced and their size (which may increase upon induction) will affect their stability after isolation. Thus far, peroxisome protein import has been reconstituted *in vitro* with peroxisomes from three different organisms: rat (liver) and two yeasts, *Candida tropicalis* and *Saccharomyces cerevisiae*. Each has advantages. Peroxisomes are naturally abundant in rat liver. There is a long history of purifying them, so the optimal conditions are well-established, and they are quite stable as isolated. It is possible to induce proliferation of peroxisomes in rat liver by the administration of hypolipidemic drugs, but this is disadvantageous because the larger peroxisomes thus induced are more fragile.

Peroxisomes are inducible in all yeast species thus far investigated, for example, by growing the yeasts on fatty acids or on methanol (reviewed by Veenhuis *et al.*, 1983; Fukui and Tanaka, 1979). Peroxisome formation is repressed by growing the yeasts on glucose. The resulting huge differences in the expression of genes encoding peroxisomal proteins have many advantages for cloning peroxisomal genes and for studying peroxisome biogenesis. We chose to study *in vitro* import into peroxisomes from *C. tropicalis* because large and abundant peroxisomes could be induced by growth on oleic acid, and the enzymatic functions of these peroxisomes resembled those of rat liver (Fujiki *et al.*, 1986). Oleate-induced peroxisomes turned out to be too fragile for use in import studies. Moderate-sized, more stable peroxisomes could be induced by growth on Brij 35, a polymer containing lauric acid residues, and these peroxisomes proved suitable for *in vitro* import (Small *et al.*, 1987).

Although it was long thought to be infeasible to induce peroxisomes in *S. cerevisiae*, recently such induction was accomplished by growth on fatty acids (Veenhuis *et al.*, 1987). These peroxisomes are quite fragile. After considerable trial and error we have been able to purify the peroxisomes and accomplish *in vitro* import (Thieringer *et al.*, 1991). Work on *in vitro* import with *S. cerevisiae* peroxisomes is worth the effort, despite the fragility problems, because of the many other advantages of *S. cerevisiae*, in particular the molecular genetics, for studying organelle assembly.

B. Cell Fractionation

1. RAT LIVER

The purification of peroxisomes from rat liver is based on the procedure of Leighton *et al.* (1968) with some simplification (Imanaka *et al.*, 1987). Peroxisomes are prepared from three female rats. Food is withdrawn the night before. The rats are anesthetized and guillotined. The livers are removed, weighed, and cut into pieces. They are suspended in five volumes of ice-cold 0.25 M sucrose, 1 mM EDTA, 0.1% ethanol, pH 7.4 (SVE), containing the following protease inhibitors at a concentration of 5 μg/ml: leupeptin, pepstatin, chymostatin, and antipain (stock in dimethyl sulfoxide). The liver is homogenized in a motor-driven Potter–Elvehjem homogenizer (one stroke) and the homogenate is centrifuged at 4500 rpm (2500 g) for 10 minutes in a Sorvall SS34 rotor. The supernatant is carefully removed, avoiding the lipid layer at the top, and the pellet resuspended in SVE (approximately one-half of the original homogenate volume). This is homogenized and centrifuged as before and the two supernatants (including the pink fluffy layer) are combined. This postnuclear supernatant fraction is then centrifuged at 12,000 rpm (17,000 g) for 20 minutes (SS34 rotor). The supernatant is removed and the pellet gently resuspended in approximately 20 ml of cold SVE using two to three strokes with a Dounce homogenizer (Kontes Blass Corporation, Vineland, NJ; type B pestle). The pellet fraction is resuspended and recentrifuged as before (17,000 g for 20 minutes) and the two 12,000-rpm supernatants are combined.

The 17,000 g pellet (containing mainly mitochondria, peroxisomes, and lysosomes) is carefully resuspended in approximately 8 ml of SVE and is subjected to isopycnic centrifugation in Nycodenz. For this step we routinely use a custom-built Beaufay zonal rotor, which gives an excellent separation of peroxisomes from other organelles and at the same time causes minimal organelle damage (Leighton *et al.*, 1968). We use a 24-ml Nycodenz gradient [1.15–1.25 g/ml; 21–39% (w/v) resting on a 6-ml cushion of dense Nycodenz (1.30 g/ml; 48%)]. The Nycodenz solutions are prepared in SVE plus 5 mM HEPES, pH 7.4 (SVEH). Centrifugation is at 35,000 rpm (90,000 g) for 40 minutes. A good alternative to the Beaufay rotor (which is not commercially available) is the use of a vertical rotor (see preparation of peroxisomes from *S. cerevisiae*; see also Alexson *et al.*, 1985).

Peroxisomes and mitochondria are located in the gradient fractions by measuring a marker enzyme for each organelle across the gradient (Leighton *et al.*, 1968). We use catalase for peroxisomes (see Lazarow *et al.*, 1988, for the assay procedure) and cytochrome oxidase for mitochondria (Leighton *et al.*, 1968). The peak peroxisome fractions are pooled and the density of the medium is lowered by the dropwise addition of five volumes of ice-cold SVEH. After 10 minutes on ice, the peroxisomes are pelleted by centrifugation at 12,000 rpm (17,000 g) for 20 minutes. The pellet is gently resuspended in import buffer (0.25 M sucrose, 0.1% ethanol, 5 mM HEPES, pH 7.4) using a Dounce homogenizer with a type B pestle. The result of this step is to concentrate the purified peroxisomes and to change the medium to one suitable for *in vitro* import.

After fractionation of the density gradient we routinely determine the refractive indices of the gradient fractions in order to calculate their densities. This allows the investigator to check whether the gradient formation was correct. The equilibrium density of peroxisomes should be 1.23 g/cm^3 and the density of mitochondria is approximately 1.19.

2.   *Candida tropicalis*

To prepare stable peroxisomes from *C. tropicalis*, we (Small *et al.*, 1987) use a growth medium consisting of 0.3% yeast extract, 0.5% Bacto-peptone, 0.5% K_2HPO_4, 0.5% KH_2PO_4, and 0.3% Brij 35 (YPB). First, a loopful of yeast cells freshly grown on agar plates containing 0.3% malt extract is precultured for 24 hours in 85 ml of 0.3% malt extract. The cells are harvested by centrifugation at 600 g for 10 minutes (2000 rpm in the Sorvall GS-3 rotor) and are washed by dilution in distilled water and recentrifugation. Half of the cells are transferred into 1 liter of the YPB medium. Cells are grown to midlogarithmic phase (10 hours at 30°C) and are then harvested and washed.

The cells are resuspended by diluting the final wet weight fivefold in a solution which consists of 50 mM phosphate buffer pH 7.5, 0.5 M KCl, 10 mM NA_2SO_3, 0.01 M β-mercaptoethanol, and 1 mg Zymolase-100T per gram of cells. The cells are incubated for 30 minutes at 35°C with continuous shaking to digest the cell walls and form spheroplasts. The spheroplasts are diluted 10-fold in 50 mM MOPS, pH 7.2, and 0.5 M KCl, and are concentrated by pelleting them through a cushion of 5% Ficoll, 0.6 M sorbitol, 1 mM EDTA, and 2.5 mM MOPS, pH 7.2 (F/S). The concentrated cells are resuspended in 3–4 ml of F/S and are homogenized using 10 strokes with a Potter–Elvehjem homogenizer. The homogenate is centrifuged at 2000 rpm (1000 g) for 10 min in a Sorvall H6000A rotor and the supernatant is carefully removed and stored on ice. The pellet is resuspended in 5 ml of F/S, homogenized, and centrifuged as before. The supernatants are combined and centrifuged at 13,000 rpm (20,000 g) for 20 minutes (SS34 rotor) and the supernatant is carefully removed.

The 20,000 g pellet is gently resuspended with a Dounce homogenizer in 8 ml of F/S containing final concentrations of 0.7 mM of the following protease inhibitors: leupeptin, chymostatin, antipain, and pepstatin, using a Dounce homogenizer. A final low-speed centrifugation at 2000 rpm (1000 g) for 5 minutes (H6000A rotor) is carried out to remove any large aggregates in this suspension. Isopycnic centrifugation is carried out in a linear sucrose gradient [density span of 1.15–1.25 g/ml; 35–54% (w/w)]. The gradient solutions contain 2.5 mM MOPS, pH 7.2, 0.5 mM EDTA, and 0.1% ethanol. The cushion is 62% sucrose in the same medium (density of 1.30 g/ml). We generally use the Beaufay zonal rotor at 35,000 rpm (90,000 g) for 40 minutes, but again, a vertical rotor is a good alternative (see below).

Peroxisomal and mitochondrial fractions are identified as described above for rat liver. Their equilibrium densities are again approximately 1.23 and 1.19 g/ml, respectively. The peroxisomes are concentrated as before except that they are diluted and resuspended in a slightly different import buffer, which consists of 0.5 M sucrose, 2.5 mM MOPS, pH 7.2, and 0.5 mM EDTA. (see Table I).

In a typical experiment, 0.4 g cells (wet weight) are harvested from each of three 85-ml precultures. Six flasks, each containing 1 liter of YPB medium, are inoculated with 0.3 g of washed cells/liter. After 10 hours, the yield is 4 g cells/liter. This is sufficient to isolate approximately 9 mg of peroxisomal protein on the sucrose gradient. The center of the peroxisome peak is combined and concentrated, for a final yield of 4.5 mg of peroxisomes.

3. Saccharomyces cerevisiae

The preparation of peroxisomes from S. cerevisiae is generally similar to that for C. tropicalis. However, there are several important differences that are described below (Thieringer et al., 1991).

A fresh loopful of cells is precultured in 200 ml of 1% yeast extract, 2% Bacto-peptone, and 2% glucose (YPD). The total cell mass from each

TABLE I

COMPARISON OF THE in Vitro IMPORT SYSTEMS

Component	Rat liver	C. tropicalis	S. cerevisiae
Import buffer			
Buffer system	5 mM HEPES	2.5 mM MOPS	5 mM MES
pH[a]	7.4	7.2	6.0
Sucrose conc.	0.25 M	0.5 M	0.5 M
EDTA[b]	—	0.5 mM	0.5 mM
Ethanol[c]	0.1%	—	0.1%
MgCl$_2$[d]	3 mM	—	—
Protease inhibitor mix[e]	-	+	+

[a] The lower pH of the import buffer for S. cerevisiae peroxisomes is essential to maintain the stability of the organelles.

[b] We suspect that EDTA is best omitted. It was included for historical reasons. It will chelate some of the Mg^{2+} coming from the translation mix; Mg^{2+} is likely to be required for ATPase function. EDTA could be useful as a protease inhibitor, but must be omitted if thermolysin will be used subsequently for the protease protection assay.

[c] Ethanol is included routinely in most of our experiments to maintain the activity of catalase (marker enzyme for peroxisomes) and does not influence the results of the import assays.

[d] Imanaka et al. (1987) described that $MgCl_2$ stimulates the import of acyl-CoA oxidase into rat liver peroxisomes.

[e] The mix consists of the following protease inhibitors (10 mg/ml of each in DMSO): antipain, chymostatin, leupeptin, and chymostatin. Of the mix, 1 μl is added to the 36-μl import assay. We found it necessary to include these protease inhibitors in order to inhibit endogenous yeast proteases.

preculture is transferred to a flask with 500 ml of 0.3% yeast extract, 0.5% Bacto-peptone, 30 mM K_2HPO_4, 0.15% oleic acid, and 0.015% Tween 40 adjusted to pH 6.0 (YPOT). Cells are grown at 30°C for 18 hours. In order to prepare spheroplasts, the cells are first incubated for 20 minutes in 100 mM Tris-HCl, pH 7.4, 50 mM EDTA, and 10 mM β-mercaptoethanol. They are then treated with Zymolyase-100T (2 mg per gram wet weight of cells) in 20 mM potassium phosphate buffer, pH 7.4, containing 1.2 M sorbitol for 45–60 minutes at 30°C. The total yield from the six flasks is typically 24 g (wet weight) of cells.

Spheroplasts are resuspended in 1 ml per gram of cells of 5 mM MES, pH 6.0, 0.6 M sorbitol, 1 mM KCl, 0.5 mM EDTA, 0.1% ethanol, and 0.7 mM of the protease inhibitor mixture described above (MSKE medium). Preparation of a large organelle fraction is the same as for C. tropicalis except that low-speed centrifugation is carried our for 5 minutes at 3500 rpm (1500 g; SS34 rotor), high-speed centrifugation is at 15,000 rpm (25,000 g) for 15 minutes (SS34 rotor), and MSKE is used throughout.

Isopycnic centrifugation is carried out using a linear Nycodenz gradient (15–36%) containing a uniform concentration of 0.25 M sucrose in 5 mM MES, pH 6, 1 mM EDTA, and 0.1% ethanol. The gradient is underlaid with a 1-ml cushion of 42% Nycodenz in the same medium. Centrifugation is performed either with a Sorvall Vt865B (using two 17-ml gradients) or a Beckman vti50 (using one 33 ml gradient and a balance tube) vertical rotor at 35,000 rpm (100,000 g, slow acceleration mode) for 60 minutes. The equilibrium density of these *S. cerevisiae* peroxisomes is approximately 1.17 g/ml and the density of the mitochondria is about 1.11. The total yield of peroxisomes from the two gradients combined is typically 5 mg.

Peroxisomes are concentrated as above except that the import buffer for *S. cerevisiae* consists of 0.5 M sucrose, 1 mM EDTA, 5 mM MES, and 0.1% ethanol, pH 6.0.

C.   Protocol for Density Gradient Centrifugation

1.   OPERATING THE BEAUFAY ROTOR

The operating of this custom-built rotor is described by Leighton *et al.* (1968).

2.   OPERATING THE VERTICAL ROTOR

The following protocol is based on the use of 15–36% Nycodenz gradients in 33-ml Easy-seal centrifuge tubes (25 × 89 mm; Sarstedt Inc.) in the vti50 vertical rotor (Beckman Instruments) for the isolation of peroxisomes from *S. cerevisiae*. However, this protocol can be adapted for the use of sucrose gradients as described for the isolation of *C. tropicalis* peroxisomes. Devices that can be used to unload various Beckman vertical rotor tubes have been manufactured by Sarstedt Inc. (Easy-seal fractionation assembly) and by Accurate Chemical and Scientific Corp. (gradient unloader). We think that these systems, in which the samples are collected from the top of the density gradient is superior to the practice of recovering the fractions from the bottom by tube puncture.

Protocol.   Place a 1-ml cushion of 42% Nycodenz solution onto the bottom of the centrifuge tube. This is done more conveniently with a syringe to which a long needle is attached. Carefully overlay the cushion with a 29-ml gradient of 15–36% Nycodenz prepared in a gradient-forming device consisting of two equal-diameter cylinders, connected at the bottom, with a magnetic stirrer in the mixing chamber (BioRad or Sigma). We use an Auto-densi Flow II loading device (Buchler Instruments), equipped with a peristaltic pump. With the help of a sensor on this device the gradient solution is continuously added to the surface of the forming density gradient, thereby minimizing the chance of turbulences. The gradients are then cooled and can be stored up to approximately 3 hours at 4°C.

Overlay the gradients with 3 ml of the sample (25,000 *g* pellet fraction). Should a single gradient be used, balance the rotor with another centrifuge tube, containing the gradient solutions and homogenization buffer (MSKE medium). The tubes are sealed and placed into the precooled rotor. Care should be taken not to disturb the gradient during loading of the sample and handling of the tube. Place the rotor gently into the centrifuge.

Run the centrifugation at 35,000 rpm (100,000 *g*) for 60 minutes at 4°C, using the slow acceleration/deceleration mode. If the centrifuge is equipped with an ω^2 dt-integrator, set it to 4.24×10^{10} rad^2/second.

After the centrifugation, remove the tube from the rotor and attach the fractionation assembly. Collect the fractions from the top of the centrifuge tube to the bottom by slowly injecting a dense solution (65% sucrose or fluorocarbon) into the bottom of the tube with the help of a peristaltic pump. The volume of the fractions should be approximately 2 ml. The fraction volumes may be determined accurately by using preweighed tubes, weighing the fractions, and dividing by the densities, measured with a refractometer (Leighton *et al.*, 1968).

III. Preparation of Proteins

The proteins tested in our *in vitro* import experiments are radioactively labeled with ^{35}S during *in vitro* translation. We translate either isolated total cell RNA or a single RNA species obtained by *in vitro* transcription of a cloned gene (Melton *et al.*, 1984). Initially, a rabbit reticulocyte lysate system (Pelham and Jackson, 1976) was used for translations of rat liver and *C. tropicalis* RNA. Later, we found that several mRNAs from *C. tropicalis* were translated more efficiently in a wheat germ system (Erickson and Blobel, 1983), and we now use this system routinely for *C. tropicalis* and *S. cerevisiae* mRNAs. However, the choice of the translation system does not influence the result of the import assay (binding and protease protection), at least for import of acyl-CoA oxidase into *C. tropicalis* peroxisomes (Small *et al.*, 1988).

A. Sources of RNA

1. ISOLATION OF TOTAL RNA

Clofibrate is administered to rats (5 g/kg chow) for two weeks; this induces mRNAs encoding peroxisomal β-oxidation proteins 10- to 50-fold (Lazarow and Fujiki, 1985). Total RNA (of which 1–2% is mRNA) is extracted from the liver with guanidinium thiocyanate/guanidinium chloride (Chirgwin *et al.*, 1979). Several peroxisomal proteins are easily recognized among the major translation products of this RNA in a fluorogram of an SDS–polyacrylamide gel. In *C. tropicalis*, mRNAs encoding peroxisomal proteins are induced by growth on oleic acid. Total RNA is isolated by the guanidinium thiocyanate method. The yield is typically 500 μg of RNA per gram wet weight of cells.

2. *In Vitro* TRANSCRIPTION

We use *in vitro* transcription to synthesize a single mRNA species. The coding region for the protein or protein fragment of interest is cloned into the multiple cloning site of a pGEM vector (Promega Biotech) that contains the SP6 and T7 promoters in opposite orientations, flanking the multiple cloning site. Plasmid DNA is prepared by one of the standard miniprep or large-scale procedures. The DNA is further purified by phenol/chloroform extraction and ethanol precipitation (Sambrook *et al.*, 1989). The plasmid is linearized downstream of the coding region by digestion with an appropriate restriction enzyme. This allows the synthesis of "run off" transcripts that are free of vector sequences. By cutting the plasmid within the cloned coding region, shorter RNAs coding for amino-terminal portions of the protein may be produced in the *in vitro* transcription reaction. The DNA is transcribed with either SP6 or T7 polymerase (depending on the orientation of the gene), and the mRNA is simultaneously capped as described by Promega Biotech based on the method of Melton *et al.* (1984). We have obtained much better yields of mRNA with SP6 than with T7 polymerase, so it is preferable to insert the gene into the multiple cloning site such that the SP6 promoter may be used.

B.   Translation of RNA *in Vitro*

We use the rabbit reticulocyte lysate system for the translation of rat mRNAs essentially as described by Clemens (1984). The translation system includes [^{35}S]methionine or Tran^{35}S label (ICN). The lysate is pretreated with ribonuclease to digest endogenous mRNAs. To increase the translation efficiency, the RNA is heated at 65°C for 5 minutes and then quickly cooled in an ice-water bath. This increases the synthesis of several peroxisome proteins 10- to 100-fold; total protein synthesis doubles (Mortenson *et al.*, 1984).

To remove polyribosomes after *in vitro* translation, the translation mixture is centrifuged in a ti50 rotor (Beckman Instruments) at 40,000 rpm (150,000 *g*) for 1 hour.

Yeast mRNA is translated either in a rabbit reticulocyte lysate system as described for rat liver RNA or in a wheat germ extract (Erickson and Blobel, 1983).

Several considerations influence the choice of the *in vitro* translation system. One important factor is the efficiency of mRNA translation. This is influenced by, among other things, the tRNA population provided, specifically by the extent of similarity in the codon usage of the organism from which the translation system is derived and the codon usage in the gene being expressed. We sometimes add bovine liver tRNA to the reticulocyte lysate when synthesizing rat liver proteins. Another criterion is the evolutionary distance between the source of the *in vitro* translation system lysate and the source of the peroxisomes used in the import assay. The lysate may contain soluble factors that influence the import into peroxisomes, e.g., by affecting protein folding. In import experiments with *S. cerevisiae* peroxisomes, a good alternative to the wheat germ system might be a cell-free translation system isolated from *S. cerevisiae* (Tuite and Plesset, 1986).

IV.   Import Assay

The design of the import assay is shown schematically in Fig. 1.

A.   Import Conditions

For *in vitro* import, freshly prepared concentrated peroxisomes are incubated with the translation product. The composition of a typical import mixture is shown below. The differences among our three import systems (using peroxisomes isolated from rat liver, *C. tropicalis*, and *S. cerevisiae*) are summarized in Table I.

1. *In vitro* import

Wheat germ *in vitro* translation system
^{35}S-labeled translation product
Concentrated peroxisomes
Import buffer containing 50 mM KCl

Incubation Mix

Incubation: 0-120 min, 26°C

2. Protease protection assay

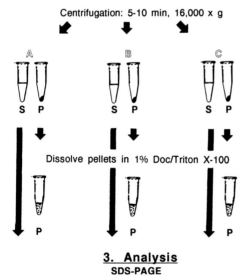

-	+	+	Protease
-	-	+	1% DOC/Triton X-100

Incubation 15-30 min, 0°C

Dilution with import buffer
containing protease inhibitor and 50 mM KCl

Centrifugation: 5-10 min, 16,000 x g

Dissolve pellets in 1% Doc/Triton X-100

3. Analysis
SDS-PAGE
Fluorography
Densitrometric scanning

FIG. 1. Schematic design of the *in vitro* assay of protein import into peroxisomes.

Protocol.   In vitro import assay mixture.[1]

12 μl translation mix[2]
  4 μl sucrose: 1 M for rat liver, 2 M for yeast peroxisomes[3]
16 μl concentrated peroxisomes in import buffer (60–240 μg of protein)[4]
  4 μl 500 mM KCl in import buffer

Incubate at 26°C for 30 minutes.[5] Cool on ice. Several replicates of this import assay mixture are prepared and used as controls. These are incubated at 0°C, or not incubated at all, or the peroxisomes are omitted or replaced with mitochondria. See Section V.

B.   Protease Protection Assay

The amount of gene product associated with the peroxisomes is assayed by pelleting the peroxisomes from an aliquot of the import mix and analyzing the pellet by SDS–PAGE and fluorography. The amount of gene product translocated inside the peroxisomes is determined by a protease protection assay as follows: Protease is added to digest all ^{35}S-labeled translation product that remains on the outside of the peroxisomes. The protease is selected such that the translation product in the import medium is completely digested, but the peroxisome membrane remains an intact barrier. Thus the translation product inside the organelle is protected from proteolysis. The choice of protease for import studies using yeast peroxisomes is more limited than for those using rat liver peroxisomes due to the necessity of adding protease inhibitors in the yeast import assays. Obviously, the subsequently added protease must be unaffected by the inhibitors already present.

---

[1] In earlier experiments, larger volumes (100–140 μl) have been used. We recommend the reduced reaction volume of 36 μl as shown. The smaller volume works just as well and more import assays can be carried out after one preparation of fresh peroxisomes.

[2] The translation mix contains all the constituents of the *in vitro* translation procedure, including ATP (approximately 1.2 mM final concentration in the import assay). In both the reticulocyte lysate and wheat germ systems. The mix can be frozen in aliquots at -80°C after translation and before import.

[3] For import experiments with rat liver peroxisomes, the final sucrose concentration of the translation mix is adjusted to 0.25 M (by addition of 4 μl of 1 M sucrose). In the case of yeast peroxisome import, we found that adjusting the translation mixture to 0.5 M sucrose gave better import results.

[4] Approximately 2 mg of peroxisomal protein per milliliter of import assay mixture appeared to be optimal for the import of acyl-CoA oxidase into rat liver peroxisomes (Imanaka *et al.*, 1987). Larger amounts of peroxisomal protein have proved useful for the import of catalase A into yeast peroxisomes.

[5] A temperature of 30°C and a longer incubation time of 2 hours were better for catalase A import into *C. tropicalis* peroxisomes.

Detergent is added to one aliquot of the import assay to disrupt the peroxisome membrane in order to demonstrate that the protease is capable of digesting all of the translation product when the product is not protected by the membrane barrier. We have used a mixture of deoxycholate and Triton X-100; however, we believe that the nonionic detergent alone would suffice and is preferable.

Protocol. Divide the mixture into three aliquots of 10 μl (A, B, and C), and save the remaining 6 μl of the import mix as a reference for analysis (see below).

Add: to aliquot A: 4 μl import buffer

Add: to aliquot B: 4 μl import buffer containing protease (See Table II)

Add: to aliquot C: 4 μl import buffer containing protease plus sodium deoxycholate and Triton X-100 to give 1% final concentrations

Incubate the samples on ice for 15–30 minutes.

Terminate the reaction by diluting the samples ninefold in import buffer containing the appropriate protease inhibitor (final volume, 125 μl).[6]

Centrifuge the sample in a refrigerated microfuge (16,000 g) for 5–10 minutes.

Transfer the supernatants to a fresh tube and resuspend the peroxisomal pellets in 25 μl of import buffer containing 1% sodium deoxycholate, 1% Triton X-100, and the appropriate protease inhibitor.

The samples are then analyzed by SDS–PAGE and fluorography (see below).

TABLE II

PROTEASES SUCCESSFULLY USED

Factor	Rat liver	C. tropicalis		S. cerevisiae
Protease	Proteinase K	Proteinase K	Thermolysin	Proteinase K
Concentration	2 μg/140 μl	2–4 μg/14 μl	0.6 μg/14 μl	5 μg/14 μl
pH	7.4	7.2	7.2	6.0
When	Digestion after 10× dilution of import assay mixture	Digestion before 10× dilution of import assay mixture		
Inhibitor added afterward	PMSF (1 mM)	PMSF (1 mM)	EDTA (1 mM)	PMSF (1 mM)

[6] For historical reasons, the rat liver import mixture was diluted prior to protease digestion. Probably it could be diluted after digestion, as for yeast peroxisomes. This difference has not been tested.

C.   Quantitation of Binding and Import

Aliquots of the supernatant and pellet fractions obtained from the protease protection assay (representing equal amounts of the input *in vitro* translation products) are analyzed by SDS–PAGE, fluorography, and densitometric scanning. In typical experiments, 4 μl of the solubilized pellets and 20 μl of the supernatants are denatured in SDS–PAGE sample buffer and subjected to gel electrophoresis. To quantitate the input radioactivity, an equivalent aliquot of the import mix (1.6 μl) is loaded on a separate lane. Fluorography is performed on preflashed films (Laskey and Mills, 1975); the exposure time varies with the translation efficiency of the mRNA, but is typically 1 day. Generally, more than one exposure of

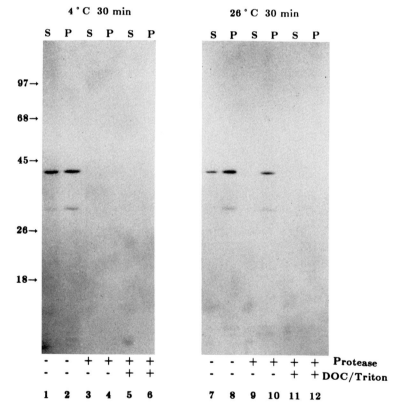

FIG. 2.   Import of a carboxy-terminal fragment of acyl-CoA oxidase into *C. tropicalis* peroxisomes *in vitro*. S, Supernatant; P, pellet. Right panel, import at 26°C; left panel, 4°C control. From Small and Lazarow (1987).

each fluorogram should be scanned to verify that the fluorographic band intensities are in the linear response range of the film. A typical experiment is illustrated in Fig. 2.

An alternative to fluorography is to excise the radioactive band from the SDS gel and determine the radioactivity in a liquid scintillation counter. When a complex mixture of translation products is used in an import assay, e.g., when total cell mRNA is translated, the protein of interest may be identified unambiguously by immunoprecipitation prior to gel electrophoresis. For example, this was done after translation of total RNA from oleate-induced *C. tropicalis* (Small *et al.*, 1987).

The radioactive protein recovered in the pellet from aliquot A (no protease) represents protein associated with peroxisomes, both bound to the outside of the peroxisomal membrane and translocated inside. The radioactive protein in the pellet from aliquot B (protease) is presumed to have been translocated into peroxisomes (see Section V). The difference between pellets A and B is translation products bound to the membrane. There should be no significant pellet from aliquot C (protease plus detergents) and there should be no intact radioactive protein in supernatants B and C.

V. Controls

The import of peroxisomal proteins is in general not accompanied by posttranslational modifications of the newly synthesized proteins. For example, cleavage of topogenic peptides or addition of sugar residues has not been observed, and thus protein import cannot be assessed by analysis of molecular size differences by SDS–PAGE alone. Analysis of import relies on the protease protection assay, in which the organelle membrane protects the imported ^{35}S-labeled proteins from the action of externally added protease.

It is necessary to control for the possibility that the ^{35}S-labeled translation product could become protease resistant for reasons other than its translocation into peroxisomes. For example, the translation product could aggregate, or could stick on the tube wall, or could bind to the organelle membrane (either specifically to a receptor or nonspecifically) such that its protease susceptibility is reduced. These events could give false-positive indications of import. Therefore, several controls are strongly recommended and are described in the next paragraphs. After that, we mention controls that are useful when adapting the import assay for your favorite gene product. Genetic engineering of protein fragments increases the risk of aggregation or sticking.

A. Import in the Absence of Organelles (Mock Import)

We observed that some proteins and protein fragments form aggregates when translated *in vitro*, which cosediment with organelles. Omitting organelles from the import mixture is an easy way to check for this. When the organelles are replaced with import buffer, translation products should not be found in the pellet.

Aggregation may also be tested by measuring the approximate sedimentation rate of the translation product in sucrose gradients (in the absence of organelles). Aggregates will sediment much more rapidly than protein monomers.

Some aggregates may not be dissolved by deoxycholate and Triton X-100. The peroxisome pellet, dissolved in these detergents at the end of the import assay, may be recentrifuged prior to adding SDS. The supernatant is then analyzed by SDS–PAGE and compared with peroxisome pellets that did not get the extra centrifugation step. A loss of radioactive band intensity in SDS–PAGE/fluorography after centrifugation indicates the existence of such aggregates. In a couple of cases, we found that certain protein fragments nearly disappeared with such centrifugation. In one case, we observed that newly synthesized catalase A of *S. cerevisiae* adhered to the wall of the microfuge tube in which the import assay was carried out. This was corrected by siliconizing the tubes.

B. Temperature Control

In our experience, peroxisomal protein translocation is a temperature-dependent event. Import is seen only at higher temperatures (typically 26°C), not at 0°C. This temperature dependence allows a dissection of the translocation process into two steps: (1) binding of a protein to the peroxisomal membrane (receptor) that occurs at 0°C, and (2) translocation of a protein into the organelle at higher temperatures (import). This temperature dependence likely reflects the requirements of the ATP-utilizing translocation machinery. Altered fluidity of the peroxisomal membrane could also be involved. Regardless of the reason, if the translocation product is bound to the outside of the peroxisome membrane (presumably to a receptor) at 0°C, this permits one to verify its protease susceptibility when it adheres to a membrane. Therefore, a control import assay is routinely carried out for 30–120 minutes at 0°C. Three aliquots are analyzed as described above for the regular 26°C assay. If the translation product sediments with the peroxisomes in the absence of protease (aliquot A), but is digested in the presence of protease (aliquot B, no detergent) and thus disappears from both pellet and supernatant, then this implies that (1) the translocation product was bound to the outside of the peroxisome and (2) such binding alone does not give artifactual protease resistance. If this control works, it greatly increases our confidence that any protease resistance observed at 26°C is due to real translocation, not to mere binding.

C.   Zero-Time Control and Time Dependence

It is important to distinguish between a time-dependent translocation event and a spontaneous aggregation or binding of protein to the organelle membrane. Therefore, we carry out a zero-time control. After adding the translation products to the peroxisomes in the import assay at 0°C, one aliquot of the incubation mixture is immediately diluted and the peroxisomes are pelleted by centrifugation. The other two aliquots immediately receive protease plus or minus detergent as described in Section IV, B. Because there has been no time for translocation to occur, no translation product should be associated with peroxisomes, except perhaps for some very rapid binding event.

In addition to zero-time controls, one may establish the kinetics of *in vitro* protein uptake by taking additional time points during incubation. The time range used in our laboratory is between 0 and 120 minutes of incubation: import of acyl-CoA oxidase ceased after 30 minutes; import of *S. cerevisiae* catalase A continued longer.

D.   Control Organelles

To demonstrate that the proteins being studied are imported specifically into peroxisomes, it is necessary to use a different cell organelle as a control. We routinely isolate mitochondria for this purpose. The mitochondrial peak fraction from a density gradient is diluted and concentrated in the same fashion as described for peroxisomes. The protein concentration of mitochondria is adjusted to the same protein concentration used for peroxisomes in the import assay. Care should be taken to minimize the contamination of the control mitochondrial fraction with peroxisomes. For the *S. cerevisiae* import system we use mitochondria isolated on a separate gradient from cells that were grown on glucose-containing medium, which represses peroxisome formation.

When analyzing protein fragments engineered by cutting or splicing genes, one should keep in mind the possibility that *in vitro* translation products may contain mitochondrial-targeting signals that were cryptic in the normal, full-length protein (Eilers and Schatz, 1986).

E. Other Experiments

1. ATP DEPLETION

It has been shown that translocation of acyl-CoA oxidase into rat liver peroxisomes requires the hydrolysis of ATP (Imanaka *et al.*, 1987). This has also been observed for the *C. tropicalis* import system (G. Small, unpublished results), and it seems reasonable to assume that ATP hydrolysis is part of the translocation process in all systems. ATP is provided to the import assay from the *in vitro* translation system, and may be depleted by two methods: (1) treatment of the translation mixture with apyrase or hexokinase plus glucose, or (2) centrifugation of the translation mixture through Sephadex G-25. In the absence of ATP, rat acyl-CoA oxidase bound efficiently to peroxisomes but remained protease digestible. This demonstrated that ATP is required for the translocation step. Moreover, it served as a further control, demonstrating that binding *per se* does not render the translocation product resistant to digestion.

2. REPORTER PROTEINS

For protein import studies, much work has been carried out that involved the use of hybrid proteins. In the case of peroxisomes, dihydrofolate reductase (DHFR) and chloramphenicol acetyltransferase (CAT) have proved useful as reporter proteins. DHFR was used for acyl-CoA oxidase from *C. tropicalis* (Small *et al.*, 1988) and rat liver (Miyazawa *et al.*, 1989), and CAT was used for acyl-CoA oxidase from rat liver (Miyazawa *et al.*, 1989).

The use of DHFR should allow further dissection of the import process. The tertiary structure of DHFR can be stabilized by addition of methotrexate, a folate analog (Eilers and Schatz, 1986). In the case of mitochondrial import studies, this alteration of the structure of hybrid proteins prevents transport into the organelle (Hurt and Schatz, 1987). One drawback to the use of DHFR is that the protein itself is quite protease resistant. We found that DHFR protease sensitivity increased whenever a DHFR–hybrid protein was constructed, indicating a conformational change of the DHFR domain. Vestweber and Schatz (1988) constructed a modified, more protease-sensitive DHFR protein by introducing a mutation into the DHFR structural gene. It is worth mentioning that (cytosolic) DHFR contains a cryptic targeting sequence for mitochondria within the first 85 amino-terminal amino acids (Hurt and Schatz, 1987).

VI.   Adapting the Peroxisome Import Assay to Your Favorite Gene Product

In order to adapt the import assays described here to study another peroxisomal protein, it is necessary to verify that the protease protection assay functions for the new protein, and, if not, to find suitable conditions. The selection of the protease is based on two criteria: (1) the ^{35}S-labeled protein has to be digestible by the protease and (2) the peroxisomes have to remain intact during the protease treatment. Protease type and concentration should be selected carefully in control experiments. The *in vitro* translation mixture containing your favorite gene product is mixed at 0°C with peroxisomes. Aliquots are incubated on ice with a variety of candidate proteases at various concentrations as described above for the protease protection assay. Digestion of the translation product is easily assessed by SDS–PAGE and fluorography. The intactness of peroxisomes can be tested by two different methods, an assay of catalase latency and analysis of peroxisomal proteins by SDS–PAGE.

1.   ANALYSIS OF PEROXISOMAL PROTEIN PATTERNS AFTER THE PROTEASE PROTECTION ASSAY

The peroxisomes are pelleted after protease treatment (in the presence of an inhibitor of the protease). The peroxisomes are analyzed by SDS–PAGE and Coomassie blue or silver staining. It is fairly easy to recognize the specific pattern of the peroxisomal proteins. There should be no shift in size or intensity of the peroxisomal protein bands after protease treatment compared to the bands of the peroxisomal pellet obtained after incubation without protease treatment.

2.   CATALASE LATENCY AFTER THE PROTEASE PROTECTION ASSAY

The intactness of the peroxisome may be assessed by measuring the latency of catalase. Because of the large activity of catalase, hydrogen peroxide entering peroxisomes is effectively decomposed by the enzyme at the periphery of the organelle. The catalase in the center of the peroxisome sees little or no substrate and hence is latent. Its activity appears when it escapes from the peroxisome, either because detergent is added in the assay, or, for instance, if the peroxisome membrane is damaged by protease such that catalase leaks out. Thus, in order to determine the intactness of the peroxisome, the catalase assay is carried out in duplicate, with the addition or omission of 0.1% Triton X-100 (Lazarow *et al.*, 1988).

Catalase latency is defined as the percentage of the total catalase activity that is not assayable when the detergent is omitted. The catalase latencies of our purified peroxisomes are typically between 70 and 90%. The latency should not decrease after protease treatment.

VII.   Comparison to Other Systems

There are alternative methods of *in vitro* analysis that may be equally dependable. Miyazawa *et al.* (1989) have also measured the import of acyl-CoA oxidase into rat liver peroxisomes. The major difference in their assay and ours is the preparation of the organelles. When purified peroxisomes are used, they do not attempt to remove the gradient media or to concentrate the peroxisomes before carrying out import. Alternatively, they use a postnuclear supernatant fraction of rat liver. In some cases they prepare a light mitochondrial fraction for the import assay; following import and protease treatment they then separate the organelles by isopycnic centrifugation. The peroxisomal pellet that is recovered at the bottom of the gradient is analyzed for the imported polypeptides. This group has successfully adapted their assay to show the *in vitro* insertion of a peroxisomal membrane protein (Fujiki *et al.*, 1989).

An *in vitro* protein import assay has also been described for glycosomes, peroxisome-like organelles found in trypanosomes and *Leishmania* (Dovey *et al.*, 1988). In this case, the import assay was performed at 37°C for 1 hour and was followed by protease digestion. Specific import of glycosomal phosphoglycerate kinase was reported. However, some doubt has been raised by Borst and Swinkels concerning the interpretation of these results, because the association of this protein with peroxisomes was not dependent on time, temperature, or ATP hydrolysis (Borst, 1989).

VIII.   Conclusions

We have spent much time developing and testing the systems outlined above, and are confident that measurements of *in vitro* import are reliable if the indicated precautions are taken. In two instances, the nature of peroxisomal targeting information has been compared *in vitro* and *in vivo*. Rat liver acyl-CoA oxidase has the carboxy-terminal Ser-Lys-Leu tripeptide shown *in vivo* by Gould *et al.* (1989) to be important for peroxisome targeting. Miyazawa *et al.* (1989) studied the same protein by *in vitro* import, and likewise found the same tripeptide to be critical. The same enzyme in the yeast, *C. tropicalis*, is targeted differently. It lacks a carboxy-terminal Ser-Lys-Leu. Small *et al.* (1988) showed by *in vitro* import studies that it contains redundant targeting sequences, one near the amino-terminus and one in the middle. Kamiryo *et al.* (1989) obtained similar results on the *C. tropicalis* oxidase with studies *in vivo*. Thus, there is a satisfying consistency in the data obtained so far *in vitro* and *in vivo*.

REFERENCES

Alexson, S. E. H., Fujiki, Y., Shio, H., and Lazarow, P. B. (1985). *J. Cell Biol.* **101**, 294–305.
Borst, P. (1989). *Biochim. Biophys. Acta* **1008**, 1–13.
Brul, S., Westerveld, A., Strijland, A., Wanders, R. J. A., Schram, A. W., Heymans, H. S. A., Schutgens, R. B. H., van den Bosch, H., and Tager, J. M. (1988). *J. Clin. Invest.* **81**, 1710.
Chirgwin, J. M., Przybyla, A. E., MacDonald, R. J., and Rutter, W. J. (1979). *Biochemistry* **18**, 5294–5299.
Clemens, M. J. (1984). *In* "Transcription and Translation: A Practical Approach" (B. D. Hames and S. Z. Higgins, eds.), pp. 231–270, IRL Press, Oxford, England.
Dovey, H. F., Parsons, M., and Wang, C. C. (1988). *Proc. Natl. Acad. Sci. U.S.A.* **85**, 2598–2602.
Eilers, M., and Schatz, G. (1986). *Nature (London)* **322**, 228–232.
Erickson, A. H., and Blobel, G. (1983). *In* "Methods in Enzymology" (S. Fleischer and B. Fleischer, eds.), Vol. **96**, pp. 38–50. Academic Press, New York.
Fujiki, Y., Rachubinski, R., Zentella-Dehesa, A., and Lazarow, P. B. (1986). *J. Biol. Chem.* **261**, 15787–15793.
Fujiki, Y., Kasuya, I., and Mori, H. (1989). *Agric. Biol. Chem.* **53**, 591–592.
Fukui, S., and Tanaka, A. (1979). *J. Appl. Biochem.* **1**, 171–201.
Gould, S. J., Keller, G.-A., and Subramani, S. (1987). *J. Cell Biol.* **105**, 2923–2931.
Gould, S. J., Keller, G.-A., Hosken, N., Wilkinson, J., and Subramani, S. (1989). *J. Cell Biol.* **108**, 1657–1664.
Gould, S. J., Keller, G.-A., Schneider, M., Howell, S. H., Garrard, L. J., Goodman, J. M., Distel, B., Tabak, H., and Subramani, S. (1990). *EMBO J.* **9**, 85–90.
Hurt, E. C., and Schatz, G. (1987). *Nature (London)* **325**, 499–503.
Imanaka, T., Small, G. M., and Lazarow, P. B. (1987). *J. Cell Biol.* **105**, 2915–2922.
Kamiryo, T., Sakasegawa, Y., and Tan, H. (1989). *Agric. Biol. Chem.* **53**, 179–186.
Laskey, R. A., and Mills, A. D. (1975). *Eur. J. Biochem.* **56**, 335–341.
Lazarow, P. B. (1989). *Curr. Opin. Cell Biol.* **1**, 630–634.
Lazarow, P. B., and Fujiki, Y. (1985). *Annu. Rev. Cell Biol.* **1**, 489–530.
Lazarow, P. B., and Moser, H. W. (1989). *In* "The Metabolic Basis of Inherited Disease" C. R. Scriver, A. L. Beaudet, W. S. Sly and D. Valle, eds.), 6th ed., Vol. **II**, pp. 1479–1508. McGraw-Hill, New York.
Lazarow, P. B., Small, G. M., Santos, M., Shio, H., Moser, A., Moser, H., Esterman, A., Black, V., and Dancis, J. (1988). *Pediatr. Res.* **24**, 63–67.

Leighton, F., Poole, B., Beaufay, H., Baudhuin, P., Coffey, J. W., Fowler, S., and de Duve, C. (1968). *J. Cell Biol.* **37**, 482–513.

Melton, D. A., Krieg, P. A., Rebagliati, M. R., Maniatis, T., Zinn, K., and Green, M. R. (1984). *Nucleic Acids Res.* **12**, 7035–7056.

Miyazawa, S., Osumi, T., Hashimoto, T., Ohno, K., Miura, S., and Fujiki, Y. (1989). *Mol. Cell. Biol.* **9**, 83–91.

Mortensen, R. M., Rachubinski, R. A., Fujiki, Y., and Lazarow, P. B. (1984). *Biochem. J.* **223**, 547–550.

Osumi, T., and Fujiki, Y. (1990). *BioEssays* **12**, 217–222.

Pelham, H. R. B., and Jackson, R. J. (1976). *Eur. J. Biochem.* **67**, 247–256.

Roscher, A., Hoefler, S., Hoefler, G., Paschke, E., Patlauf, F., Moser, A., and Moser, H. (1989). *Pediatr. Res.* **26**, 67–72.

Sambrook, J., Fritsch, E. F., and Maniatis, T. (1989). "Molecular Cloning: A Laboratory Manual," 2nd ed., Vols. 1–3. Cold Spring Harbor Lab., Cold Spring Harbor, New York.

Santos, M. J., Imanaka, T., Shio, H., and Lazarow, P. B. (1988a). *J. Biol. Chem.* **263**, 10502–10509.

Santos, M. J., Imanaka, T., Shio, H., Small, G. M., and Lazarow, P. B. (1988b). *Science* **239**, 1536–1538.

Small, G. M., and Lazarow, P. B. (1987). *J. Cell Biol.* **105**, 247–250.

Small, G. M., and Lewin, A. S. (1990). *Biochem. Soc. Trans.* **18**, 85–87.

Small, G. M., Imanaka, T., Shio, H., and Lazarow, P. B. (1987). *Mol. Cell. Biol.* **7**, 1848–1855.

Small, G. M., Szabo, L. J., and Lazarow, P. B. (1988). *EMBO J.* **7**, 1167–1173.

Thieringer, R., Shio, H., Han, Y., Cohen, G., and Lazarow, P. B. (1991). *Mol. Cell. Biol.* **11** 510–522.

Tuite, M. F., and Plesset, J. (1986). *Yeast* **2**, 35–52.

Veenhuis, M., van Dijken, J. P., and Harder, W. (1983). *Microbi. Physiol.* **24**, 1–82.

Veenhuis, M., Mateblowski, M., Kunau, W. H., and Harder, W. (1987). *Yeast* **3**, 77–84.

Vestweber, D., and Schatz, G. (1988). *EMBO J.* **7**, 1147–1151.

Chapter 15

In Vitro Reconstitution of Protein Transport into Chloroplasts

SHARYN E. PERRY, HSOU-MIN LI, AND KENNETH KEEGSTRA

Department of Botany
University of Wisconsin
Madison, Wisconsin 53706

I. Introduction

In vitro reconstitution of protein transport into chloroplasts requires several different steps. The first is production of precursor proteins. This is normally accomplished with an *in vitro* translation system, but other

327

methods are also possible. The second step is isolation and purification of intact chloroplasts. Next, the precursor proteins and chloroplasts are incubated together under conditions that will allow transport to occur. The transport reaction is terminated by separating the chloroplasts, containing imported proteins, from the reaction mixture, containing residual precursors. If desired, it is possible to fractionate the recovered chloroplasts to determine which chloroplastic compartment contains the imported protein. Finally, the extent of transport is determined by measuring the amount of mature protein that has accumulated inside the chloroplasts.

Transport has been reconstituted with many different precursors and with chloroplasts from several different species (Keegstra, 1989; Keegstra *et al.*, 1989; Mishkind *et al.*, 1987; Schmidt and Mishkind, 1986). What is described herein are the procedures that we use to study the transport of precursors into pea chloroplasts (Cline *et al.*, 1985; Lubben and Keegstra, 1986; Olsen *et al.*, 1989; Smeekens *et al.*, 1986; Theg *et al.*, 1989).

II.   Preparation of Precursor Proteins

An *in vitro* translation system is commonly used for the production of precursor proteins. We generally use a translation system isolated from wheat germ, but a rabbit reticulocyte translation system also works (Mishkind *et al.*, 1987). The source of RNA for directing the synthesis of precursors can be either poly(A^+) RNA isolated from plant tissues or an *in vitro*-synthesized RNA produced by transcription of a cloned gene. Translation of poly(A^+) RNA yields a mixture of precursors as well as other proteins, making quantitative analysis of the products more difficult. Translation of an *in vitro* transcript derived from a cloned gene yields a single precursor protein that is radiochemically pure. This procedure has many advantages that make it the method of choice for most studies. In either case, it is important to remember that the newly synthesized proteins are present in very small quantities compared to the large amount of protein present in the wheat germ extract.

Future studies will be greatly enhanced by the ability to produce precursor proteins by expression in bacterial or yeast cells. Such expression systems allow the production of significant quantities of precursor proteins, which can then be purified. Some success with this approach has already been reported (Pilon *et al.*, 1990), but one major limitation at the present time is the inability to keep the precursors in a soluble state. Solutions containing urea or other detergents are required to solubilize the precursors (Pilon *et al.*, 1990; Waegemann *et al.*, 1990). After removal of the urea, the precursors have limited ability to be transported and one group has reported that other factors are required (Waegemann *et al.*, 1990).

A. *In Vitro* Transcription of Cloned Precursor Genes

The cloned precursor gene (without any introns) must be present in a vector with SP6 or some other suitable promoter that allows transcription of the insert using commercially available RNA polymerases. The plasmid should first be linearized by cutting with an appropriate restriction enzyme that will cut only in the 3' flanking region of the insert. The linearized plasmid should be extracted with phenol:chloroform:isoamyl alcohol (25:24:1), precipitated, and resuspended at 1 mg/ml. Hydroxyquinoline is added to the phenol (to 0.1%), which then must be equilibrated to a pH greater than 7.8 before use (Sambrook *et al.*, 1989).

The procedure described below is used in our laboratory for conducting transcription reactions. It is slightly modified from the procedure described by Krieg and Melton (1987). Capping of the transcript is accomplished by lowering the GTP concentration [relative to the other nucleoside triphosphates (NTPs)] and including a cap structure (diguanosine triphosphate, GpppG) in the reaction mixture. This allows the cap structure to compete with GTP during initiation of transcription.

In a small microfuge tube, mix together 10 μl 5× transcription buffer (200 mM Tris, pH 7.5, at 37°C, 30 mM MgCl$_2$, 10 mM spermidine), 10 μl rNTP stock (2.5 mM each UTP, CTP, and ATP in 50 mM Tris, adjusted to a pH near 7.0 by addition of Na carbonate), 1 μl 2.5 mM GTP stock in 50 mM Tris, pH 7.0, 5 μl 5 mM GpppG in water, 5 μl 0.1 M dithiothreitol (DTT; made fresh), 5 μl 0.5 mg/ml bovine serum albumin (BSA), 5 μg linearized DNA, 0.5–1 μl RNasin (Promega), and water to a final volume of 49 μl. Add 1 μl SP6 polymerase (10–20 U) to initiate the transcription reaction. Incubate for 30 minutes at 40°C, add an additional 1 μl of GTP stock, and incubate for an additional 30 minutes at 40°C. Precipitate the RNA with 2.5 volumes of ethanol plus 0.1 volume of 3 M potassium acetate, 15–30 minutes on ice. Always use acetate salts because KCl or NaCl may inhibit translation of the RNA. The RNA can be extracted with phenol:chloroform:isoamyl alcohol before precipitation, but our experience indicates that this extraction step is not necessary. Pellet the RNA by centrifugation for 15–20 minutes, at top speed, in a microfuge. Remove as much of the supernatant as possible. Wash the pellet twice with 75% ethanol. Dry the pellet completely (Speedvac, without heat, works best). Dissolve the dry pellet in 10–20 μl water containing 10 mM DTT and some RNasin. These RNA preparations can be stored for 1–2 months at −80°C. This RNA stock is added to the *in vitro* translation system to 10% of the final volume.

B. *In Vitro* Translation

The capped RNA generated by the *in vitro* transcription reaction can be used to direct synthesis of the corresponding precursor proteins using an *in vitro* translation system. The translation systems derived from rabbit reticulocytes and from wheat germ both produce precursors that can be imported into chloroplasts (Mishkind *et al.*, 1987). The size of the precursor protein is an important factor in deciding which translation system to use. The wheat germ system provides excellent translation of proteins under 50,000 Da, but often gives poor translation of larger proteins. The reticulocyte system gives good translation of all proteins, but some preparations of reticulocyte lysate are reported to cause lysis of chloroplasts (Mishkind *et al.*, 1987).

It is possible to include any radioactive amino acid in the translation reaction to label the nascent proteins. Those most frequently used are [^{3}H]leucine or [^{35}S]methionine. An advantage of [^{35}S]methionine is the very high specific activity that is available. On the other hand, leucine is more abundant in most proteins, thereby partially compensating for the lower specific activity of [^{3}H]leucine when total incorporation is measured. Moreover, [^{3}H]leucine has the advantage of extended storage life and is less expensive. Consequently, we use [^{3}H]leucine for most studies, but use [^{35}S]methionine when it is particularly important to have highly labeled precursors or when a particular precursor is poorly labeled with leucine.

Kits containing all the materials needed to conduct translation reactions are available from several suppliers. If these are used, the protocols provided with the kit should be followed. In our lab we prepare wheat germ extracts using published procedures (Anderson *et al.*, 1983). To conduct protein synthesis with this extract, a volume of [^{3}H]leucine equal to the final volume of the translation reaction is dried down in a microfuge tube. The final concentration of [^{3}H]leucine during translation will be 1 mCi/ml. If [^{35}S]methionine is used, it is not dried down but is added directly to the reaction. In a normal translation reaction of 50 μl, 30 μl of wheat germ extract is added to the tube containing the dried label, followed by appropriate levels of energy mixture and amino acids, salts, 1 μl RNasin, and water to a final volume of 45 μl. Initiate the translation reaction by adding 5 μl of transcript. Mix well to make sure the label has redissolved. Incubate for 1–1.5 hours at 23–25°C (it is acceptable to leave the reaction on the bench top). To terminate the translation, add an equal volume of 60 mM leucine (or 50 mM methionine if [^{35}S]methionine was used as the radiolabel) in 2× import buffer (see below for the recipe). The unlabeled amino acid dilutes the radiolabel, so that it does not enter chloroplasts and generate radiolabeled proteins inside chloroplasts via *in organello* protein synthesis during the import reaction. The $2\times$ import buffer brings the translation products to the appropriate osmotic concentration so that they can be added directly to chloroplasts without causing lysis.

The diluted translation products are stored on ice until use. Alternatively, they can be prepared ahead and stored frozen in small aliquots until needed. The precursor preparation should not be frozen more than one time, as repeated cycles of freezing and thawing often lead to a reduction in import competence of precursors. Incorporation of the radiolabel into precursor proteins can be measured by TCA precipitation of a sample of the translation products onto filter paper and then counting in a scintillation counter. This will give a measure of dpm incorporated into protein. If a binding reaction is to be performed with low levels of ATP, the translation should be desalted to remove the ATP present in the energy mixture (see Section V).

III. Preparation of Intact Chloroplasts

Intact chloroplasts have been isolated from a wide range of plant tissues (Walker *et al.*, 1987). Not all of these chloroplast preparations have been tested for their ability to import precursor proteins, but is seems likely that any chloroplast preparation capable of CO_2-dependent oxygen evolution will be capable of protein transport. Spinach and pea tissue have been used extensively for isolation of intact chloroplasts, and in both cases, these chloroplasts are capable of importing precursor proteins (Keegstra *et al.*, 1989).

The most common method for isolation of intact chloroplasts is to grind the plant tissue in buffer containing an osmoticum to prevent lysis of the released organelles. A crude preparation of chloroplasts can be recovered from the homogenate by differential centrifugation (Walker *et al.*, 1987). The best way to purify intact chloroplasts from this crude preparation is by density gradient centrifugation on Percoll gradients (Price *et al.*, 1987). Intact chloroplasts are resolved from broken chloroplasts and other cellular debris. They can be separated from Percoll by dilution and differential centrifugation.

The method we use for isolation of pea chloroplasts is summarized below. Before beginning the tissue homogenization, it is best to prepare the Percoll gradients that will be used for purification of the isolated chloroplasts. Add 15 ml of $2\times$ grinding buffer (see below) to 15 ml of Percoll and mix. Centrifuge for 30 minutes at 40,000 g (18,000 rpm in a Sorvall SS34 rotor) to form the gradient. Allow the rotor to stop without a brake. Store the gradient on ice until use.

Harvest 50–75 g of pea tissue by cutting the expanded top leaves of plants that are 10–15 days old. Tissue should be harvested no later than 2 hours after lights come on to prevent accumulation of starch grains inside the plastids. Store leaves in a covered ice bucket until use. Grind the tissue in 200 ml of ice-cold grinding buffer ($1\times$ grinding buffer is 50 mM HEPES-KOH, pH 7.3, containing 0.33 M sorbitol, 0.1% bovine serum albumin, 1 mM MgCl$_2$, 1 mM MnCl$_2$, and 2 mM Na$_2$EDTA) using a polytron homogenizer on setting #7: three bursts of around 5 seconds each should effectively pulverize the leaves. Filter the homogenate through two layers of Miracloth (Calbiochem) into a 500-ml Erlenmeyer flask on ice. Transfer the filtrate to polycarbonate tubes and centrifuge for 5 minutes at 2600 g (4000 rpm in a Sorvall HB-4 rotor). Remove the supernatant and resuspend each pellet of crude chloroplasts in 2 ml of grinding buffer. Carefully layer the resuspended chloroplasts on top of the Percoll gradient formed earlier. Centrifuge in a swinging bucket rotor for 10 minutes at 8,000 g (7,000 rpm in a Sorvall HB-4 rotor) or for 5 minutes at 13,000 g (9,000 rpm in an HB-4 rotor). Allow the rotor to stop without a brake.

Remove the upper band of broken chloroplasts and discard. Then collect the lowest green band, representing the intact chloroplast fraction, using a Pasteur pipet. Transfer to a clean polycarbonate tube and dilute with three volumes of import buffer (import buffer is 50 mM HEPES-KOH, pH 8.0, containing 0.33 M sorbitol). Centrifuge for 5 minutes at 2600 g (4000 rpm in an HB-4 rotor). Resuspend the chloroplast pellet in 3 ml of import buffer with a Pasteur pipet. Then adjust the volume to 20 ml with import buffer. Remove fractions at this point for chlorophyll determination (see below). Pellet the plastids again for 5 minutes at 2600 g. Resuspend the final pellet at a concentration of 1 mg chlorophyll/ml. Store in a covered ice bucket until use.

If quantitation is to be performed, an aliquot of resuspended chloroplasts must be saved for determination of protein concentration. Chloroplast counts must be performed if the results are to be expressed on a "per chloroplast" basis. Generally there are (1–2×10^9 chloroplasts per mg of chlorophyll. (See Section VII for additional information on quantitation.)

Chlorophyll assays are performed as described (Arnon, 1949). Briefly, a small aliquot of chloroplasts is diluted to 1 ml with water and brought to 80% acetone by the addition of 4 ml of acetone. Insoluble material is removed by centrifugation in a clinical centrifuge and the optical density of the supernatant is recorded at 652 nm. The level of chlorophyll in each aliquot can be determined by the following conversion:

$$\text{chlorophyll} \quad (\text{mg/ml}) = 5 \times \text{OD}_{652} \times 0.02899/\text{sample volume}$$

where the sample volume is the volume of chloroplast suspension added to the acetone

$$\text{total mg chlorophyll} = \text{chloropyll} \quad (\text{mg/ml}) \times 20 \text{ ml}$$

IV. Transport Reactions

Protein transport occurs when precursors and chloroplasts are incubated together under the proper conditions. Transport requires adequate temperatures and an energy source. Transport occurs readily in the range of 20–30°C, but is inhibited at low temperatures. Transport requires energy in the form of ATP (Grossman *et al.*, 1980; Theg *et al.*, 1989). ATP can be provided by incubating chloroplasts in the light, which causes the synthesis of ATP via photophosphorylation. ATP can also be added exogenously, which is the procedure that we normally use. In this case, an external ATP concentration of 1 m*M* or greater is needed to support transport (Olsen *et al.*, 1989; Theg *et al.*, 1989).

A. Transport Conditions

In a standard tranport reaction, we generally use chloroplasts containing 50 μg chlorophyll (about 1×10^8 chloroplasts) in a reaction volume of 150 μl. It is possible to scale up or down proportionally, to suit the experimental design. Keep in mind, however, that it is difficult to work with less chloroplasts than are represented by 50 μg chlorophyll, especially if they are to be treated with protease after import. To a 1.5-ml microfuge tube add 3 μl of 100 m*M* MgATP in import buffer (the final concentration will be 2 m*M*), an aliquot of the translation mixture that contains $(1-2) \times 10^6$ dpm of precursor protein, and import buffer to bring the volume to 100 μl. The transport reaction is initiated by the addition of 50 μl of chloroplast suspension (suspended at 1 mg chlorophyll/ml). The reaction is incubated at 25°C. It is best to control the temperature in a waterbath. However, it is acceptable to run transport reactions at room temperature on a lab bench. The time of the transport reaction can be varied. Times less than 1 or 2 minutes are not convenient and times longer than 30 minutes are probably not useful. Generally, transport proceeds linearly for about 10 to 20 minutes, then slows and very little transport occurs after 30 minutes. It is a good idea to gently agitate the reactions every 5 minutes.

B. Termination

Transport reactions can be terminated in several different ways, depend-ing upon the purpose of the experiment. For time course experiments, where it is important to control the length of incubation precisely, one of the easiest termination methods is to rapidly sediment the intact chloro-plasts through a layer of silicone oil into a quench solution. One disadvan-tage of this method is that it precludes any posttreatment or fractionation of the chloroplasts used in the transport reaction. If posttreatment or fractionation is needed, an alternative is to terminate the reaction by dilution in 10 volumes or more of ice-cold import buffer. Intact chloro-plasts can then be recovered and treated or fractionated as needed. If timing is not critical, it is acceptable to simply proceed with repurification of the chloroplasts or with the protease treatment.

1.   SILICONE OIL TERMINATION

The silicone oil method is a quick and relatively easy way to stop a transport reaction (Flugge and Hinz, 1986). When the incubation time for a transport reaction is almost complete, a sample of the reaction (about 20–50 μl) is layered on top of 100 μl of silicone oil (Wacker AR 200), which is layered on top of 100 μl of 1 M perchloric acid (PCA) in a 400-μl polyethylene tube. [Alternatively, a quench solution of SDS and sucrose can be used (Flugge and Hinz, 1986).] When the desired incubation time is up, this tube is spun at 9200 g for 20 seconds at speed in a microfuge. Only the intact chloroplasts will spin through the oil. Broken chloroplasts will be left at the transport reaction/oil interface, and free radiolabeled protein will be left in the aqueous supernatant. When the chloroplasts contact the PCA, lysis and precipitation occur, stopping all reactions. The tube is then frozen until final work-up. To recover the chloroplast pellet, the frozen tube is cut with a razor blade or scalpel below the oil/PCA interface, but well above the chloroplast pellet. For ease of cutting, polyethylene rather than polypropylene tubes must be used. It is important not to have any oil in the PCA phase because oil will make subsequent steps difficult. Once the PCA has thawed, it is removed from the pellet. It is easier to handle the sample if the tube section is placed inside a 1.5-ml microfuge tube. The pellet is resuspended in 20 μl of 2× SDS sample buffer and 20 μl of 0.5 M Tris, pH 9.5 [2× SDS sample buffer is prepared by mixing 25% (v/v) 0.5 M Tris, pH 6.8, 20% (v/v) glycerol, 10% (v/v) 2-mercaptoethanol, 44% (v/v) 10% SDS stock, and 0.01% bromophenol blue for tracking dye]. The resuspension should be blue to blue–green, but not yellow. If it is yellow, the PCA has not been neutralized. In this case, a few microliters of 0.5 M NaOH can be added to raise the pH. The bottom of the 400-μl tube is inverted in the microfuge tube and then spun to transfer the contents of the polyethylene tube to the microfuge tube. Dissolution of the pellet can be facilitated by mixing with a pipet and using a sonicating water bath. Samples are then boiled for 2 minutes before analysis by SDS–PAGE.

If quantitation is to be performed, the pellet must be resuspended in 1× modified SDS sample buffer/0.5 M Tris, pH 8.0. Modified sample buffer is the same as sample buffer except that it lacks 2-mercaptoethanol and bromophenol blue. These chemicals interfere with the BCA protein assay (Pierce). A sample of 5 μl is removed for the protein assay. Bromophenol blue and 2-mercaptoethanol must be added to the rest of the sample and the sample is then boiled for 2 minutes before running on a gel.

The radiolabeled protein remaining in the supernatant may also be analyzed by the PCA/oil method. The frozen polyethylene tube is cut just below the transport reaction/oil interface. All of the aqueous transport reaction must remain in the top, even if some silicone oil is included. The top of the 400-μl tube is uncapped and put into a 1.5-ml microfuge tube. An equal volume of 2× SDS sample buffer is layered on top of the aqueous layer. This is then centrifuged to transfer the sample to the microfuge tube. The sample buffer on top of the aqueous layer washes the 400-μl tube, giving a more complete recovery of the aqueous supernatant. The samples are mixed, boiled for 2 minutes, and analyzed by SDS–PAGE.

The silicone oil method is convenient for time course experiments, in that it allows for a rapid and more complete arrest of the transport reaction than does the Percoll method (see below). Also, smaller reactions are used, conserving precursor and chloroplasts. However, it does not include a step to wash the chloroplasts, as does the Percoll method. Because the chloroplasts sediment with a shell of mixture reactions surrounding them, some quantity of radiolabeled precursor protein will be nonspecifically included with the chloroplasts. This impedes the use of this method for the study of binding of precursor proteins to chloroplasts. Controls to distinguish specific from nonspecific association of precursors with chloroplasts must be done. Also, it is not possible to do any posttreatment of the chloroplast, such as fractionation.

2. RECOVERY OF INTACT CHLOROPLASTS VIA A PERCOLL CUSHION

Use of a Percoll cushion to recover chloroplasts is slower and does not provide for an immediate termination of transport reactions. However, it does allow for the posttreatment of chloroplasts before the Percoll step and for fractionation of the intact chloroplasts recovered from the Percoll step.

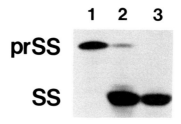

Fɪɢ. 1. Transport of prSS into chloroplasts. Intact chloroplasts and radiolabeled prSS were incubated for 12 minutes at room temperature in the presence of 5 m*M* ATP. Intact chloroplasts were reisolated after transport (lane 2) or were treated with thermolysin and then reisolated (lane 3). Lane 1 shows an aliquot of *in vitro* translation containing [³H]prSS. See the text for experimental details.

Load the reaction mixture (it can first be diluted with cold buffer if timing is important) onto a 1.0-ml cushion of 40% Percoll in import buffer. Centrifuge at 3000 *g* for 6 minutes in a swinging bucket rotor with low brake. The intact chloroplasts should pellet through the Percoll cushion, whereas broken chloroplasts will remain on top of the cushion. Remove the broken chloroplasts and the Percoll solution and resuspend the intact chloroplasts in 1 ml of import buffer. If quantitation is to be performed, remove a 50-μl aliquot for protein determination. Pellet the chloroplasts by centrifugation at 1500 *g* for 3 min. Dissolve the pellet of chloroplasts in 50–100 μl of SDS sample buffer and boil for 2 minutes. Analyze by running 10–30 μl on SDS–PAGE.

An example of transport of the precursor of the small subunit of ribu-lose-1,5-bisphosphate carboxylase (prSS) is shown in Fig. 1. Lane 1 is a sample of the translation reaction containing [³H]prSS. In the presence of chloroplasts and 5 m*M* ATP to support translocation, the precursor has been transported into chloroplasts, where it was processed to its mature form (SS) (lane 2). The processed form was protected from thermolysin digestion, but the prSS associated with the chloroplasts (lane 2) was ther-molysin sensitive, indicating an external location (lane 3) (see Section VI,A for a discussion of protease treatment).

V. Binding Reactions

Often some precursor protein is found associated with the repurified chloroplasts in a transport experiment (see Fig. 1, lane 2). This precursor is protease sensitive, implying an external location (Fig. 1, lane 3). However, it does not necessarily represent specific binding. If translocation is not blocked, specifically bound precursors should be transported into the chloroplast, leaving only nonspecifically bound precursors associated with the chloroplastic surface. Specific binding can be measured reliably only when translocation is blocked, separating the binding step of transport from the rest of the process.

Protein translocation across the chloroplastic envelope is inhibited by low temperatures, but binding of the precursor to the chloroplastic surface occurs readily on ice (Friedman and Keegstra, 1989). This provides a method for separating binding from the rest of the transport process. Binding reactions can be run the same way as transport reactions (see above), but the samples must be kept cold at all times to prevent translocation of the bound precursor across the envelope.

Another method to separate binding from the rest of the transport process is to take advantage of different energy requirements for these steps. Transport requires high levels of ATP, but binding requires only low levels of ATP (50–100 μM) (Olsen *et al.*, 1989). Thus, binding and translocation can be separated by adjusting the level of ATP. This allows a binding experiment to be performed at room temperature, but the reaction must be done in the dark to prevent ATP from being generated by photophosphorylation. Also, the *in vitro* translation must be treated to remove the ATP present in the translation mixture. This can be done by passing the *in vitro* translation through a Sephadex G-25 column (Olsen *et al.*, 1989). Small molecules, such as ATP, will be retained by the column, whereas precursor protein passes through the column in the void volume.

The G-25 columns are prepared in 1-ml disposable syringes that are hung in 15-ml Falcon tubes. The G-25 beads are swollen in import buffer, at 4°C, overnight. Place two glass fiber filters in the bottom of the syringe and pack the G-25 beads by centrifugation at 1700 g for 2 minutes at speed. Place a glass fiber filter on top of the packed bed of G-25 to help ensure even loading of sample. Wash the columns three times with import buffer by centrifugation at 1700 g for 15 seconds at speed. Treat the columns with 2% bovine serum albumin in import buffer by centrifugation for 2 minutes at speed, 1700 g. Load a sample of *in vitro* translation product of not more than 200 μl on top of the column, centrifuge at 1700 g for 2 minutes at speed, and collect the void volume in a microcentrifuge tube. Almost all of the ATP (98–99%) is removed by this procedure (Oslen *et al.*, 1989) as well as free radiolabeled amino acid and other small molecules. The columns can be washed three times with import buffer, stored in import buffer with 0.02% azide, and reused.

The physiological significance of binding in the above assays can be shown by importing the bound precursor. The reaction mixture is diluted with 1 ml of cold import buffer and the chloroplasts are pelleted by centrifugation at 1000 g for 3 minutes to separate bound precursors from free precursors. The pellet is resuspended and transport conditions are restored by raising the temperature or by the addition of sufficient ATP to support transport. The reaction is then treated as a transport reaction (see above). The majority of prSS (50–85%) bound when transport is blocked can be imported into chloroplasts when the block is removed (Cline *et al.*, 1985; Friedman and Keegstra, 1989).

VI.   Posttreatment and Fractionation of Chloroplasts

Protein translocation into chloroplasts can usually be monitored by the appearance of mature-sized protein inside chloroplasts. However, processing protease from broken plastids can generate mature proteins outside of chloroplasts. Also, when working with a new precursor or new reaction conditions, it is important to verify the locations of each protein species. Translocation can be confirmed by the resistance of proteins to exogenously added proteases. Thermolysin has been shown to be an ideal probe for the chloroplast surface (Cline *et al.*, 1984). It is thus the enzyme of choice to distinguish externally bound from translocated proteins (Fig. 1).

A.   Protease Treatment to Remove Externally Bound Proteins

Add 1 ml of cold import buffer to an import reaction mixture. Pellet the plastids by centrifugation at 1000 g for 3 minutes. Resuspend the chloroplasts in 0.25 ml of import buffer containing 0.2 mg/ml thermolysin plus 1 mM $CaCl_2$. Prepare the thermolysin as a stock solution of 2 mg/ml in import buffer/10 mM $CaCl_2$ immediately before use. It is convenient to weigh out the thermolysin powder ahead of time and dissolve it just before use. Incubate the protease digestion reaction at 4°C for 30 minutes with occasional agitation. Terminate proteolysis by adding 50 μl of import buffer containing 50 mM EDTA. Reisolate intact chloroplasts through a 40% Percoll cushion and analyze by SDS–PAGE as described in Section IV,B,2. The only modification is that all solutions, including the 40% Percoll, should contain 5 mM EDTA to ensure that any residual thermolysin is not active.

B.   Fractionation of Chloroplasts

For more detailed localization of translocated proteins, a small-scale fractionation protocol can be used. It provides a method to separate the major compartments of chloroplasts (envelope, stroma, thylakoid membrane, and thylakoid lumen) and determine where the imported proteins are located. For fractionation of chloroplasts, the import reaction should be scaled up threefold. This gives an adequate yield of each fraction, but still allows most of the procedures to be performed in a microfuge tube. In the procedure described here, chloroplasts are first hypotonically lysed and different fractions are then separated on a sucrose step gradient. The stromal fraction remains on top of the 0.46 M sucrose. Envelope membranes (a mixture of inner and outer membranes) fractionate to the interface of 0.46 and 1 M sucrose. Thylakoids form a green pellet at the bottom of the tube. A scheme for the fractionation procedure is illustrated in Fig. 2.

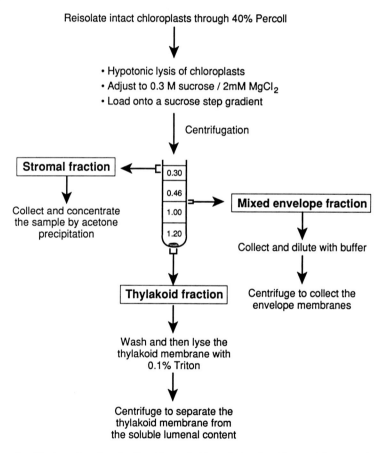

FIG. 2. Strategy for the fractionation of chloroplasts. See the text for experimental details.

Before fractionation, intact chloroplasts are reisolated through 1 ml of 40% Percoll as described in Section IV,B,2. The pellet of chloroplasts is resuspended in 900 μl of import buffer. Remove a 100-μl aliquot to be analyzed for total transport. Dilute this aliquot with 1 ml of import buffer and pellet the chloroplasts by centrifugation at 1500 g for 3 min. This pellet is resuspended in 40 μl of sample buffer and the total transport products analyzed by SDS–PAGE. The remaining 800 μl of resuspended intact chloroplasts is centrifuged at 1500 g for 3 minutes to recover the chloroplasts. The chloroplast pellet is resuspended in 450 μl of 25 mM HEPES buffer (pH 8.0) and incubated on ice for 10 minutes to cause lysis of the chloroplasts. To check for complete lysis, remove a 50-μl aliquot of the resuspended chloroplasts and spin it through 40% Percoll as described in Section IV,B,2. There should not be a green pellet (intact chloroplasts) at the bottom of the tube. To the remaining 400 μl of resuspended chloroplasts, add 400 μl of 0.6 M sucrose and 4 mM MgCl$_2$ in 25 mM HEPES buffer, pH 8.0. The sucrose and Mg^{2+} keep the thylakoids intact and stacked so that they will not break into small fragments and contaminate the envelope fraction (Douce and Joyard, 1982).

Make a sucrose step gradient in a 5-ml SW 50.1 tube by adding 1.2 ml of 1.2 M sucrose, 1.5 ml of 1 M sucrose, and then 1.5 ml of 0.46 M sucrose. All sucrose solutions are in 25 mM HEPES buffer, pH 8.0. Mark the interface of the 0.46 and 1 M sucrose on the tube wall to help identify the position of the envelope fraction after the centrifugation, because there will be too little envelope at that point to form a visible yellow band. The step gradients can be made before the transport experiments starts.

Load the lysed chloroplast suspension onto the step gradient. Centrifuge in a Beckman SW 50.1 rotor at 47,000 rpm for 1 hour. Collect 200 μl of the supernatant into a microfuge tube. This represents the stromal fraction. Carefully remove the remaining supernatant and most of the 0.46 M sucrose solutions. Collect the envelope fraction from the interface of 0.46 and 1 M sucrose solution. As much as 500 μl can be collected without risk of thylakoid contamination. Remove the remaining sucrose solution. The green pellet at the bottom of the tube is the thylakoid fraction.

To the stromal sample, add 800 μl of ice-cold acetone and precipitate at $-20°$C for 1 hour. Spin down the precipitant in a microfuge for 15 minutes. Dry and then resuspend the pellet in 40 μl of sample buffer. To the envelope sample, add at least 1 ml of 25 mM HEPES buffer, pH 8.0, and mix well to ensure the sucrose concentration is sufficiently diluted. Centrifuge at 48,000 g for 45 minutes (20,000 rpm in a SS34 rotor). The pellet should be easily visible and yellow in color. A greenish pellet indicates contamination by the thyakoids. To the thylakoid pellet, add 1 ml of import buffer to resuspend the pellet. Transfer the resuspension to a microfuge tube and spin in a microfuge at 7500 rpm for 5 minutes.

If further fractionation of thylakoids into thylakoid membrane and lumenal content is not necessary, the pellet can simply be resuspended in 200 μl of sample buffer. If separation of the thylakoid membrane from the lumenal content is desired, the pellet is resuspended in 100 μl of import buffer containing 0.1% Triton X-100 and incubated on ice for 5 minutes. This concentration of Triton releases all of the thylakoid lumen protein plastocyanin, without release of any chlorophyll from the thylakoid membranes (S. Theg, personal communication). Thylakoid membrane is then separated from the soluble lumenal content by spinning the tube in a microfuge at top speed for 15 minutes. Remove the supernatant to a new microfuge tube and add 100 μl of 2× sample buffer. This represents the soluble lumenal fraction. To the thylakoid membrane pellet, add import buffer to wash the membrane. Resuspend the final membrane pellet in 200 μl of 1× sample buffer. Analyze the samples on SDS–PAGE by loading 20 μl of each fraction.

This fractionation procedure gives very clean stromal and envelope fractions. However, some envelope proteins will remain in the thylakoid fraction. The amount of envelope contamination in the thylakoid fraction varies from 5 to 33% of the total envelope membrane (H.-M. Li, personal observation). As a result, an envelope protein will show up in both the envelope and the thylakoid membrane fractions. Nonetheless, a thylakoid membrane protein should only be present in the thylakoid fraction. It is thus possible to determine the membranous compartment in which an imported protein is located.

VII. Analysis of Products

A. Electrophoretic Separation of Products

We normally use SDS–PAGE for analysis of the import reactions. We sometimes use a minigel apparatus, but usually full-sized gels are used for publication-quality work. We use the discontinuous buffer system described by Laemmli (1970). There are several sources that provide detailed descriptions of these techniques (for example, Hames and Rickwood, 1981). For analysis of the Rubisco small subunit and other small proteins, a 15% running gel is most useful. For separation of larger imported proteins, lower percentage gels are recommended. Gradient gels are useful if a mixture of import products of widely different molecular weights is to be analyzed.

Detection of radiolabeled proteins is accomplished by fluorography. Even if [^{35}S]methionine is used to label the proteins, detection is much more rapid if fluorography is used rather than autoradiography. Several solutions are commercially available for conducting fluorography. We normally use homemade solutions following the protocol described below.

Place the gel into once-used dimethyl sulfoxide (DMSO) for 10 minutes. Remove the used DMSO and replace it with fresh DMSO for 10 minutes. Remove the fresh DMSO and save it as once-used DMSO. Add 20% (w/v) 2,5-diphenyloxazole (PPO)/DMSO (20 g PPO + 80 ml DMSO; store in darkness or in an amber bottle). Leave the gel in this solution for 30 minutes. Remove the PPO/DMSO and save it (it can be used many times). Wash the gel in running tap water for 30 minutes to remove the DMSO and to precipitate the PPO. Dry the gel onto filter paper and expose it to X-ray film at $-80°C$. If quantitation is to be done, radioactive markers must be placed on the gel before exposure to the film (see below).

Protein import results are often presented in a qualitative fashion by showing the SDS gel patterns of precursors and imported proteins (Fig. 1). This is acceptable for some purposes, but many studies require quantitative analysis of the results.

B.   Quantitative Analysis of Import Products

Quantitation allows the expression of experimental results in terms of molecules of protein bound or imported per chloroplast. This value is easy to understand and is useful for comparison of experiments and among laboratories. The number of molecules of protein bound or imported is determined by excision of the radioactive proteins from the dried gel and counting in a liquid scintillation counter. The number of chloroplasts loaded on the gel lane is determined by a protein assay.

If quantitation is to be performed, radioactive markers must be placed on the dried gel before exposure to X-ray film to provide a means of lining up fluorogram and gel for excision of gel bands. We find that writing on tape with ink that has ^{14}C added (about 10 μCi in 5 ml ink) and then placing three to four of these markers at the corners of the gels works well. A light box is needed for viewing the fluorogram through the dried gel. Regions of the gel corresponding to bands on the fluorogram are excised with a razor blade. The dried gel slice is pulled away from the backing filter paper with forceps and is placed in a scintillation vial. A sample is cut from an area without a band of protein to provide a background value. Rehydration of the gel and extraction of the counts are achieved by adding 1 ml of 30% hydrogen peroxide and incubating at $55-60°C$ overnight. Scintillation cocktail is added the next day, and the vials are mixed and incubated overnight at $55-60°C$. The samples are counted in a liquid scintillation counter to determine the amount of radioactivity in the gel slice. Background radioactivity is subtracted from this value. We have found that only about 85% of the counts are extracted from the gel slice by the above procedure, thus the dpm/gel slice must be divided by 0.85 to obtain the total dpm/gel slice (Cline *et al.,* 1985).

To determine the number of molecules of protein in the gel slice, first calculate the number of protein molecules/dpm. This is obtained from the specific activity of the radiolabeled amino acid and the number of labeled amino acids per protein molecule (obtained from the deduced sequences of the precursor and mature protein). By multiplying total dpm/gel slice by the number of protein molecules/dpm, the number of protein molecules/ gel slice is obtained.

This value can be normalized by dividing by the number of chloroplasts in the sample. The number of chloroplasts/milliliter was determined during the chloroplast preparation (see Section III). A sample of the chloroplast preparation was saved for a protein assay. When compared to a standard curve, micrograms of protein/milliliter of chloroplasts can be determined. The two above values allow calculation of the number of chloroplasts/microgram of protein. Likewise, protein determination is done on the samples saved from the binding or transport reactions (see Section IV,B). When all dilution factors are taken into account, micrograms of protein recovered through Percoll from the transport reaction can be determined. From this and the value of number of chloroplasts/ micrograms of protein, the number of chloroplasts recovered can be calculated. Taking into consideration the fraction of the binding or transport reaction that was run on the gel, the number of chloroplasts run on the gel lane can be determined. Molecules bound or transported per chloroplast can be calculated from the number of molecules in the gel slice (above) and the number of chloroplasts run on the gel lane.

REFERENCES

Anderson, C. W., Straus, J. W., and Dudock, B. S. (1983). Preparation of a cell-free protein-synthesizing system from wheat germ. *In* "Methods in Enzymology" (R. Wu, L. Grossman, and K. Moldave, eds.), Vol. 101, pp 635–644. Academic Press, New York.

Arnon, D. I. (1949). Copper enzymes in isolated chloroplasts. Polyphenoloxidase in *Beta vulgaris. Plant Physiol.* **24,** 1–15.

Cline, K., Werner-Washburne, M., Andrews, J., and Keegstra, K. (1984). Thermolysin is a suitable protease for probing the surface of intact pea chloroplasts. *Plant Physiol.* **75,** 675–678.

Cline, K., Werner-Washburne, M., Lubben, T. H., and Keegstra, K. (1985). Precursors to two nuclear-encoded chloroplast proteins bind to the outer envelope membrane before being imported into chloroplasts. *J. Biol. Chem.* **260,** 3691–3696.

Douce, R., and Joyard, J. (1982). Purification of the chloroplast envelope. *In* "Methods in Chloroplast Molecular Biology" (M. Edelman, R. B. Hallick, and N.-H. Chua, eds.) pp. 239–256. Elsevier Biomedical, New York.

Flügge, U. I., and Hinz, G. (1986). Energy dependence of protein translocation into chloroplasts. *Eur. J. Biochem.* **160**, 563–570.

Friedman, A. L., and Keegstra, K. (1989). Chloroplast protein import: Quantitative analysis of precursor. *Plant Physiol.* **89**, 993–999.

Grossman, A., Bartlett, S., and Chua, N.-H. (1980). Energy-dependent uptake of cytoplasmically snynthesized polypeptides by chloroplasts. *Nature (London)* **285**, 625–628.

Hames, B. D., and Rickwood, D. (1981). "Gel Electrophoresis of Proteins." IRL Press, Washington, D.C.

Keegstra, K. (1989). Transport and routing of proteins into chloroplasts. *Cell (Cambridge, Mass.)* **56**, 247–253.

Keegstra, K., Olsen, L. J., and Theg, S. M. (1989). Chloroplastic precursors and their transport across the envelope membranes. *Annu. Rev. Plant Physiol. Plant Mol. Biol.* **40**, 471–501.

Krieg, P. A., and Melton, D. A. (1987). *In vitro* RNA synthesis with SP6 RNA polymerase. *In* "Methods in Enzymology" (R. Wu, ed.), Vol. 155, pp. 397–415, Academic Press, San Diego, California.

Laemmli, U. K. (1970). Cleavage of structural proteins during the assembly of the head of bacteriophage T4. *Nature (London)* **227**, 680–685.

Lubben, T. H., and Keegstra, K. (1986). Efficient *in vitro* import of a cytosolic heat shock protein into pea chloroplasts. *Proc. Natl. Acad. Sci. U.S.A.* **83**, 5502–5506.

Mishkind, M. L., Greer, K. L., and Schmidt, G. W. (1987). Cell-free reconstitution of protein transport into chloroplasts. *In* "Methods in Enzymology" (L. Packer and R. Douce, eds.), Vol. 148, pp. 274–294. Academic Press, San Diego, California.

Olsen, L. J., Theg, S. M., Selman, B. R., and Keegstra, K. (1989). ATP is required for the binding of precursor proteins to chloroplasts. *J. Biol. Chem.* **264**, 6724–6729.

Pilon, M., De Boer, A. D., Knols, S. L., Koppelman, M. H. G. M., Van der Graaf, R. M., de Kruigff, B., and Weisbeek, P. J. (1990). Expression in *Escherichia coli* and purification of a translocation-competent precursor of the chloroplast protein ferredoxin. *J. Biol. Chem.* **265**, 3358–3361.

Price, C. A., Cushman, J. C., Mendiola-Morgenthaler, L. R., and Reardon, E. M. (1987). Isolation of plastids in density gradients of Percoll and other silica sols. *In* "Methods in Enzymology" (L. Packer and R. Douce, eds.), Vol. 148, pp. 157–179. Academic Press, San Diego, California.

Sambrook, J., Fritsch, E. F., and Maniatis, T. (1989). "Molecular Cloning: A Laboratory Manual," 2nd ed. Cold Spring Harbor Lab., Cold Spring Harbor, New York.

Schmidt, G. W., and Mishkind, M. L. (1986). The transport of proteins into chloroplasts. *Annu. Rev. Biochem.* **55**, 879–912.

Smeekens, S., Bauerle, C., Hageman, J., Keegstra, K., and Weisbeek, P. (1986). The role of the transit peptide in the routing of precursors toward different chloroplast compartments. *Cell (Cambridge, Mass.)* **46**, 365–375.

Theg, S. M., Bauerle, C., Olsen, L. J., Selman, B. R., and Keegstra, K. (1989). Internal ATP is the only energy requirement for the translocation of precursor proteins across chloroplastic membranes. *J. Biol. Chem.* **264**, 6730–6736.

Waegemann, K., Paulsen, H., and Soll, J. (1990). Translocation of proteins into islated chloroplasts requires cytosolic factors to obtain import competence. *FEBS Lett.* **261**, 89–92.

Walker, D. A., Cerovic, Z. G., and Robinson, S. P. (1987). Isolation of intact chloroplasts: General principles and criteria of integrity. *In* "Methods in Enzymology" (L. Packer and R. Douce, eds.), Vol. 148, pp. 145–157. Academic Press, San Diego, California.

Chapter 16

Analysis of Mitochondrial Protein Import Using Translocation Intermediates and Specific Antibodies

THOMAS SÖLLNER, JOACHIM RASSOW,
AND NIKOLAUS PFANNER

Institut für Physiologische Chemie
Universität München
D-8000 München 2, Germany

I. Introduction

Most mitochondrial proteins are synthesized as precursor proteins on cytosolic polysomes and are posttranslationally translocated into and across the mitochondrial membranes (Attardi and Schatz, 1988; Hartl and

345

Neupert, 1990; Horwich, 1990; Pfanner and Neupert, 1990). The precursor proteins carry positively charged targeting sequences that are either located in amino-terminal presequences or in the mature part of the proteins. The precursors are recognized by specific receptors, MOM19 and MOM72, on the mitochondrial surface and are inserted into the outer membrane. Further translocation predominantly occurs through contact sites between both mitochondrial membranes. Import requires energy in the form of the electrical potential $\Delta\Psi$ across the inner membrane and ATP in the cytosol and in the mitochondrial matrix. The presequences are proteolytically cleaved off during or after translocation by the processing peptidase in the matrix.

The characterization of mitochondrial protein uptake was greatly advanced by the accumulation of precursor proteins at distinct stages of their import pathway (translocation intermediates). Here we focus on three translocation intermediates: (1) accumulation of the precursor of ADP/ATP carrier at its receptor MOM72 on the mitochondrial surface; (2) accumulation of the precursor of ADP/ATP carrier in the mitochondrial outer membrane at the common membrane insertion site "general insertion protein" (GIP); and (3) arrest of various precursor proteins in contact sites between both mitochondrial membranes. A basic requirement for a true translocation intermediate is the reversibility of the transport arrest, i.e., upon release of the block, the precursor protein has to be fully imported into mitochondria and correctly assembled.

We first will describe the currently used standard procedures for in vitro synthesis and import of precursor proteins. The generation of various translocation intermediates will then be achieved by modifications of the standard protocol. Where appropriate, we indicate the possible range of concentrations of various components that are routinely used. With new precursor proteins it may be useful to test import conditions also outside the given range.

II. Synthesis of Mitochondrial Precursor Proteins in Rabbit Reticulocyte Lysates

In most cases, pGEM4 plasmids (Promega Biotec) containing DNA inserts that code for mitochondrial precursor proteins are linearized with restriction enzymes that cleave downstream (3') of the coding region for the precursor. After extraction with phenol/chloroform, precipitation with ethanol, and resuspension of the DNA, the transcription is performed with

SP6 RNA polymerase for 60 minutes at 37–40°C in the presence of 7 mGpppG (Melton *et al.*, 1984; Stueber *et al.*, 1984; Krieg and Melton, 1984; Sambrook *et al.*, 1989). The samples are again extracted with phenol/chloroform. After precipitation in presence of 70% ethanol and 250–300 mM LiCl, the RNA and DNA are dissolved in H_2O containing RNAsin (0.5 U/μl) and stored in aliquots at -80°C. The translation is performed in rabbit reticulocyte lysates (Pelham and Jackson, 1976) (e.g., from Amersham Buchler) in the presence of [^{35}S]methionine (specific radioactivity 1000 Ci/mmol; Amersham Buchler) for 60 minutes at 30°C. For an efficient synthesis of various mitochondrial precursor proteins, the optimal concentrations of RNA, potassium ions (100–200 mM), and magnesium ions (1–3 mM) in the translation reaction have to be determined. A postribosomal supernatant is prepared by centrifugation for 60 minutes at 226,000 g and unlabeled methionine (5 mM final concentration) is added. The reticulocyte lysates are made isotonic (for mitochondria) by addition of sucrose (250 mM final concentration), are frozen in liquid nitrogen, and are stored in aliquots at -80°C.

III.   Standard Protocol for Protein Import into Isolated Mitochondria

The import reactions contain mitochondria (10–50 μg of mitochondrial protein) that were isolated from *Neurospora crassa* (Pfanner and Neupert, 1985, 1986; Hartl *et al.*, 1986) or the yeast *Saccharomyces cerevisiae* (Daum *et al.*, 1982; Hartl *et al.*, 1987) as described, and are resuspended in SEM buffer [250 mM sucrose, 1 mM ethylenediaminetetraacetic acid (EDTA), 10–20 mM 3-(N-morpholino)propanesulfonic acid (MOPS) adjusted to pH 7.2 with KOH or NaOH] at a protein concentration of 1–5 mg/ml; rabbit reticulocyte lysate [1–30% (v/v)] containing [^{35}S]methionine-labeled precursor protein(s); bovine serum albumin (BSA) buffer [3% (w/v) BSA, 250 mM sucrose, 50–120 mM KCl, 0–5 mM $MgCl_2$, 10–40 mM MOPS adjusted to pH 7.2 by KOH or NaOH]; and 5 mM unlabeled methionine (leading to a strong reduction of "unspecifically" labeled protein bands that are associated with the mitochondria isolated after the import reaction); as energy substrate either 2 mM NADH or 8 mM potassium ascorbate (pH 7) plus 0.2 mM N,N,N',N'-tetramethylphenylenediamine (TMPD) are included (the ATP levels present in reticulocyte lysate are usually sufficient for the import reaction). The final volume of the import reaction is usually 200 μl (with a variation from 50 to 600 μl). The amount of BSA buffer added is variable, depending on the volume of the other additions; the final concentration of BSA in the import reaction thus is about 2–2.5% (w/v). Most precursor proteins (e.g., ADP/ATP carrier, cytochrome c_1, porin, cytochrome b_2) are efficiently imported at low concentrations of reticulocyte lysate [1–10% (v/v)]. Import of a few precursor proteins, in particular the β-subunit of F_1-ATPase, is enhanced by increasing the concentration of reticulocyte lysate in the import reaction up to 30% (probably because of a requirement for cytosolic cofactors.)

All additions except of reticulocyte lysate and mitochondria are mixed in a 1.5-ml tube. Then the reticulocyte lysate (thawed immediately before use) is added. After mixing, the isolated mitochondria [freshly isolated (*N. crassa*) or from aliquots that were stored at $-80°C$ (yeast)] are added and the samples are gently mixed and incubated at 25°C. The incubation time is from 3 minutes up to 25 minutes. With most precursor proteins, the rate of import is linear depending on the incubation time, up to 8–10 min (with F_1-ATPase β-subunit at least up to 20 minutes).

The samples are cooled to 0°C and the mitochondria are reisolated by centrifugation at 10,000–20,000 g for 12–20 minutes. Protease treatment of mitochondria is performed either before the reisolation of mitochondria by adding proteinase K (10–200 μg/ml) or trypsin (20–100 μg/ml) to the mixture containing reticulocyte lysate, BSA buffer, and mitochondria or after the reisolation of mitochondria by resuspending the mitochondria in SEM buffer or BSA buffer and adding protease (in SEM buffer roughly three times less protease is needed than in BSA buffer). Protease treatment is performed for 10–20 minutes at 0°C and stopped by the addition of 1–2 mM phenylmethylsulfonyl fluoride (PMSF) in the case of proteinase K or a 10- to 30-fold weight excess of soybean trypsin inhibitor in the case of trypsin, followed by a further incubation for 5–10 minutes at 0°C. The reisolated mitochondria are then analyzed by SDS–polyacrylamide gel electrophoresis (Laemmli, 1970) and fluorography of the dried gel (Chamberlain, 1979).

For dissipation of the mitochondrial membrane potential, NADH and potassium ascorbate plus TMPD are omitted and 0.1–1 μM valinomycin (a potassium ionophore; final concentration of KCl in the import reaction is 70–80 mM or higher), 8 μM antimycin A, and 20 μM oligomycin are included. With *N. crassa* mitochondria, a sequential dissipation and reestablishment of the membrane potential is possible (a convenient method for accumulation and chase of translocation intermediates) (Zwizinski *et al.*, 1983; Pfanner and Neupert, 1985, 1987). For this purpose a membrane potential is generated by addition of NADH (via complex I of the respiratory chain), then oligomycin (block of the F_0F_1-ATPase) and antimycin A (virtually complete block of complex III of the respiratory chain in *Neurospora* mitochondria) are added, leading to dissipation of the membrane potential. By addition of potassium ascorbate plus TMPD, a membrane potential is generated via complex IV of the respiratory chain that can finally be dissipated by addition of potassium cyanide or (conveniently) valinomycin.

IV. Translocation Intermediates of the ADP/ATP Carrier

A. Receptor-Bound Intermediate

Accumulation of the precursor of the ADP/ATP carrier at its surface receptor (MOM72) is performed by incubation of reticulocyte lysate and mitochondria at low levels of ATP and in the absence of a membrane potential (Fig. 1) (Pfanner *et al.*, 1987; Pfaller *et al.*, 1988; Söllner *et al.*, 1988, 1990). Mitochondria (1 mg protein/ml SEM buffer) and reticulocyte lysate containing the radiolabeled precursor are separately treated with 5 U apyrase/ml for 20 minutes at 0 or 25°C, respectively (apyrase is an ATPase and ADPase from potato). The reticulocyte lysate is added to the mitochondria in BSA buffer in the presence of valinomycin, antimycin A, and oligomycin (to dissipate the membrane potential). The mixture is incubated for 5–10 minutes at 25°C and then cooled to 0°C. The mitochondria are reisolated, washed with 200 μl of SEM buffer, and the proteins are separated by SDS–polyacrylamide gel electrophoresis and analyzed as described above.

The reticulocyte lysate should be used freshly or after only one cycle of freeze-thawing to avoid a high unspecific (nonproductive) binding of the precursor of ADP/ATP carrier [the amount of unspecifically bound precursor is determined by using mitochondria pretreated with trypsin (10–20 μg/ml) in a parallel sample (Pfaller *et al.*, 1988; Söllner *et al.*, 1989, 1990)]. To test for the accumulation of the ADP/ATP carrier on the mitochondrial surface, an aliquot is treated with low concentrations of protease (20 μg trypsin/ml) before the reisolation of mitochondria (Fig. 1).

The productivity of the binding can be examined by the further transport ("chase") of the receptor-bound ADP/ATP carrier to the GIP site. For this purpose, the samples are treated as described before except that 1 U apyrase/ml is used. By addition of nucleoside triphosphates (e.g., 6 mM GTP that can be used to generate ATP by nucleoside diphosphate kinases present in the *in vitro* system) and incubation for 20 minutes at 25°C, the ADP/ATP carrier is translocated from its receptor to GIP. With *N. crassa*

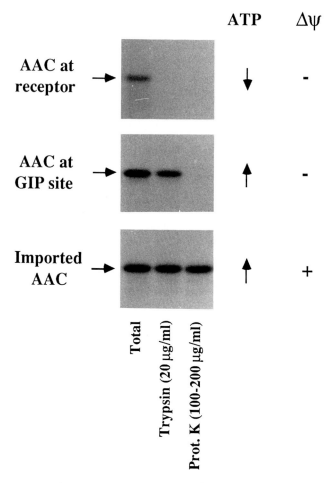

FIG. 1. Translocation intermediates of the ADP/ATP carrier. Reticulocyte lysates containing radiolabeled precursor of the ADP/ATP carrier (AAC) and isolated *Neurospora crassa* mitochondria were incubated at 25°C, and aliquots were then treated with protease as described in the text. The reisolated mitochondria were analyzed by SDS–polyacrylamide gel electrophoresis and fluorography of the dried gel. Arrows indicate high or low levels of ATP; $\Delta\Psi$ is the electrical potential across the mitochondrial inner membrane; AAC, ADP/ATP carrier; GIP, general insertion protein in the mitochondrial outer membrane; Prot. K, proteinase K.

mitochondria, the completion of transport into the inner membrane can be analyzed: valinomycin is omitted in the binding reaction, then GTP and potassium ascorbate plus TMPD are added (to reestablish a membrane potential), leading to transfer of the ADP/ATP carrier into the inner membrane and its assembly into the functionally active dimeric form (Pfanner *et al.*, 1987; Klingenberg, 1985).

B. GIP Intermediate

The ADP/ATP carrier is accumulated at the general insertion site GIP in the outer membrane by incubation of reticulocyte lysate (containing radiolabeled precursor) with mitochondria in the absence of a membrane potential, but in the presence of ATP (Pfanner and Neupert, 1987; Pfaller *et al.*, 1988). The incubation is performed in BSA buffer in the presence of valinomycin, antimycin A, and oligomycin for 20 minutes at 25°C. The samples are cooled to 0°C and treated with 20 μg trypsin/ml to remove receptor-bound and unspecifically bound ADP/ATP carrier. To exclude transport of the precursor into the inner membrane, an aliquot is treated with 200 μg proteinase K/ml (in BSA buffer) or (after reisolation of mitochondria) with 100 mM Na$_2$CO$_3$ (pH 11.5) (Fujiki *et al.*, 1982a,b; Pfanner and Neupert, 1987). In contrast to completely imported ADP/ATP carrier, the precursor accumulated at the GIP site is digested by high concentrations of protease and can be extracted at alkaline pH.

To complete the translocation of ADP/ATP carrier from the GIP site into the inner membrane (*N. crassa* mitochondria), a membrane potential is generated. The ADP/ATP carrier is accumulated at the GIP site as described before, but without addition of valinomycin. Then an aliquot is incubated in the presence of potassium ascorbate plus TMPD for 25 minutes at 20°C. Treatment with proteinase K (200 μg/ml) or at alkaline pH allows the examination of the translocation event. Furthermore, the assembly of ADP/ATP carrier to the dimeric form in the inner membrane can be tested by using the specific inhibitor carboxyatractyloside that only binds to the assembled dimer. This ADP/ATP carrier–inhibitor complex will pass through hydroxylapatite columns at low ionic strength (Klingenberg *et al.*, 1979; Schleyer and Neupert, 1984; Pfanner *et al.*, 1987).

V. Translocation Intermediates in Contact Sites

Most mitochondrial precursor proteins are proteolytically processed during or after translocation across the mitochondrial membranes. This provides a convenient assay for the translocation of the presequence into the mitochondrial matrix. Under special import conditions, specifically processed precursor proteins remain accessible to externally added proteases [the intactness of mitochondrial outer and inner membranes is controlled by the analysis of marker proteins for the mitochondrial subcompartments, e.g, adenylate kinase or cytochrome c for the intermembrane space and fumarase or isocitrate dehydrogenase for the matrix (Schleyer and Neupert, 1985; Hartl et al., 1986, 1987; Schwaiger et al., 1987)]. The susceptibility of the precursor protein to proteolytic cleavage both on the matrix side (processing peptidase) and on the cytosolic side [usually proteinase K (20–50 μg/ml) for 10–25 minutes at 0°C] demonstrates the accumulation of the precursor as a translocation intermediate in contact sites between both mitochondrial membranes (Fig. 2) (Schleyer and Neupert, 1985; Schwaiger et al., 1987). Such contact site intermediates can be created by different means.

A. Low-Temperature Intermediate

The original and basic procedure to create a translocation intermediate is to perform the import reaction at low temperature (2–15°C) (Schleyer and Neupert, 1985). The translocation block is reversible (by raising the temperature to 25°C), and the approach has the advantage that a chemical manipulation of the import system is avoided. In a typical experiment, precursors of the F_1-ATPase β-subunit (Fig. 2), of cytochrome c_1, or a fusion protein between the 167 amino-terminal amino acid residues of the precursor of cytochrome b_2 and the entire dihydrofolate reductase [cytochrome b_2 (1–167)–DHFR] are imported into N. crassa or yeast mitochondria for 5–15 minutes at 8°C. Thereby, more than 90% of the processed precursors accumulate as contact site intermediates, i.e., are accessible to externally added protease. Experiments with contact site intermediates should include a control sample with deenergized mitochondria (addition of valinomycin, oligomycin, and antimycin A) to exclude processing of the precursors outside the mitochondria (e.g., by released processing peptidase) (Fig. 2).

The low-temperature intermediates are reasonably stable for at least 30 minutes at 0°C. However, because these intermediates are kinetic intermediates (probably due to a retardation of the unfolding of the precursors), they are not stable enough to allow extended manipulations (e.g., subfractionation) of the mitochondria carrying the intermediate.

B. Low-ATP Intermediate

With many precursor proteins, complete translocation requires the hydrolysis of nucleoside triphosphates (NTPs) in the cytosol, probably to allow release of the precursor proteins from cytosolic cofactors such as the

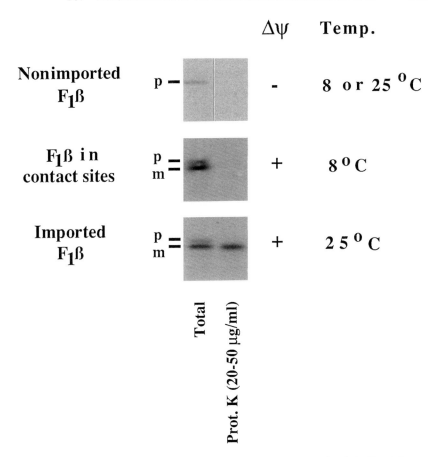

FIG. 2. Contact site intermediate (low-temperature intermediate) of the F_1-ATPaseβ-subunit. Reticulocyte lysates containing radiolabeled precursor (p) of the F_1-ATPase β-subunit ($F_1\beta$) and isolated *Neurospora crassa* mitochondria were incubated at different temperatures (in the presence of ATP), and aliquots were then treated with protease as described in the text. The reisolated mitochondria were analyzed by SDS–polyacrylamide gel electrophoresis and fluorography of the dried gel; $\Delta\Psi$ is the electrical potential across the mitochondrial inner membrane; m, mature-sized (processed) $F_1\beta$; Prot. K, proteinase K.

70-kDa heat-shock proteins (hsp70) (Rothman, 1989; Pfanner *et al.*, 1990). It is thus possible to control the translocation by depletion and readdition of NTPs (Pfanner and Neupert, 1986; Pfanner *et al.*, 1987; Chen and Douglas, 1987a; Eilers *et al.*, 1988).

To accumulate the precursor of the F_1-ATPase β-subunit in contact sites of *N. crassa* mitochondria, the reticulocyte lysate and the mitochondria are pretreated with apyrase (0.5 U/ml) at 25°C for 25 or 15 minutes, respectively (Pfanner et al., 1987). Upon mixing of mitochondria and reticulocyte lysate, the import reaction is carried out in the presence of 20 μM oligomycin and 6 mM ADP for 10–25 minutes at 25°C. Under these conditions the ATP levels (generated from ADP) are sufficient to promote the translocation of an amino-terminal portion of the precursor (including the presequence), but a complete translocation of the precursor protein is prevented. This contact site intermediate is probably more stable than the low-temperature intermediate (Pon et al., 1989).

The precursor proteins can be fully imported into the matrix by adding nucleoside triphosphates (e.g., 6 mM GTP) and performing a second incubation for 15 minutes at 25°C (with or without reisolation of the mitochondria). The completion of translocation does not require an energized inner membrane, i.e., it can occur in the presence of valinomycin (the same $\Delta\Psi$ independence is found with the low-temperature and the reversible-folding intermediate).

C. Reversible-Folding Intermediate

Protein domains cannot traverse the mitochondrial membranes in a fixed tertiary structure (Eilers and Schatz, 1986) The best characterized system for this purpose is mouse DHFR fused to amino-terminal parts of mitochondrial precursor proteins. Reticulocyte lysates containing the fusion protein cytochrome b_2 (1–167)–DHFR are preincubated with 100–1000 nM methotrexate (a specific ligand that stabilizes the tertiary structure of DHFR) for 5 minutes at 0°C (Rassow et al., 1989). Then isolated mitochondria are added and the import reaction is performed for 5–15 minutes at 25°C. The folded DHFR exhibits a high endogenous resistance to proteases and can be cleaved off from the translocation intermediate by proteinase K (1–50 $\mu g/ml$). With this assay, the localization as well as the folding state of the DHFR domain can be determined (Rassow et al., 1989; Ostermann et al., 1989). The reversible-folding intermediate is more stable than the low-temperature and the low-ATP intermediates (it is stable for at least several hours).

The folding of DHFR can be reversed and the translocation of fusion proteins can be completed by dilution of methotrexate, i.e., reisolation of the mitochondria, followed by a second incubation for 5–40 minutes at 25°C in the absence of methotrexate. A similar reversible-folding intermediate can be generated by use of yeast copper metallothionein in a fusion protein (with copper as the specific ligand) (Chen and Douglas, 1987b). To become processed in the mitochondrial matrix, fusion proteins need at least 50 amino acid residues (that span both mitochondrial membranes) between the site of proteolytic cleavage by processing peptidase (i.e., the carboxyl-terminal end of the presequence) and the folded domain on the cytosolic side (Rassow *et al.*, 1990).

D. Antibody Intermediate

The complete translocation of a precursor protein through contact sites can be stopped directly by the addition of antibodies directed against non-amino-terminal (carboxyl-terminal) portions of the precursor (Schleyer and Neupert, 1985; Schwaiger *et al.*, 1987). The reticulocyte lysates containing the radiolabeled precursors are incubated with the corresponding antiserum for 15 minutes at 25°C. Large precursor–antibody complexes are removed by a centrifugation for 15 minutes at 24,000 g. To the supernatant, isolated mitochondria are added and the import reaction is performed. The optimal ratio of reticulocyte lysate and antiserum has to be determined. In case of the F_1-ATPase β-subunit we found the optimal ratio to be roughly 100:1 (v/v) reticulocyte lysate to rabbit polyclonal antiserum (Schleyer and Neupert, 1985; Schwaiger *et al.*, 1987).

The antibody intermediate is not a true translocation intermediate because it is irreversible (i.e., the completion of translocation can not be analyzed). However, the completely stable arrest of a precursor in contact sites is a very helpful tool for analysis of contact sites, e.g., by subfractionation of mitochondria (Schwaiger *et al.*, 1987). With a hybrid protein containing the tightly folded bovine pancreatic trypsin inhibitor at the carboxyl terminus, a similar contact site intermediate could be generated (Vestweber and Schatz, 1988).

VI.   Identification of Surface Components of the Mitochondrial Import Machinery by Use of Inhibitory Antibodies

To test the effect of antibodies on import of precursor proteins into isolated mitochondria (Söllner *et al.*, 1989, 1990), immunoglobulins G (IgGs) are isolated from polyclonal rabbit antisera by protein A– Superose chromatography (Pharmacia). Bound IgGs are eluted with 0.1 *M* citrate (pH 3), immediately neutralized with 2 *M* Tris (pH 8), and concentrated by lyophilization. The lyophilized IgGs are solubilized in SEM buffer and stored as aliquots at −20°C. IgGs (5–150 μg) are prebound to mitochondria (10 μg of protein) in BSA buffer in the presence of protease inhibitors [100 μg/ml of cytosolic protease inhibitor fraction from

N. crassa (Schmidt *et al.*, 1984) and 100 μg α_2-macroglobulin/ml] in a final volume of 50–150 μl. As controls, IgGs prepared from preimmune serum that was obtained from rabbits before injection with mitochondrial antigens and IgGs directed against the major mitochondrial outer membrane protein (porin) are used. The incubation is performed for 30–45 minutes at 0°C. Then the mitochondria are reisolated by centrifugation and incubated with reticulocyte lysate containing radiolabeled precursor proteins in the presence of different agents (the reisolation of mitochondria can be omitted). The choice of the agents and the further treatment of the samples depend on the distinct translocation intermediates that will be generated (see above).

To avoid unspecific effects caused by divalent IgGs, Fab fragments are isolated. The preparation is performed by digestion of purified IgGs with soluble papain according to Mage (1980) or with immobilized papain (Pierce). Fab fragments are separated from intact IgGs and Fc fragments by protein A–Superose chromatography, dialyzed against water, and lyophylized. The Fab fragments are solubilized in SEM buffer and are stored as aliquots at − 20°C or for shorter periods at 4°C. The Fab fragment are used freshly or after only one cycle of freeze-thawing. Prebinding of Fab fragments to mitochondria is performed as described for IgGs.

VII. Conclusions

Most translocation intermediates of mitochondrial precursor proteins are obtained by variations of a basic import scheme. Thereby, receptor sites, the GIP site, and contact sites can be titrated and their properties can be analyzed (Pfaller and Neupert, 1987; Pfaller *et al.*, 1988; Vestweber and Schatz, 1988; Rassow *et al.*, 1989). Translocation intermediates are an important tool to analyze the function of putative components of the import apparatus that were or will be identified by biochemical means (e.g., inhibitory antibodies) (Söllner *et al.*, 1989, 1990) or by genetic means (such as mutants defective in import) (Kang *et al.*, 1990). In fact, the use of translocation intermediates is one of the most important controls to define a certain inhibitory (or stimulatory) condition and characterize its specificity by showing that only a distinct import step is affected. The analysis of subreactions of mitochondrial protein import, such as binding of precursor proteins to receptors, membrane insertion, or membrane translocation of precursors, should provide the basis for a reconstitution of these reactions with purified components.

ACKNOWLEDGMENTS

We thank Dr. Walter Neupert for helpful comments. Studies in the authors' laboratory were supported by the Deutsche Forschungsgemeinschaft (Sonderforschungsbereich 184).

REFERENCES

Attardi, G., and Schatz, G. (1988). *Annu. Rev. Cell Biol.* **4,** 289–333.

Chamberlain, J. P. (1979). *Anal. Biochem.* **98,** 132–135.

Chen, W.-J., and Douglas, M. G. (1987a). *Cell (Cambridge, Mass.)* **49,** 651–658.

Chen, W.-J., and Douglas, M. G. (1987b). *J. Biol. Chem.* **262,** 15605–15609.

Daum, G., Gasser, S. M., and Schatz, G. (1982). *J. Biol. Chem.* **257,** 13075–13080.

Eilers, M., and Schatz, G. (1986). *Nature (London)* **322,** 228–232.

Eilers, M., Hwang, S., and Schatz, G. (1988). *EMBO J.* **7,** 1139–1145.

Fujiki, Y., Hubbard, A. L., Fowler, S., and Lazarow, P. B. (1982a). *J. Cell Biol.* **93,** 97–102.

Fujiki, Y., Fowler, S., Shio, H., Hubbard, A. L., and Lazarow, P. B. (1982b). *J. Cell Biol.* **93,** 103–110.

Hartl, F.-U., and Neupert, W. (1990). *Science* **247,** 930–938.

Hartl, F.-U., Schmidt, B., Weiss, H., Wachter, E., and Neupert, W. (1986). *Cell (Cambridge, Mass.)* **47,** 939–951.

Hartl, F.-U., Ostermann, J., Guiard, B., and Neupert, W. (1987). *Cell (Cambridge, Mass.)* **51,** 1027–1037.

Horwich, A. (1990). *Curr. Opin. Cell Biol.* **2,** 625–633.

Kang, P.-J., Ostermann, J., Shilling, J., Neupert, W., Craig, E. A., and Pfanner, N. (1990). *Nature (London)* **348,** 137–143.

Klingenberg, M. (1985). *Ann. N. Y. Acad. Sci.* **456,** 279–288.

Klingenberg, M., Aquila, H., and Riccio, P. (1979). *In* "Methods in Enzymology" (S. Fleischer and L. Packer, eds.) Vol. 56, pp. 407–414. Academic Press, New York.

Krieg, P. A., and Melton, D. A. (1984). *Nucleic Acids Res.* **12,** 7057–7070.

Laemmli, U. K. (1970). *Nature (London)* **227,** 680–685.

Mage, M. G. (1980). *In* "Methods in Enzymology" (H., Van Vunakis and J. J. Langone, eds.), Vol. **70,** pp. 142–150. Academic Press, New York.

Melton, D. A., Krieg, P. A., Rebagliati, M. R., Maniatis, T., Zinn, K., and Green, M. R. (1984). *Nucleic Acids Res.* **12,** 7035–7056.

Ostermann, J., Horwich, A. L., Neupert, W., and Hartl, F.-U. (1989). *Nature (London)* **341,** 125–130.

Pelham, H. R. B., and Jackson, R. J. (1976). *Eur. J. Biochem.* **67,** 247–256.

Pfaller, R., and Neupert, W. (1987). *EMBO J.* **6,** 2635–2642.

Pfaller, R., Steger, H. F., Rassow, J., Pfanner, N., and Neupert, W. (1988). *J. Cell Biol.* **107,** 2483–2490.

Pfanner, N., and Neupert, W. (1985). EMBO J. **4,** 2819–2825.

Pfanner, N., and Neupert, W. (1986). *FEBS Lett.* **209,** 152–156.

Pfanner, N., and Neupert, W. (1987). *J. Biol. Chem.* **262,** 7528–7536.

Pfanner, N., and Neupert, W. (1990). *Annu. Rev. Biochem.* **59,** 331–353.

Pfanner, N., Tropschug, M., and Neupert, W. (1987). *Cell (Cambridge, Mass.)* **49,** 815–823.

Pfanner, N., Rassow, J., Guiard, B., Söllner, T., Hartl, F.-U., and Neupert, W. (1990). *J. Biol. Chem.* **265,** 16324–16329.

Pon, L., Moll, T., Vestweber, D., Marshallsay, B., and Schatz, G. (1989). *J. Cell Biol.* **109,** 2603–2616.

Rassow, J., Guiard, B., Wienhues, U., Herzog, V., Hartl, F.-U., and Neupert, W. (1989). *J. Cell Biol.* **109,** 1421–1428.

Rassow, J., Hartl, F.-U., Guiard, B., Pfanner, N., and Neupert, W. (1990). *FEBS Lett.* **275,** 190–194.

Rothman, J. E. (1989). *Cell (Cambridge, Mass.)* **59,** 591–601.

Sambrook, J., Fritsch, E. F., and Maniatis, T. (1989). "Molecular Cloning: A Laboratory Manual." 2nd ed. Cold Spring Harbor Lab., Cold Spring Harbor, New York.

Schleyer, M., and Neupert, W. (1984). *J. Biol. Chem.* **259,** 3487–3491.

Schleyer, M., and Neupert, W. (1985). *Cell (Cambridge, Mass.)* **43,** 339–350.

Schmidt, B., Wachter, E., Sebald, W., and Neupert, W. (1984). *Eur. J. Biochem.* **144,** 581–588.

Schwaiger, M., Herzog, V., and Neupert, W. (1987). *J. Cell Biol.* **105,** 235–246.

Söllner, T., Pfanner, N., and Neupert, W. (1988). *FEBS Lett.* **229,** 25–29.

Söllner, T., Griffiths, G., Pfaller, R., Pfanner, N., and Neupert, W. (1989). *Cell (Cambridge, Mass).* **59,** 1061–1070.

Söllner, T., Pfaller, R., Griffiths, G., Pfanner, N., and Neupert, W. (1990). *Cell (Cambridge, Mass.)* **62,** 107–115.

Stueber, D., Ibrahimi, I., Cutler, D., Dobberstein, B., and Bujard, H. (1984). *EMBO J.* **3,** 3143–3148.

Vestweber, D., and Schatz, G. (1988). *J. Cell Biol.* **107,** 2037–2043.

Zwizinski, C., Schleyer, M., and Neupert, W. (1983). *J. Biol. Chem.* **258,** 4071–4074.

Chapter 17

Import of Precursor Proteins into Yeast Submitochondrial Particles

THOMAS JASCUR

Department of Biochemistry
Biocenter, University of Basel
CH-4056 Basel, Switzerland

I. Introduction

Import of precursor proteins into the mitochondrial matrix involves translocation across both the outer and the inner membranes. Early observations had suggested that this process occurs at sites of close contact between the two membranes (Kellems *et al.*, 1975; Ades and Butow, 1980; Suissa, and Schatz, 1982). Recently, it was directly demonstrated that precursor proteins that were stuck in the import sites of isolated mitochondria were located at membrane contact sites (Schleyer and Neupert, 1985; Schwaiger *et al.*, 1987; Pon *et al.*, 1989).

359

However, recent studies imply that the two mitochondrial membranes contain two separate translocation systems that operate in tandem, contact sites representing the point where precursors are handed over from the outer membrane to the inner membrane import machinery (Hwang *et al.,* 1991; Jascur *et al.,* 1991). This model was first suggested by the observation that import into intact mitochondria can be blocked either by antibodies against certain outer membrane proteins (Ohba and Schatz, 1987a), by treating intact mitochondria with trypsin (Ohba and Schatz, 1987b), or by jamming import sites with an artificial precursor protein (Vestweber and Schatz, 1988). Import into the matrix is restored after each block by disrupting the outer membrane with an osmotic shock; this suggests that the inner membrane alone contains a complete protein translocation system that can import precursors into the matrix (Ohba and Schatz, 1987b; Hwang *et al.,* 1989). Import into mitoplasts closely resembles import into intact mitochondria: it requires a mitochondrial-targeting signal at the N-terminus of the precursor protein, unfolding of the precursor, an electrochemical potential ($\Delta\Psi$) across the inner membrane, and ATP hydrolysis in the matrix. However, it is apparently independent of import receptors that function in the outer membrane.

Everything that is true for import into mitoplasts also holds for import into inner membrane vesicles. Additionally, inner membrane vesicles have the advantage of being free from contaminating outer membrane fragments that are attached to mitoplasts. These vesicles represent a purer system, which is essential for biochemical work (Hwang *et al.,* 1989).

Inner membrane vesicles import both authentic and artificial precursors with an efficiency comparable to that of intact mitochondria (based on equal amounts of inner membrane marker proteins in all samples). Inner membrane vesicles from yeast are mainly in a right-side-out orientation (matrix side inside), in contrast to submitochondrial particles from mammalian cells, which tend to be inside-out. So far, it has not been possible to isolate inside-out inner membrane vesicles from yeast mitochondria.

Import into inner membrane vesicles, like import into intact mitochondria, requires an electrochemical potential across the inner membrane. In contrast to mitoplasts, however, $\Delta\Psi$ in vesicles can only be generated by respiration, e.g., with L-ascorbate and cytochrome *c* as substrates. It is not possible to set up a potential via the F_0F_1-ATPase, probably because the ATPase is damaged by the high levels of EDTA in the preparation. The vesicles contain significant amounts of matrix proteins, such as heat-shock proteins involved in import of precursors, and matrix protease, which cleaves the presequence from precursors. However, the activity of the matrix protease is less than in intact mitochondria, which results in incomplete processing of imported precursors.

The following sections outline procedures for preparing mitoplasts and inner membrane vesicles and for assaying the import of precursors into these preparations.

II.   Import into Mitoplasts

A.   Inactivation of Outer Membrane Components

1.   TRYPSIN TREATMENT OF MITOCHONDRIA

1. Suspend mitochondria in 0.6 M mannitol, 20 mM HEPES–KOH, pH 7.4, at 10 mg/ml.
2. Add L-1-p-tosylamino-2-phenylethyl chloromethyl ketone (TPCK)-treated trypsin to 1 mg/ml and incubate for 30 minutes at 0°C.
3. Stop proteolysis by adding a 10-fold excess (w/w) of soybean trypsin inhibitor (STI) and incubation for an additional 10 minutes at 0°C.
4. Reisolate the mitochondria by centrifugation at 12,000 g for 5 minutes.
5. Resuspend in 0.6 M mannitol, 20 mM HEPES–KOH, pH 7.4, 1 mg/ml STI.
6. Resediment through 0.6 ml of 1 M mannitol, 20 mM HEPES–KOH, pH 7.4, 1 mg/ml STI.
7. Resuspend in 0.6M mannitol, 20 mM HEPES–KOH, pH 7.4, 1 mg/ml STI at 10 mg/ml.
8. Treat control mitochondria in exactly the same manner, but omit trypsin.

2.   ANTIBODY TREATMENT OF MITOCHONDRIA

1. Incubate mitochondria in 0.6 M mannitol, 20 mM HEPES–KOH, pH 7.4, with an appropriate amount of IgGs or Fab fragments for 1 hour at 0°C. For every IgG or Fab preparation, optimal conditions must be determined. Generally, a 2- to 20-fold excess of IgG over mitochondria (on a protein basis) will be sufficient (Ohba and Schatz, 1987a; Vestweber et al., 1989; Söllner et al., 1989, 1990; Hines et al., 1990). As a control, IgGs directed against outer membrane proteins that are not involved in import of precursors (such as porin) can be used.
2. Reisolate mitochondria by sedimentation through 0.6 ml of 1 M mannitol, 20 mM HEPES–KOH, pH 7.4.

3. Resuspend in 0.6 M mannitol, 20 mM HEPES–KOH, pH 7.4, at 10 mg/ml.

B. Preparation of Mitoplasts by Osmotic Shock

1. Dilute mitochondria (untreated, trypsin treated, or antibody treated) with 9 volumes of 20 mM HEPES–KOH, pH 7.4, and incubate for 20 minutes at 0°C with occasional agitation on a Vortex mixer to disrupt the outer membrane.
2. Reisolate the resulting mitoplasts by centrifugation at 12,000 g for 10 minutes.
3. Resuspend in 0.6 M mannitol, 20 mM HEPES–KOH, pH 7.4.
   Note: 1 mg/ml STI should be added to all buffers if the mitochondria have been trypsin-treated.

C. Assaying the Effectiveness of Conversion of Mitochondria to Mitoplasts

1. Subject 50-μg samples of mitochondria and mitoplasts to SDS–12% PAGE.
2. Transfer the separated proteins to a nitrocellulose membrane by electroblotting.
3. Assay the amounts of a matrix marker (e.g., citrate synthase) and a soluble intermembrane space marker (e.g., cytochrome b_2) by quantitative immune blotting.
4. Calculate the efficiency of conversion into mitoplasts as follows:

$$\% \text{ conversion} = \left[1 - \frac{b_2/\text{CS} \quad (\text{mtp})}{b_2/\text{CS} \quad (\text{mito})}\right]100$$

where b_2 is cytochrome b_2, CS is citrate synthase, mtp is mitoplasts, and mito is mitochondria. Usually, 90% of the mitochondria are converted to mitoplasts.

D. Import of Precursor Proteins into Mitoplasts

Import into mitoplasts is performed exactly as import into intact mitochondria (Ohba and Schatz, 1987b; van Loon and Schatz, 1987; Hwang et al., 1989). Mitoplasts, like mitochondria, can set up a membrane potential across the inner membrane by respiration or by hydrolyzing externally added ATP via the F_0F_1-ATPase (Eilers et al., 1987). (For further information, follow the protocol of Glick, Chapter 20.)

III. Import into Inner Membrane Vesicles

A. Isolation of Right-Side-Out Inner Membrane Vesicles from Yeast Mitochondria

1. Isolate mitochondria using breaking buffer supplemented with 10 mM EDTA for the homogenization steps and suspend them in 0.6 M sorbitol, 20 mM HEPES–KOH, pH 7.4, at 20 mg/ml. [We usually grow 10- or 40-liter cultures in semisynthetic lactate medium. A 10-liter culture usually yields approximately 80 g of cells (wet weight), 300 mg of mitochondrial protein, and 10 mg of inner membrane vesicles (protein)]. EDTA serves to strip the mitochondria of cytoplasmic ribosomes that are bound to contact sites; this results in a better separation of inner membrane vesicles from contact site vesicles (Pon et al., 1989).
2. Add 9 volumes of 20 mM HEPES–KOH, pH 7.4, 1 mM phenylmethylsulfonyl fluoride (PMSF), and incubate for 30 minutes at 0°C.
3. Add one-third volume of 1.8 M sucrose (to give a final sucrose concentration of 0.45 M) and incubate for another 10 minutes at 0°C.
4. Sonicate in 100- to 150-ml aliquots on ice. We use the following conditions: a model W-375 sonicator by Heat Systems-Ultrasonics, Inc., Farmingdale, NY, equipped with macrotip. Sonication is at full power, 80% duty cycle for a total of 3 minutes, including 15-second on/off intervals (i.e., a total of 1.5 minutes of sonication).
5. Remove unbroken mitochondria and large fragments by centrifugation at 30,000 g for 20 minutes at 4°C; collect the supernatant.
6. Pellet all membrane vesicles by ultracentrifugation at 200,000 g for 1 hour at 4°C.
7. Resuspend the pellet by homogenization in approximately 0.5 ml of 5 mM HEPES–KOH, 7.4, 10 mM KCl per 100 mg of starting mitochondria.
8. Load sample onto linear sucrose gradients (0.85–1.6 M sucrose in 5 mM HEPES–KOH, pH 7.4, 10 mM KCl); we use one 11-ml gradient for 100 mg of mitochondrial starting material.
9. Spin for 16 hours at 100,000 g 4°C, in a swing-out rotor (Kontron TST 41.14 or comparable).
10. Collect the densest of the three bands (containing the inner membrane vesicles) with a syringe (Fig. 1).
11. Dialyze against 0.6 M sorbitol, 20 mM KP$_i$, pH 7.4, for 4–6 hours with one change of buffer.

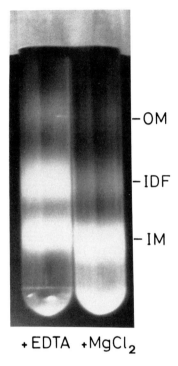

FIG. 1. The effect of EDTA on preparation of yeast mitochondrial inner membrane vesicles. Submitochondrial membrane vesicles were prepared from isolated yeast mitochondria and separated on a sucrose density gradient as described by Pon *et al.* (1989). (Left) Vesicles prepared from EDTA-treated mitochondria in $MgCl_2$-free buffers; (right) vesicles prepared from untreated mitochondria in buffers containing 2 mM $MgCl_2$. OM, Outer membrane; IDF, intermediate-density fraction; IM, inner membrane. Reproduced from *J. Cell Biol.* (1989), **109**, p. 2604 by copyright permission of the Rockefeller University Press.

12. Determine the protein concentration colorimetrically or by OD_{280} measurement in the presence of 0.6% SDS; 3 OD_{280} units correspond to 1 mg/ml of protein.
13. Add fatty acid-free bovine serum albumin to 1 mg/ml and freeze in aliquots in liquid nitrogen. Samples can be stored at −80°C.

B. Measurement of Membrane Potential

Semiquantitative measurement of $\Delta\Psi$ can be performed with a fluorescence spectrophotometric assay and the potential-sensitive fluorescent dye, 3,3′-dipropylthiocarbocyanine iodide (diS-C$_3$-(5); Molecular Probes, Inc., Junction City, OR) (Sims *et al.*, 1974). Excitation is at 620 nm, emission at 670 nm. The measurement is done at room temperature with stirring after addition of reagents. The reaction buffer is as follows:

> 0.6 M sorbitol
> 20 mM KP$_i$, pH 7.4
> 10 mM MgCl$_2$
> 0.5 mM EDTA
> 1 mg/ml bovine serum albumin

1. Add 2.5 ml of aerated reaction buffer to a 3-ml fluorescence cuvette.
2. Calibrate blank signal.
3. Add 5–10 μl of dye from a 1 mM stock solution in ethanol; this will give the maximum signal.
4. Add 10–50 μg of inner membrane vesicles; a small decrease of the signal is observed due to interaction of the dye with the membranes.
5. Add Na-ascorbate or Na-succinate to 5 mM.
6. Add cytochrome c (Sigma, type VI, from horse heart) to 40 μg/ml. Within a few seconds, a decrease in the fluorescence signal indicates establishment of a membrane potential (positive outside) if the vesicles are intact.
7. Uncouple the vesicles by adding valinomycin to 1 μg/ml. The signal reverses to the initial level (Fig. 2).

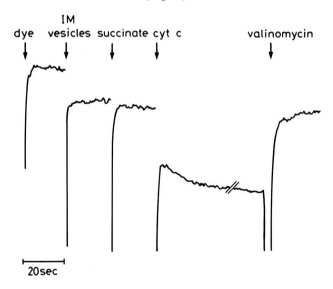

FIG. 2. Fluorescence measurement of a membrane potential generated by yeast mitochondrial inner membrane (IM) vesicles. The experiment was done as described in this article. Where indicated, dye was added to 4 μM, succinate to 5 mM, horse heart cytochrome c to 50 μg/ml, and valinomycin to 1 μg/ml. Fluorescence is given in arbitrary units; an increase in potential is represented by a downward deflection. The break in the fluorescence measurement represents 1.5 min. Reproduced from *J. Cell Biol.* (1989), **109,** p. 491 by copyright permission of the Rockefeller University Press.

Note: Carbonyl cyanide m-chlorophenylhydrazone (CCCP) is not useful for uncoupling because it interferes with the fluorescence assay. Ascorbate should be prepared freshly as a 0.2 M stock solution in 20 mM HEPES–KOH, pH 7.4.

C. Import of Precursors into Inner Membrane Vesicles

Import into inner membrane vesicles follows a protocol similar to that of import into intact mitochondria, with the modification that the import buffer is supplemented with ascorbate and cytochrome c (Hwang et al., 1989). Typically, we use 160-μl import reactions containing 10–20 μl of precursor solution and 10–40 μg of inner membrane vesicles. The precursor can either be synthesized by coupled in vitro transcription/translation or purified after overexpression and ^{35}S-labeling in Escherichia coli. We carry out import reactions in shallow 3-ml scintillation vials with vigorous shaking in a water bath in order to ensure good aeration. For every precursor, the linear range of import must be determined; 10 minutes at 30°C can serve as a starting point. We use the following import buffer (final concentrations in an import reaction after addition of the vesicles):

0.6 M sorbitol	10 mM L-malate
20 mM KP$_i$	1 mg/ml bovine serum albumin
5 mM MgCl$_2$	0.625 mg/ml cytochrome c
20 mM KCl	3.8 mM L-ascorbate
0.3 mM EDTA	0.9 mM ATP
0.6 mM DTT	8 mM creatine phosphate
10 mM succinate	12.5 units/ml creatine kinase

The samples are prewarmed in a shaking waterbath and the import reaction is then started by adding the precursor. It is stopped either by chilling on ice or by adding valinomycin to 10 μM in order to collapse the membrane potential.

Imported precursor is assayed by measuring protease-protected precursor: vesicles are treated with 100 μg/ml of proteinase K for 30 minutes at 0°C. However, depending on the precursor used, these conditions may have to be optimized. Proteinase K is best stopped by the procedure suggested by Glick (Chapter 20).

Inner membrane vesicles can be pelleted in an Airfuge (Beckman Instruments, Inc., Fullerton, CA) at 130,000 g for 20 minutes at 4°C. Analysis of the reaction products by SDS–PAGE and fluorography is done as described for intact mitochondria (Glick, Chapter 20).

IV. Outlook

Current research on mitochondrial protein import aims at the identification of mitochondrial proteins that mediate import of precursor proteins into the organelle. Several such components have already been identified: some are localized in the outer membrane (Vestweber *et al.*, 1989; Söllner *et al.*, 1989, 1990; Hines *et al.*, 1990; Murakami *et al.*, 1990; Pain *et al.*, 1990), others in the matrix (Hawlitschek *et al.*, 1988; Yang *et al.*, 1988; Ostermann *et al.*, 1989; Kang *et al.*, 1990; Scherer *et al.*, 1990). However, no inner membrane protein has so far been shown to be involved in precursor import. In order to identify such proteins, biochemical systems are needed that allow one to study the inner membrane alone and to manipulate it with biochemical methods. Reconstitution of protein translocation activity into proteoliposomes has been achieved with the *E. coli* export system (Driessen and Wickner, 1990) and with microsomes from endoplasmic reticulum (Nicchitta and Blobel, 1990). Yeast mitochondrial inner membrane vesicles will provide a starting point to make such a system available for studying mitochondrial protein import in more detail.

REFERENCES

Ades, I. Z., and Butow, R. A. 1980. The products of mitochondria-bound cytoplasmic polysomes in yeast. *J. Biol. Chem.* **255**, 9918–9924.

Driessen, A. J. M., and Wickner, W. (1990). Solubilization and functional reconstitution of the protein-translocation enzymes of *Escherichia coli*. *Proc. Natl. Acad. Sci. U.S.A.* **87**, 3107–3111.

Eilers, M., Oppliger, W., and Schatz, G. (1987). Both ATP and an energized inner membrane are required to import a purified precursor protein into mitochondria. *EMBO J.* **6**, 1073–1077.

Hawlitschek, G., Schneider, H., Schmidt, B., Tropschug, M., Hartl, F.-U., and Neupert, W. (1988). Mitochondrial protein import: Identification of processing peptidase and of PEP, a processing enhancing protein. *Cell (Cambridge, Mass.)* **53**, 795–806.

Hines, V., Brandt, A., Griffiths, G., Horstmann, H., Brütsch, H. and Schatz, G. (1990). Protein import into yeast mitochondria is accelerated by the outer membrane protein MAS70. *EMBO J.* **9**, 3191–3200.

Hwang, S., Jascur, T., Vestweber, D., Pon, L., and Schatz, G. (1989). Disrupted yeast mitochondria can import precursor proteins directly through their inner membrane. *J. Cell Biol.* **109**, 487–493.

Hwang, S. *et al.* (1991). In preparation.

Jascur, T. *et al.* (1991). In preparation.

Kang, P.-J., Ostermann, J., Shilling, J., Neupert, W., Craig, E. A., and Pfanner, N. (1990). Requirement for hsp70 in the mitochondrial matrix for translocation and folding of precursor proteins. *Nature (London)* **348**, 137–143.

Kellems, R. E., Allison, V. F., and Butow, R. A. (1975). Cytoplasmic type 80S ribosomes associated with yeast mitochondria. IV. Attachment of ribosomes to the outer membrane of isolated mitochondria. *J. Cell Biol.* **65**, 1–14.

Mukarami, H., Blobel, G., and Pain, D. (1990). Isolation and characterization of the gene for a yeast mitochondrial import receptor. *Nature (London)* **347**, 488–491.

Nicchitta, C. V., and Blobel, G. (1990). Assembly of translocation-competent proteolipo-somes from detergent-solubilized rough microsomes. *Cell (Cambridge, Mass.)* **60**, 259–269.

Ohba, M., and Schatz, G. (1987a). Protein import into yeast mitochondria is inhibited by antibodies raised against 45-kd proteins of the outer membrane. *EMBO J.* **6**, 2109–2115.

Ohba, M., and Schatz, G. (1987b). Disruption of the outer membrane restores protein import to trypsin-treated yeast mitochondria. *EMBO J.* **6**, 2117–2122.

Ostermann, J., Horwich, A. L., Neupert, W., and Hartl, F.-U. (1989). Protein folding in mitochondria requires complex formation with hsp60 and ATP hydrolysis. *Nature (London)* **341**, 125–130.

Pain, D., Murakami, H., and Blobel, G. (1990). Identification of a receptor for protein import into mitochondria. *Nature (London)* **347**, 444–449.

Pon, L., Moll, T., Vestweber, D., Marshallsay, B., and Schatz, G. (1989). Protein import into mitochondria: ATP-dependent protein translocation activity in a submitochondrial fraction enriched in membrane contact sites and specific proteins. *J. Cell Biol.* **109**, 2603–2616.

Scherer, P. E., Krieg, U. C., Hwang, S. T., Vestweber, D., and Schatz, G. (1990). A precursor protein partly translocated into yeast mitochondria is bound to a 70 kDa mitochondrial stress protein. *EMBO J.* **9**, 4315–4322.

Schleyer, M., and Neupert, W. (1985). Transport of proteins into mitochondria: Transloca-tion intermediates spanning contact sites between outer and inner membranes. *Cell (Cambridge, Mass.)* **43**, 339–350.

Schwaiger, M., Herzog, V., and Neupert, W. (1987). Characterization of translocation contact sites involved in the import of mitochondrial proteins. *J. Cell Biol.* **105**, 235–246.

Sims, P. J., Waggoner, A. S., Wang, C.-H., and Hoffman, J. F. (1974). Studies on the mechanism by which cyanine dyes measure membrane potential in red blood cells and phosphatidylcholine vesicles. *Biochemistry* **13**, 3315–3330.

Söllner, T., Griffiths, G., Pfaller, R., Pfanner, N., and Neupert, W. (1989). MOM19, an import receptor for mitochondrial precursor proteins. *Cell (Cambridge, Mass.)* **59**, 1061–1070.

Söllner, T., Pfaller, R., Griffiths, G., Pfanner, N., and Neupert, W. (1990). A mitochondrial import receptor for the ADP/ATP carrier. *Cell (Cambridge, Mass.)* **62**, 107–115.

Suissa, M., and Schatz, G. (1982). Import of proteins into mitochondria: Translatable mRNAs for imported mitochondrial proteins are present in free as well as mitochondria-bound cytoplasmic polysomes. *J. Biol. Chem.* **257**, 13048–13055.

van Loon, A. P. G. M., and Schatz, G. (1987). Transport of proteins to the mitochondrial intermembrane space: The 'sorting' domain of the cytochrome c_1 presequence is a stop-transfer sequence specific for the mitochondrial inner membrane. *EMBO J.* **6**, 2441–2448.

Vestweber, D., and Schatz, G. (1988). A chimeric mitochondrial precursor protein with internal disulfide bridges blocks import of authentic precursors into mitochondria and allows quantitation of import sites. *J. Cell Biol.* **107**, 2037–2043.

Vestweber, D., Brunner, J., Baker, A., and Schatz, G. (1989). A 42K outer-membrane protein is a component of the yeast mitochondrial protein import site. *Nature (London)* **341**, 205–209.

Yang, M., Jensen, R. E., Yaffe, M. P., Oppliger, W., and Schatz, G. (1988). Import of proteins into yeast mitochondria: The purified matrix processing protease contains two subunits which are encoded by the nuclear *MAS1* and *MAS2* genes. *EMBO J.* **7**, 3857–3862.

Chapter 18

Pulse Labeling of Yeast Cells as a Tool to Study Mitochondrial Protein Import

ANDERS BRANDT[1]

Department of Physiology
Biocenter, University of Basel
CH-4056 Basel, Switzerland

I. Introduction

In this chapter, a simple labeling protocol is presented that measures protein import into mitochondria of intact yeast cells. The method exploits the fact that most nuclear-encoded mitochondrial proteins are synthesized in the cytoplasm as larger precursors. The amino-terminal presequence is cleaved off by the matrix protease during import into the mitochondria, resulting in a smaller protein inside the mitochondria. Thus, by pulse labeling yeast cells with a radioactive amino acid, the newly synthesized precursor as well as the newly imported mature form can each be detected

---

[1] Present address: Department of Physiology, Carlsberg Laboratory, DK-2500 Copenhagen, Denmark.

as labeled polypeptides after immunoprecipitation, SDS–gel electrophoresis, and fluorography. This assay depends on the availability of monospecific antibodies that recognize both the precursor and the mature form of the protein. Most polyclonal antibodies raised against mature proteins fulfill this criterion.

In vivo labeling has proved to be a valuable part of the methodology to analyze mitochondrial protein import. Even though intact yeast cells not are amenable to the detailed manipulations that can be applied to isolated mitochondria, labeling of whole cells has the advantage that the results reflects the conditions *in vivo*. As it is the classical method to study mitochondrial protein import, pulse-labeling experiments with intact cells have contributed significantly to the identification of important features and components of the mitochondrial protein import machinery. Thus, they showed that most nuclear-encoded mitochondrial proteins contain presequences, which upon import are removed by the matrix-localized protease (Maccecchini *et al.*, 1979; Reid and Schatz, 1982a, b) and that some nuclear encoded mitochondrial polypeptides contain complex presequences that are cleaved twice, first by the matrix protease and then by other proteases (Reid *et al.*, 1982; Ohashi *et al.*, 1982). These latter proteins are targeted either to the inner membrane (cytochrome c_1, subunit IV of cytochrome oxidase, the nonheme iron-carrying subunit of the cytochrome b/c_1 complex) or to the intermembrane space (cytochrome b_2, cytochrome c peroxidase). Pulse-labeling experiments have also provided the first indications that ATP is required in the mitochondrial matrix for presequence cleavage and/or import to occur (Nelson and Schatz, 1979) and that mutations affecting components in the protein import machinery cause accumulation of mitochondrial precursor polypeptides, e.g., mutations in the two subunits of the matrix protease (Yaffe and Schatz, 1984), in mitochondrial hsp70 (Kang *et al.*, 1990), and in MAS70, an import regulator in the outer membrane (Hines *et al.*, 1990). Depletion of ISP42, a component of the import machinery in the outer membrane, likewise leads to the accumulation of precursors (Baker *et al.*, 1990). In contrast, mutations in mitochondrial hsp60 do not result in accumulation of precursor polypeptides, even though *in vivo* labeling experiments were important in showing this protein to be essential for assembly of oligomeric complexes (Cheng *et al.*, 1989).

As a supplement to the more detailed mitochondrial protein import analyses *in vitro*, labeling of intact yeast cells allows a simple and convenient evaluation of mitochondrial protein import. Even though the method is simple, a few important points must be followed to ensure success. The protocol given here is an updated version of the methods published by

Maccecchini *et al.* (1979), Reid and Schatz (1982a,b), and Yaffe and Schatz (1984).

II.   Procedure

A.   Labeling

1. Grow an overnight culture of yeast in 20 ml of semisynthetic lactate medium in a 200-ml Erlenmeyer flask with vigorous shaking to ensure good aeration. When the OD_{600} of the culture is between 0.5–1.0, harvest the cells by centrifugation at 3000 rpm for 5 minutes. Galactose or glucose can also be used as a carbon source in the growth medium, but should then also be used in the subsequent labeling medium.

2. Resuspend the cell pellet to 5 OD_{600}/ml in 40 mM KP_i, pH 6.0, 2% lactate. Add [^{35}S]methionine to 50–100 μCi/ml (1000 Ci/mmol). Translation-grade [^{35}S]methionine is excellent, but methionine with 10-fold lower specific activity can be used as well, provided that the radioactive concentration is increased to 200–400 μCi/ml. Label the cells for 2–5 minutes with vigorous shaking. Use a 25-ml Erlenmeyer flask for cell suspensions up to 5 ml. Labeling of the cells can also be performed in the semisynthetic growth medium, but the level of incorporation is usually only one-half to one-fourth of the level achieved in 40 mM KP_i (Fig. 1).

3. Stop the incorporation of label into protein by adding unlabeled methionine to 10 mM and cycloheximide to 100 μg/ml.

4. Chase with vigorous shaking. Convenient chase times are 0, 1, 2, 5, and 10 minutes.

5. Stop the chase as indicated below.

B.   Measuring Incorporation of [^{35}S]Methionine into Protein

Prepare 1 cm^2 Whatman 3MM paper squares prewetted in 10% trichloracetic acid (TCA) and then dried. The filters can be marked with a pencil. Spot 5 μl of the cell suspension on the filters and boil the filters in 10% TCA for 10 minutes in a beaker. Add ice so the filters sink to the bottom of the beaker and wash the filters twice in water, twice in ethanol, and twice in acetone and then air dry. Count in a liquid scintillation counter. Sample at the start and at the end of the labeling period. This method gives an accurate measure of the incorporation of [^{35}S]methionine into protein. A

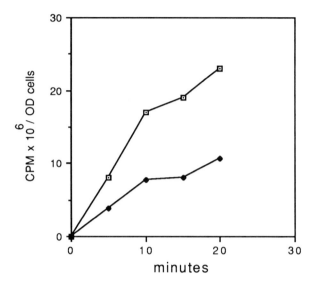

Fig. 1. The level of incorporation of [^{35}S]methionine into protein is dependent on the composition of the labeling medium. Yeast cells (1 OD$_{600}$) were labeled in a final volume of 0.2 ml. Samples were withdrawn at different times after the start of the labeling and incorporation of [^{35}S]methionine into protein was determined as described in the text. Open squares, labeling in 40 mM KP$_i$, pH 6.0, and 2% lactate; filled squares, labeling in semisynthetic lactate growth medium.

reasonable labeling results in the incorporation of 10^{7}cpm/OD$_{600}$ of cells in 10 minutes, corresponding to an incorporation of 5–10% of the added label. Incorporation of [^{35}S]methionine into protein is linear up to 10 minutes and it then levels off (Fig. 1). The following factors influence the level of incorporation: (1) Exponentially growing cells incorporate high levels of label into protein, whereas stationary-phase cells poorly incorporate label due to the reduced protein synthesis. (2) Even though the labeled amino acid is usually rapidly taken up by the cells, incorporation into protein can be low due to substantial intracellular pools of amino acid. The semisynthetic growth medium used in this protocol is adequate for keeping the pool sizes at a given level, which allows high incorporation levels. If the incorporation of label into protein is not satisfactory, reduce the amount of yeast extract in the growth media two- to threefold, in an attempt to lower the amino acid pools prior to the labeling experiment. (3) The quality of the labeled amino acid is crucial. Oxidation of [^{35}S]methionine can be minimized by storing the labeled amino acid in aliquots at −70°C and avoiding repeated freezing and thrawing.

C.   Extraction of Total Yeast Proteins and Immune Precipitation of Specific Polypeptides

1. Transfer 200 μl of the labeled cell suspension corresponding to 1 OD_{600} directly into 200 μl 0.4 N NaOH and 5% β-mercaptoethanol in an Eppendorf centrifuge tube. Incubate on ice for 10 minutes. Add 400 μl acetone and 80 μl 100% TCA and leave on ice for 10 minutes. Centrifuge for 10 minutes and remove the supernatant by aspiration. Wash the pellet in 500 μl of acetone. Resuspend the pellet in 50 μl 2% SDS, 0.1 M Tris-HCl, pH 7.5, 5 mM EDTA, and 0.005% bromophenol blue. If the solution turns yellow due to remaining TCA, add 1 M Tris base in 1-μl aliquots until the solution turns blue. Boil the sample for 10 minutes and spin in an Eppendorf microfuge for 10 minutes at room temperature. Remove the supernatant into 1 ml TNTE, 1 mM PMSF, and mix well.

2. Add 5 μl of preimmune serum and incubate at 4°C for 1 hour by slowly rotating the tube. Spin for 10 minutes in a microfuge, recover the supernatant, and add 25 μl of a protein A–Sepharose slurry. To prepare the slurry, swell the protein A–Sepharose beads overnight in TNTE, collect the swollen beads by a brief centrifugation, and add a volume of TNTE equal to the volume of the sedimented packed beads. Incubate at 4°C for 0.5 hours and collect the protein A–Sepharose particles by a 10-second centrifugation. Save the supernatant. Step 2 is optional; if no preimmune serum is available, pretreat the extract with protein A–Sepharose only.

3. To the supernatant from step 2, add 10 μl of the desired antiserum. Crude rabbit serum is adequate. Incubate for 2 hours or overnight at 4°C while gently rotating the tube. As some aggregates usually form during the incubation with crude serum, spin 10 minutes in the microfuge and recover the supernatant. Add 50 μl of the protein A–Sepharose slurry to the supernatant and incubate for 1 hour at 4°C with gentle shaking. Collect the protein A–Sepharose beads by a 10-second spin in the microfuge. The supernatant can be used for at least five subsequent immune precipitations without significant contamination from one immune precipitation to the next, provided that the extract is treated with 25 μl protein A–Sepharose between each immune precipitation step.

4. Wash the protein A–Sepharose beads twice in 1 ml TNTE and twice in 1 ml NTE for 15 minutes.

5. Resuspend the beads in 30 μl 5% SDS, 0.1 M Tris-HCl, pH 8.8, 5 mM EDTA, 0.005% bromophenolblue, and 20% glycerol; boil for

10 minutes and subject the supernatant to SDS–PAGE. After electrophoresis, incubate the gel in 1 M Na-salicylate, 20% methanol, and 2% glycerol for 20 minutes at room temperature, dry it on a vacuum gel dryer, and expose it to Kodak X-omat AR film or any X-ray film of similar sensitivity with an intensifying screen at −70°C. The fluorogram can either be evaluated by eye or with a scanner to assess the distribution of label in the precursor and the mature forms of the polypeptide under study.

D.   Preparation of Spheroplasts and Mitochondria

Spheroplasts can be labeled nearly as efficiently as yeast cells. Labeling of spheroplasts has the advantage that mitochondria can be readily isolated after labeling and the submitochondrial localization of newly imported polypeptides can be analyzed.

1. Preparation of spheroplasts. Collect 30 OD_{600} of cells by centrifugation at 3500 rpm for 5 minutes. Resuspend the cells in 10 ml of 0.1 M $TrisSO_4$, pH 9.4, 10 mM DTT, and incubate 10 minutes at 30°C. Pellet the cells by centrifugation as above and resuspend them in 6 ml 1.2 M sorbitol, 20 mM KP_i, pH 7.4, 0.5% yeast extract, 0.05% glucose, 2% lactate, pH 7.4, and 1 mg/ml Zymolase-20T (Kirin Brewery Co., Ltd). Incubate for 30 minutes at 30°C and check for spheroplast formation by adding 100-μl aliquots of the cell suspension to 1 ml of water. The solution should clear within a few minutes if spheroplasts have been produced. Recover the spheroplasts by centrifugation as above and resuspend in 3 ml of labeling buffer composed of 1.2 M sorbitol, 20 mM KP_i, pH 6.0, and 2% lactate. Label an aliquot of spheroplasts corresponding to 10 OD_{600} by adding [^{35}S]methionine to 50–100 μCi/ml. Stop labeling by adding unlabeled methionine to 10 mM, cycloheximide to 100 μg/ml, and CCCP to 40 μM.

2. Isolation of mitochondria. All steps should be performed on ice. Collect the labeled spheroplasts by a 30-second spin in a microfuge and resuspend them to an OD_{600} of 10 in 0.6 M sorbitol, 20 mM HEPES/KOH, pH 7.4, and 1 mM PMSF. The amount of mitochondria obtained from 10 OD_{600} of spheroplasts is sufficient for the subsequent analysis. Disrupt the spheroplasts with 10 strokes in a small Dounce homogenizer. Spin at 1000 g for 10 minutes, recover the supernatant, and spin it again at 15,000 g for 10 minutes. The pellet is a crude mitochondrial pellet; it can be analyzed directly or washed once by resuspension in buffer and centrifugation as above.

The mitochondria can be subfractionated by the osmotic shock procedure outlined by Jascur in Chapter 17.

E. Pulse Chase Conditions

Most precursor polypeptides are quickly processed and imported into mitochondria *in vivo*. If pulse chase experiments are to be performed with wild-type cells or spheroplasts, the pulse should be limited to 2–3 minutes and the chase to 0, 0.5, 1, or 2 minutes (Fig. 2A). Alternatively, the temperature during the labeling and chase can be lowered to 20°C, for example. Some precursors (e.g., cytochrome b_2, cytochrome c_1, and cytochrome c peroxidase) are processed and imported relatively slowly; for these precursors the pulse and the chase periods can be extended. Import and processing *in vivo* can be prevented by 40 μM CCCP. CCCP uncouples the mitochondria, preventing the insertion of the precursor polypeptide into the inner membrane (Fig. 2B). CCCP can subsequently be inactivated by treating the cell suspension with 7 mM β-mercaptoethanol. Inactivation of CCCP and the reestablishment of the membrane potential usually require 10–15 minutes.

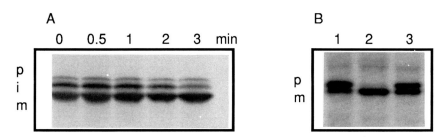

FIG. 2.   (A) Pulse labeling of yeast cells at 30°C. Yeast cells were labeled at for 2.5 minutes and chased for 0, 0.5, 1, 2, and 3 minutes as described in the text. Cytochrome b_2 was immunoprecipitated from the samples and analyzed by SDS–PAGE and fluorography; p, precursor; i, intermediate; and m, mature forms of cytochrome b_2. (B) A temperature-sensitive MAS2 subunit of the matrix-localized protease causes accumulation of unprocessed precursor of the β-subunit of the F_1-ATPase. A *mas2* mutant was grown at room temperature, shifted for 15 minutes to the nonpermissive temperature (37°C), and pulse labeled for 5 minutes at 37°C (lane 1). Incorporation of label into protein was stopped by addition of unlabeled methionine and cycloheximide; one-half of the cells was chased in the absence of and the other half in the presence of CCCP for 15 minutes (lanes 2 and 3, respectively). CCCP prevents the processing of the precursor polypeptide, indicating that the defective matrix protease leads to the accumulation of precursors outside the inner membrane. See also Yaffe and Schatz (1984).

F.   Media and Stock Solutions

Semisynthetic lactate medium: 0.3% (w/v) yeast extract, 0.1% KH_2PO_4, 0.1% NH_4Cl, 0.04% $CaCl_2$, 0.05% NaCl, 0.06% $MgCl_2$, 6 H_2O, 2% sodium lactate, pH 6.0.

CCCP (carbonyl cyanide m-chlorophenylhydrazone), 4 mM in ethanol.

TNTE (1% Triton X-100, 0.05 M Tris-HCl, pH 7.5, 0.15 M NaCl, 0.005 M EDTA).

NTE (0.01 M Tris-HCl, pH 7.5, 0.15 M NaCl, 0.005 M EDTA).

PMSF (phenylmethylsulfonyl fluoride; 1 M in dimethyl sulfoxide).

References

Baker, K. P., Schaniel, A., Vestweber, D., and Schatz, G. (1990). A yeast mitochondrial outer membrane protein essential for protein import and cell viability. *Nature (London)* **348,** 605–609.

Cheng, M. Y., Hartl, F.-U., Martin, J., Pollock, R. A., Kalousek, F., Neupert, W., Hallberg, E. M., Hallberg, R. L., and Horwich, A. L. (1989). Mitochondrial heat-shock protein hsp60 is essential for assembly of proteins imported into yeast mitochondria. *Nature (London)* **337,** 620–625.

Hines, V., Brandt, A., Griffiths, G., Horstmann, H., Brutsch, H., and Schatz, G. (1990). Protein import into yeast mitochondria is accelerated by the outer membrane protein MAS70 *EMBO J.* **9,** 3191–3200.

Kang, P.-J., Ostermann, J., Shilling, J., Neupert, W., Craig, E. A., and Pfanner, N. (1990). Requirement for hsp70 in the mitochondrial matrix for translocation and folding of precursor proteins. *Nature (London)* **348,** 137–143.

Maccecchini, M. L., Rudin, Y., Blobel, G., and Schatz, G. (1979). Import of proteins into mitochondria: Precursor forms of the extra mitochondrially made F_1-ATPase subunits of yeast. *Proc. Natl. Acad. Sci. U.S.A.* **76,** 343–347.

Nelson, N., and Schatz, G. (1979). Energy-dependent processing of cytoplasmically made precursors to mitochondrial proteins. *Proc. Natl. Acad. Sci. U.S.A.* **76,** 4365–4369.

Ohashi, A., Gibson, J., Gegor, I., and Schatz, G. (1982). Import of proteins into mitochondria. The precursor of cytochrome c_1 is processed in two steps, one of them heme-dependent. *J. Biol. Chem.* **257,** 13042–13047.

Reid, G. A., and Schatz, G. (1982a). Import of proteins into mitochondria. Yeast cells grown in the presence of carbonyl cyanide m-chlorophenylhydrazone accumulate massive amounts of some mitochondrial precursor polypeptides. *J. Biol. Chem.* **257,** 13056–13061.

Reid, G. A., and Schatz, G. (1982b). Import of proteins into mitochondria. Extramitochondrial pools and post-translational import of mitochondrial protein precursors *in vivo.* *J. Biol. Chem.* **257,** 13062–13067.

Reid, G. A., Yonetani, T., and Schatz, G. (1982). Import of protein into mitochondria. Import and maturation of the mitochondrial intermembrane space enzymes cytochrome b_2 and cytochrome c peroxidase in intact yeast cells. *J. Biol. Chem.* **257,** 13068–13074. 1982.

Yaffe, M. P., and Schatz, G. (1984). Two nuclear mutations that block mitochondrial protein import in yeast. *Proc. Natl. Acad. Sci. U.S.A.* **81,** 4819–4823.

Chapter 19

The Protein Import Machinery of Yeast Mitochondria

VICTORIA HINES AND KEVIN P. BAKER

Department of Biochemistry
Biocenter, University of Basel
CH-4056 Basel, Switzerland

I. Introduction

In recent years, considerable progress has been made in identifying components involved in yeast mitochondrial protein import. Mitochondrial protein import can be divided mechanistically into three separate stages: precursor binding to the mitochondrial outer membrane, translocation across the outer and inner membranes of the mitochondria, and subsequent processing, folding, and assembly of the newly imported protein inside the mitochondria. Using a variety of biochemical and genetic techniques, components involved in each of these three processes have been identified. Although we cannot, in this article, cover all the approaches used to identify components of the import machinery, we will review some

METHODS IN CELL BIOLOGY, VOL. 34

of the basic strategies that have been successfully employed in the past few years. For a more extensive review of this topic, see Pon and Schatz (1991) and Baker and Schatz (1991).

One of the advantages of using the yeast *Saccharomyces cerevisiae* as an experimental system is the ease with which this organism can be genetically manipulated (Botstein and Fink, 1988). The organism has a well-characterized genome and can exist in both haploid and diploid states that can be readily interconverted, permitting the study of null alleles created by gene disruption. Because the mitochondria of yeast are well characterized, *S. cerevisiae* is an organism of choice in which to study the genes involved in mitochondrial import. Both classical yeast genetics and reverse genetics can be applied to yeast and both have proved invaluable in identifying the components known to be involved in mitochondrial import.

Table I summarizes the components that have been identified and characterized to date. Components of the outer membrane involved in precursor binding and protein translocation have been identified by treating mitochondria with proteases, antibodies, and chemical cross-linkers. Components of the mitochondrial matrix involved in processing and refolding have been identified using chemical cross-linking studies and genetic screens. As the list of components expands, it has become evident that some components are functionally unique, whereas other components appear to be functionally overlapping. Many of the unique components have been identified first by genetic means and appear to be essential for mitochondrial function and cell viability. In other cases, the function of one protein can also be performed by other, functionally related proteins. These components are not essential for life and therefore, have been difficult to identify by genetic means alone. Many of the import components have also been identified in organisms other than *S. cerevisiae*. These homologous proteins are listed in Table II.

We will first discuss the different biochemical approaches that have been used to identify mitochondrial import components, and then turn to the various genetic approaches that have identified many of the components essential in protein import.

II.   Biochemical Approaches

A.   Protease Studies

The first indication that outer membrane components were involved in protein import came from treating intact mitochondria or outer membrane vesicles with different proteases. Pretreatment of mitochondria with

TABLE I

ESSENTIAL AND NONESSENTIAL COMPONENTS OF THE YEAST MITOCHONDRIAL IMPORT SYSTEM[a]

Protein	Gene	Location	Essential?	Function	Ref.
MAS1	MAS1	Matrix	Yes	Removal of matrix-targeting signal	Witte et al., (1988); Pollock et al., (1988)
MAS2	MAS2	Matrix	Yes	Removal of matrix-targeting signal	Jensen and Yaffe (1988); Pollock et al., (1988)
hsp60	MIF4	Matrix	Yes	Refolding and assembly of precursors	Cheng et al., (1989)
mhsp70	SSC1	Matrix	Yes	Refolding and transmembranous movement of precursors	Scherer et al., (1990); Kang et al., (1990)
ISP42	ISP42	Outer membrane	Yes	Translocation across outer membrane	Vestweber et al., (1989); Baker et al., (1990)
MAS70	MAS70	Outer membrane	No	Import receptor	Hines et al., (1990)
p32	MIR1	Outer membrane	No	Import receptor	Pain et al., (1990); Murakami et al., (1990)

[a] The table does not include extramitochondrial proteins, such as cytosolic 70-kDa heat shock proteins, or components specific for only one or a few components, such as cytochrome c heme lyase.

TABLE II

COMPONENTS OF THE MITOCHONDRIAL PROTEIN IMPORT MACHINERY IN YEAST AND THEIR
PROBABLE HOMOLOGUES IN OTHER ORGANISMS

Component	Location	Homologue	Organism	Ref.
MAS1	Matrix	PEP	*Neurospora crassa*	Pollock *et al.*, (1988); Hawlitschek *et al.*, (1988)
MAS2	Matrix	MPP	*Neurospora crassa*	Pollock *et al.*, (1988); Hawlitschek *et al.*, (1988)
MAS70	Outer membrane	MOM72	*Neurospora crassa*	Söllner *et al.*, (1990)
ISP42	Outer membrane	MOM38	*Neurospora crassa*	Baker *et al.*, (1990); Kiebler *et al.*, (1990)
hsp60	Matrix	hsp60	Maize	Prasad *et al.*, (1990)

varying amounts of protease, followed by quantification of protein import, showed that protease-sensitive components were involved in the import of several precursors (Ohba and Schatz, 1987b). Similar studies on isolated outer membrane vesicles demonstrated that precursor binding to these vesicles was also abolished by protease treatment (Riezman *et al.*, 1983a). More recent work by Hines *et al.* (1990) indicates that the import inhibition seen with trypsinization reflects digestion of at least two different components of the outer membrane, both of them involved in protein import. At low trypsin levels, the import of some precursors, such as the β-subunit of the F_1-ATPase, is reduced by 50% due to the specific removal of MAS70, a 70-kDa outer membrane protein (see below). Higher concentrations of trypsin further reduce the import of the β-subunit of the F_1-ATPase as well as inhibiting the import of several other precursors. This suggests that two components, which might overlap in function, are used for the import of some precursors. However, protease studies are limited for several reasons. First, they can detect only those components that are both accessible and inherently susceptible to proteases. Second, most proteases are not specific for a single protein; identification of the protease-sensitive component involved in protein import thus requires additional biochemical studies.

B. Antibody Studies

An effective tool in the identification of protein components of the import machinery has been the use of antibodies that inhibit mitochondrial import (used to identify ISP42 and MAS70) or the use of antiidiotypic antibodies (used to identify p32).

Ohba and Schatz (1987a) first characterized an antiserum raised against total outer membrane proteins that inhibited the import of several precursors into mitochondria. Both total IgGs or Fab fragments inhibited protein import up to 60%. Vestweber *et al.* (1989) showed that the active component involved in this serum was an antibody against a minor 42-kDa protein of the outer membrane (termed ISP42, for *I*mport *S*ite *P*rotein), even though the predominant antibody recognized a 45-kDa protein. Additional biochemical studies showed that ISP42 is closely associated with a partially translocated precursor.

Another outer membrane component that has been characterized using both protease and antibody studies is MAS70. This protein appears to function in the early stages of precursor binding to the mitochondrial membrane. However, *in vitro*, it is not involved in the import of all precursors (Hines *et al.*, 1990). Purified IgGs were allowed to bind for 30 minutes at 0°C and then a standard import assay was performed. Anti-MAS70 IgGs inhibited the import of the ADP/ATP translocator by 50%, but only showed minor inhibition of the import of the COXIV–DHFR fusion protein (composed of the presequence of the subunit IV of cytochrome *c* oxidase fused to mouse dihydrofolate reductase). When the same experiment was carried out with mitochondria isolated from a strain whose *MAS70* gene had been disrupted, the antibody showed no inhibitory activity with either precursor, demonstrating that the inhibition was only due to IgGs directed against the MAS70 protein.

Although antibody inhibition studies can be useful, two considerations must be kept in mind. First, one must exclude that the antibody effect is an artifact of antigen clustering or steric hindrance. An effective control for this is usually to use Fab fragments instead of total IgGs: when the Fab fragments show the same biological activity, steric hindrance is unlikely. Second, one must identify the active antibodies in a polyclonal serum. The easiest solution is to use monoclonal antibodies that are usually only directed against a unique protein. However, if a monoclonal antibody is not available, one must exclude that a minor component of the serum is responsible for the biological activity. In the identification of ISP42, Vestweber *et al.* (1989) used two approaches to identify the active antibodies. In one approach, the total serum was depleted of antibodies to ISP42 and retested for import inhibition. In the second approach, antibodies to ISP42 were affinity purified from the total serum and restested for import inhibition. Import inhibition was not seen with the depleted IgGs but was seen with the affinity-purified IgGs, thereby identifying ISP42 antibodies as the inhibitory species.

Antibodies have also been used in a completely different manner to identify another receptor component of the outer membrane. In these studies, antiidiotypic antibodies were generated against the presequence of subunit IV of yeast cytochrome *c* oxidase. These antiidiotypic antibodies recognized a 32-kDa outer membrane protein, a potential signal sequence receptor termed p32. Antibodies raised against p32 inhibited import *in vitro*, and, in a detergent extract, coprecipitated a precursor that had been bound to the mitochondrial surface (Pain *et al.*, 1990).

C. Cross-Linking Studies

If a discrete import intermediate can be generated, it is possible to probe for nearby components with chemical cross-linkers. Using a partially translocated intermediate with a covalently attached photoactivatable cross-linker, Vestweber *et al.* (1989) showed that ISP42 is in close proximity to this stuck translocation intermediate. Because antibodies to ISP42 inhibit the import of several precursors (see above) and because depletion of ISP42 in yeast cells causes the accumulation of uncleaved mitochondrial precursor proteins outside the mitochondria (see below), ISP42 appears to participate in the translocation of precursors across the outer membrane.

Using the same stuck intermediate, but a soluble chemical cross-linker, Scherer *et al.* (1990) identified the mitochondrial hsp70 in association with the precursor within the mitochondrial matrix. (For further discussions on cross-linking studies to identify import components, see Scherer and Krieg, Chapter 23.)

III. Genetic Approaches

A. Classical Genetics

We use the term classical genetics to describe approaches in which a gene is initially defined by a mutation resulting in a distinct phenotype. This phenotype may then be explored biochemically, or used to clone the gene.

The earliest successful attempt to isolate mutants in the mitochondrial import pathway relied on two assumptions: that mitochondria were essential for life, and that a block of the import pathway would result in the stable accumulation of mitochondrial precursor proteins (Yaffe and Schatz, 1984a). A direct consequence of the first assumption is that import mutants should be present in a bank of conditional lethal mutants. Starting from these assumptions, Yaffe and Schatz (1984b) screened a bank of temperature-sensitive lethal mutants for cells that accumulated the precursor to the F_1-ATPase β-subunit at the nonpermissive temperature. Two such mutants were isolated. They were shown to be due to single nuclear mutations in two different nuclear genes whose wild-type alleles were termed *MAS1* and *MAS2* (for *M*itochondrial *AS*sembly).

Cloning of these genes was accomplished by screening plasmid banks of yeast genomic DNA for genes that suppressed the temperature-sensitive growth defects of the mutants (Jensen and Yaffe, 1988; Witte *et al.*, 1988). In such screens it is important to verify that any complementing genes cure the temperature-sensitive defect if present as single copies in the cells; this makes it unlikely that the temperature-sensitive phenotype is suppressed by some other gene that is present in many copies.

Once the *MAS1* and *MAS2* genes were isolated, the construction of null alleles (gene disruption) demonstrated that these genes were indeed essential for cell viability, and that the original temperature-sensitive phenotype was not an allele-specific effect. Availability of the genes has allowed the demonstration of gene–polypeptide relationships: antibodies against peptides predicted from the sequenced genes were used to identify the polypeptides in mitochondria. *MAS1* and *MAS2* encode nonidentical subunits of the matrix protease, the "MAS-protease," which cleaves presequences from imported precursors. By placing the *MAS1* and *MAS2* genes under the control of a *GAL10–CYC1* hybrid promoter, both subunits were overexpressed in yeast. This has allowed purification of milligram amounts of the protease from yeast, leading to a biochemical characterization of this component of the import system (Geli *et al.*, 1990; Yang *et al.*, 1991).

Using a similar "brute force" approach, a second bank of temperature-sensitive yeast mutants was screened primarily for the inability to assemble active human ornithine transcarbamylase in the mitochondrial matrix. Such mutants were then screened for mitochondrial precursor accumulation at the nonpermissive temperature, as described above (Pollock *et al.*, 1988). This approach identified new alleles of *mas1* and *mas2* as well as mutations in a new complementation group, *mif4*. Using methods similar to those described above, the wild-type *MIF4* allele was cloned. It proved to be identical with the gene encoding a mitochondrial heat-shock protein, hsp 60, a close relative of bacterial groEL. hsp 60 is responsible for the stabilization of unfolded conformers and their proper assembly in the mitochondrial matrix (Cheng *et al.*, 1989).

B. Reverse Genetics

By the term reverse genetics we define approaches involving the isolation and manipulation of a gene after its gene product has been characterized. In most instances, the first step in reverse genetics is the cloning of the gene, usually from expression libraries screened with antibodies directed against the gene product. Alternatively, if the protein sequence is available, an oligonucleotide may be synthesized that can be used in hybridization screening of gene libraries.

A fruitful approach for identifying mitochondrial import components has been the characterization of mitochondrial outer membrane proteins of yeast. Outer membranes were purified and used for the production of antibodies that proved useful for the identification of import components (as discussed above) and for the cloning of the corresponding genes.

The first outer membrane component to be identified was a major 70-kDa protein (MAS70). Its gene was cloned using hybrid selection techniques (Riezman *et al.*, 1983b). Antibodies against the protein demonstrated that the right gene had been cloned: overproduction of antigen in yeast was observed in yeast cells bearing multiple copies of the cloned gene. Strains bearing a null allele were viable on fermentable carbon sources, but were unable to grow at high temperatures on nonfermentable carbon sources. Although the phenotype of *mas70* null strains initially proved difficult to characterize, recent studies with these *mas70* strains indicate that MAS70 enhances correct binding of some mitochondrial precursors to the mitochondrial surface (Hines *et al.*, 1990).

As discussed above, antiidiotypic antibodies against a mitochondrial-targeting sequence have been used to identify a receptor-like molecule of the outer mitochondrial membrane (Pain *et al.*, 1990). Antibodies were used to clone the gene from a λgt11 expression library. The gene was cloned, sequenced, and a null allele of the gene was constructed (Murakami *et al.*, 1990). The gene is not essential for growth on fermentable carbon sources, but mitochondria from a *mir1* null mutant contain reduced levels of some mitochondrial proteins.

Reverse genetics has also been instrumental in showing that the mitochondrial hsp 70 protein (encoded by the *SSC1* gene) participates in mitochondrial import. The cloned gene was used to create a temperature-sensitive allele, the phenotype of which was amenable to biochemical analysis (Kang *et al.*, 1990). The approach used to create the temperature-sensitive allele was similar to the plasmid-shuffling technique devised by Botstein and co-workers (Schatz *et al.*, 1988). One chromosomal copy of the gene was disrupted in a diploid strain; the strain was transformed with a plasmid containing the wild-type *SSC1* gene under the control of a *GAL* promoter and was induced to sporulate. An ascospore in which the plasmid supplied the only functional copy of the *SSC1* gene was identified and transformed with single-copy plasmids containing a bank of mutagenized *SSC1* genes. The resulting double transformants were screened for temperature sensitivity on glucose-containing media. The glucose in the media repressed the *GAL*-regulated wild-type copy of *SSC1*, thus allowing the detection of temperature-sensitive alleles in the double transformants. A similar approach has allowed the isolation of new temperature-sensitive alleles of *mif4* (R. Hallberg, unpublished).

Reverse genetics has also highlighted the role of ISP42 *in vivo*. This protein is located in close proximity to a partially translocated precursor protein, as shown by cross-linking studies (Section II,C). Antibodies against this protein were used to clone the gene from a λgt11 expression library and the gene was shown to be essential for yeast cell viability by gene disruption. To explore the role of this protein in mitochondrial import further, the gene was placed under a controllable *GAL10–CYC1* promoter. When this promoter was "turned off" by growing the cells in glucose, a strong arrest of mitochondrial import was observed as judged by the steady-state accumulation of mitochondrial precursors (Baker *et al.*, 1990).

IV. Prospects

It is clear from the above discussions that studies on mitochondrial import in yeast permit identification and functional characterization of many components of the import machinery. Some of the most promising biochemical tools include antibodies and cross-linkers. Antibodies provide the specificity necessary for characterization. The use of antiidiotypic antibodies seems to be particularly promising, though, at present, their general applicability is not clear. Once a component has been identified, it is usually possible to detect associated components by coimmune precipitation, chemical cross-linkers, or genetic studies. Because several components are now known, future studies should soon identify other proteins participating in mitochondrial import.

In order to complement biochemical studies on the function(s) of a particular component, a mutant allele is always of great value. Mutant alleles can also be of help to define interacting components. Such studies involve the isolation of extragenic suppressor mutations to a particular phenotype. As a rapidly growing number of temperature-sensitive alleles for import components are available, a search for second-site suppressors of such alleles should be a valuable tool in identifying novel components.

REFERENCES

Baker, K. P., and Schatz, G. (1991). Mitochondrial proteins essential for viability mediate protein import into yeast mitochondria. *Nature (London)* **349**, 205–208.
Baker, K. P., Schaniel, A., Vestweber, D., and Schatz, G. (1990). A yeast mitochondrial

outer membrane protein essential for protein import and cell viability. *Nature (London)* **348**, 605–609.

Botstein, D., and Fink, G. R. (1988). Yeast; an experimental organism for modern biology. *Science* **240**, 1439–1443.

Cheng, M. Y., Hartl, F.-U., Martin, J., Pollock, R. A., Kalousek, F., Neupert, W., Hallberg, E. M., Hallberg, R. L., and Horwich A. L. (1989). Mitochondrial heat shock protein hsp60 is essential for assembly of proteins imported into yeast mitochondria. *Nature (London)* **337**, 620–625.

Geli, V., Yang, M, Suda, K., Lustig, A., and Schatz, G. (1990). The *MAS*-encoded processing protease of yeast mitochondria. Overproduction and characterization of its two non-identical subunits. *J. Biol. Chem.* **265**, 19216–19222.

Hawlitschek, G., Schneider, H., Schmidt, B., Tropschug, M., Hartl, F.-U., and Neupert, W. (1988). Mitochondrial protein import: Identification of processing peptidase and of PEP, a processing enhancing protein. *Cell (Cambridge, Mass.)* **53**, 795–806.

Hines, V., Brandt, A., Griffiths, G., Horstmann, H., Brütsch, H., and Schatz, G. (1990). Protein import into yeast mitochondria is accelerated by the outer membrane protein MAS70. *EMBO J.* **9**, 3191–3200.

Jensen, R. E., and Yaffe, M. P. (1988). Import of proteins into yeast mitochondria: The nuclear gene *MAS2* encodes a component of the processing protease that is homologous to the *MAS1*-encoded subunit. *EMBO J.* **7**, 3863–3871.

Kang, P. J., Ostermann, J., Shilling, J., Neupert, W., Craig, E. A., and Pfanner, N. (1990). Requirement for hsp70 in the mitochondrial matrix for translocation and folding of precursor proteins. *Nature (London)* **348**, 137–142.

Kiebler, M., Pfaller, R., Söllner, T., Griffiths, G., Horstmann, H., Pfanner, N., and Neupert, W. (1990). Identification of a mitochondrial receptor complex required for recognition and membrane insertion of precursor proteins. *Nature (London)* **348**, 610–616.

Murakami, H., Blobel, G., and Pain, D. (1990). Isolation and characterization of the gene for a yeast mitochondrial import receptor. *Nature (London)* **347**, 488–491.

Ohba, M., and Schatz, G. (1987a). Protein import into yeast mitochondria is inhibited by antibodies raised against 45 kd proteins of the outer membrane. *EMBO J.* **6**, 2109–2116.

Ohba, M., and Schatz, G. (1987b). Disruption of the outer membrane restores protein import to trypsin-treated yeast mitochondria. *EMBO J.* **6**, 2117–2122.

Pain, D., Murakami, H., and Blobel, G. (1990). Identification of a receptor for protein import into mitochondria. *Nature (London)* **347**, 444–449.

Pollock, R. A., Hartl, F.-U., Cheng. M. Y., Ostermann, J., Horwich, A., and Neupert, W. (1988). The processing peptidase of yeast mitochondria: The two co-operating components MPP and PEP are structurally related. *EMBO J.* **7**, 3493–3500.

Pon, L., and Schatz, G. (1991). Biogenesis of yeast mitochondria. *In* "The Molecular Biology of the Yeast *S. cerevisiae*" (J. R. Pringle, J. Broach, and E. Jones, eds.) Vol. I. Cold Spring Harbor Press, Cold Spring Harbor, New York (in press).

Prasad, T. K., Hack , E., and Hallberg, R. L. (1990). Function of the maize mitochondrial chaperonin hsp60: Specific association between hsp60 and newly synthesized F_1-ATPase alpha subunits. *Mol. Cell. Biol.* **10**, 3979–3986.

Riezman, H., Hay, R., Witte, C., Nelson, N., and Schatz, G. (1983a). Yeast mitochondrial outer membrane specifically binds cytoplasmically-synthesized precursors of mitochondrial proteins. *EMBO J.* **2**, 1113–1118.

Riezman, H., Hase, T., van Loon, A. P. G. M., Grivell, L. A., Suda, K., and Schatz, G. (1983b). Import of proteins into mitochondria; a 70kd outer membrane protein with a large carboxy-terminal deletion is still transported to the outer membrane. *EMBO J.* **2**, 2161–2168.

Schatz, P. J., Solomon, F., and Botstein, D. (1988). Isolation and characterization of conditional lethal mutations in the *TUB1* alpha-tubulin gene of the yeast *Saccharomyces cerevisiae*. *Genetics* **120**, 681–695.

Scherer, P. E., Krieg, U. C., Hwang, S. T., Vestweber, D., and Schatz, G. (1990). A precursor protein partly translocated into yeast mitochondria is bound to a 70kd mitochondrial stress protein. *EMBO J.* **9**, 4315–4322.

Söllner, T., Pfaller, R., Griffiths, G., Pfanner, N., and Neupert, W. (1990). A mitochondrial import receptor for the ADP/ATP carrier. *Cell (Cambridge, Mass.)* **62**, 107–115.

Vestweber, D., Brunner, J., Baker, A., and Schatz, G. (1989). A 42 K outer-membrane protein is a component of the yeast mitochondrial protein import site. *Nature (London)* **341**, 205–209.

Witte, C., Jensen, R. E., Yaffe, M. P., and Schatz, G. (1988). *MAS1*, a gene essential for yeast mitochondrial assembly, encodes a subunit of the mitochondrial processing protease. *EMBO J.* **7**, 1439–1447.

Yaffe, M. P., and Schatz, G. (1984a). The future of mitochondrial research. *Trends Biochem. Sci.* **9**, 179–181.

Yaffe, M. P., and Schatz, G. (1984b). Two nuclear mutations that block mitochondrial import in yeast. *Proc. Natl. Acad. Sci. U.S.A.* **81**, 4819–4823.

Yang, M., Geli, V., Oppliger, W., Suda, K., James, P., and Schatz, G. (1991). The MAS processing protease of yeast mitochondria: Interaction of the purified enzyme with signal peptides and a purified precursor protein. *J. Biol. Chem.* (in press).

Chapter 20

Protein Import into Isolated Yeast Mitochondria

BENJAMIN S. GLICK

Biocenter, University of Basel
CH-4056 Basel, Switzerland

I. Introduction

Mitochondria import nearly all of their polypeptide components from the cytosol, and they contain an efficient machinery for the translocation and sorting of precursor proteins. The past decade has yielded major advances in our understanding of these events (for recent reviews, see Hartl *et al.*, 1989; Geli and Glick, 1990). Import can occur posttranslationally (Hallermeyer *et al.*, 1977; Reid and Schatz, 1982), and can be demonstrated *in vitro* by adding precursor proteins to isolated mitochondria in an appropriate buffer (Maccecchini *et al.*, 1979; Gasser *et al.*, 1982). As with many other biological processes, these cell-free systems permit a detailed biochemical analysis of mitochondrial protein import.

This article describes methods for the import of precursor proteins into mitochondria from the yeast *Saccharomyces cerevisiae*. Import-competent

METHODS IN CELL BIOLOGY, VOL. 34

mitochondria are prepared by gentle homogenization of spheroplasts (Daum *et al.*, 1982). Mitochondria can be stored in aliquots at $-80°C$, after being frozen in liquid nitrogen as a concentrated suspension (Kozlowski and Zagorski, 1988). Additional methods for the characterization of isolated yeast mitochondria are given by Yaffe (1990).

II.   Precursor Proteins

Import is measured by adding a radioactive precursor protein to mitochondria and analyzing the reaction products by gel electrophoresis and autoradiography (see below). In principle, any natural or artificial precursor protein can be synthesized using a cell-free translation system. The gene for the precursor must be inserted into an appropriate transcription vector, preferably one that allows for the *in vitro* synthesis of only a single major mRNA. Vectors containing the T5 (Stueber *et al.*, 1984), SP6 (Krieg and Melton, 1984), and T7 (Chen and Douglas, 1987) promoters have been used. Translation is carried out in the presence of a radioactive amino acid, usually $[^{35}S]$methionine. Ribosomes are then removed by centrifugation (15 minutes at 100,000 g) and the supernatant containing the radioactive precursor protein is added to mitochondria suspended in "import buffer" (see below). The translation mixture typically comprises between 5 and 20% of the final reaction volume.

For mitochondrial import studies, most researchers use precursors synthesized in a reticulocyte lysate (Pelham and Jackson, 1976). Translation systems derived from wheat germ do not work well, perhaps because they lack chaperone proteins of the hsp70 class, which interact with at least some mitochondrial precursors (Deshaies *et al.*, 1988; Murakami *et al.*, 1988). Some precursor proteins, such as precytochrome b_2, can become oxidized and form disulfide bridges during the synthesis reaction. This problem is avoided by adding 20 mM dithiothreitol (DTT) to the translation mixture and shielding it from light.

Protein import into isolated mitochondria can also be studied with artificial precursors containing mitochondrial-targeting sequences joined to the mouse cytosolic enzyme dihydrofolate reductase (DHFR) (Hurt *et al.*, 1984). These fusion proteins are purified after overexpression in *Escherichia coli* (Eilers and Schatz, 1986). In many cases, DHFR-containing precursors are imported into mitochondria in the absence of added soluble proteins. However, in other respects, these fusion proteins seem to utilize the same import machinery as natural precursors.

Precursors can be denatured by treatment with urea before presenting them to mitochondria. For purified precursors containing a DHFR moiety, incubation for 3 minutes at 30°C with 5 M urea/25 mM Tris-HCl, pH 7.4, is adequate (Endo and Schatz, 1988). With proteins synthesized in a reticulocyte lysate, 10–30% of the radioactive precursor may remain associated with the ribosomal pellet after the 100,000 g spin. This material can be solubilized by resuspending in a volume of 8 M urea/25 mM Tris-HCl, pH 7.4/5 mM DTT equal to the original volume of translation mixture, and incubating for 3 minutes at 30°C. At least in some cases, this solubilized precursor is imported efficiently. Alternatively, the precursor can be precipitated from the lysate with ammonium sulfate and resolubilized in 8 M urea containing Tris and DTT (Ostermann $et\ al.$, 1989). The ammonium sulfate can be removed by centrifuging the resuspended pellet through a 1-ml "spin column" of Sephadex G-25 (Sambrook $et\ al.$, 1989) equilibrated in 8 M urea/Tris/DTT (V. Hines, unpublished observations). The reaction mixture should be mixed immediately upon addition of the denatured precursor, as proteins quickly refold or aggregate after dilution of the urea. The final urea concentration in the import mixture should be 0.5 M or less.

III.   Import Conditions

A typical import reaction contains 25–200 μg of mitochondrial protein in a final volume of 100–200μl. All components except mitochondria and the precursor protein are mixed beforehand. Mitochondria are then added from a concentrated suspension, and the reaction is initiated by adding the precursor protein and transferring the mixture to a waterbath at the desired temperature. The reaction can be terminated by returning the mixture to ice. However, as some precursors import even at very low temperatures, it may be desirable to dissipate the electrochemical potential across the inner membrane with valinomycin plus K^+ or an uncoupler such as carbonylcyanide p-trifluoromethoxyphenylhydrazone (FCCP). The amounts of these compounds needed to block import vary depending upon experimental conditions, such as the concentrations of mitochondria and bovine serum albumin (BSA) in the reaction. We routinely use 1.0 μg/ml valinomycin plus 50 mM KCl, or 25 μM FCCP.

Precursor import is efficient under a variety of buffer conditions. A good general import buffer is made as follows:

0.6 M sorbitol (deionized)
50 mM HEPES

50 mM KCl
10 mM MgCl$_2$
2.5 mM Na$_2$-EDTA
2 mM KH$_2$PO$_4$
1 mg/ml fatty acid-free BSA
pH adjusted to 7.0 with KOH

Import buffer can be prepared as a 2× concentrate and stored frozen at −20°C. The optimal salt concentration for import may vary for different precursors; for example, import of the adenine nucleotide translocator is most efficient at about 200 mM KCl (Hines *et al.*, 1990). Increasing the concentration of BSA to 25 mg/ml or higher may reduce nonspecific binding of some precursors to the membranes.

The rate of import depends strongly upon temperature and the concentrations of mitochondria and precursor protein. Different precursors vary considerably in their import rates. Time courses should be performed at several temperatures, usually between 10 and 30°C, to determine the linear range of the import reaction.

IV. Energy Requirements

The translocation of precursor proteins into or through the inner membrane requires an electrochemical potential across that membrane. As only a weak potential is required, it is convenient to take advantage of the electrogenic proton ATPase in mitochondria. If ATP (0.5–2 mM) is added to the import buffer, it reaches the matrix by way of the adenine nucleotide translocator. 10 mM creatine phosphate and 0.1 mg/ml creatine kinase can be included as an ATP-regenerating system.

Alternatively, a membrane potential can be generated by providing O$_2$ and substrates of the electron transport chain. This method must be used when ATP levels in the matrix are to be kept low (see below). In order to maintain adequate levels of dissolved O$_2$, the reaction mixture should be shaken vigorously in a container with a flat bottom. We use small plastic scintillation vials for reactions of 0.5 ml or less, and normal plastic scintillation vials for larger volumes. To drive respiration, a combination of succinate and malate (5 mM each) is effective, as is 2 mM NADH. For experiments involving depletion of ATP in the matrix, NADH is the preferred substrate, because its oxidation does not involve substrate-level phosphorylation. Yeast mitochondria can generate a membrane potential by the oxidation of externally added NADH (von Jagow and Klingenberg, 1970).

In addition to maintaining a membrane potential, nucleoside triphosphates (NTPs) in the matrix are required for transporting precursors across the inner membrane (Hwang and Schatz, 1989). Mitochondria can be depleted of NTPs by treatment for 10 minutes at 25°C with 10 U/ml apyrase (Sigma, Grade VIII), plus 25 μM oligomycin to inhibit the proton-driven ATP synthase. Respiratory substrates and O_2 must then be added to generate a potential. For precursors synthesized in a reticulocyte lysate, the translation mixture can be depleted of ATP and GTP by incubating with 50 U/ml apyrase for 10 minutes at 30°C. In order to reduce NTP levels selectively outside the inner membrane, 200 μg/ml carboxyatractyloside (CAT, an inhibitor of the adenine nucleotide translocator) can be used in conjunction with apyrase and respiratory substrates. Under these conditions, the membrane potential drives ATP synthesis inside the mitochondria via ATP synthase, and CAT prevents this ATP from leaving the matrix. Any ATP outside the inner membrane is consumed by the apyrase. Conversely, NTP can be selectively depleted inside the inner membrane by incubating the mitochondria with oligomycin, and then adding CAT, respiratory substrates, and ATP. The presence of oligomycin prevents ATP synthesis in the matrix, and CAT blocks the entry of ATP via the adenine nucleotide translocator. As both oligomycin and CAT give only partial inhibition, the reduction of matrix NTP levels may be less complete than under conditions (oligomycin plus apyrase) in which the entire reaction mixture is depleted of NTP.

V. Analysis of Protein Import

Several criteria are used to monitor protein import into mitochondria. As a first step, most experimenters centrifuge the reaction mixture to remove BSA and other buffer components, unbound precursor proteins, proteases, etc. A 5-minute spin in a microfuge ($\sim$12,000 g) is sufficient. Sedimenting the mitochondria through a cushion of 0.5 ml 1.0 M sorbitol/20 mM K^+-HEPES, pH 7.4, efficiently removes soluble molecules. Whereas reisolating the mitochondria simplifies the analysis, the supernatant may also contain information, especially if mitoplasts are present (Kaput et al., 1989; K. Cunningham, unpublished observations). In most cases, the membrane pellet can be solubilized directly in sample buffer for SDS–PAGE and then heated for 3 minutes at 95°C.

Standard methods are used for SDS–PAGE. To enhance the signal from the radioactive label, gels are soaked in 1.0 M sodium salicylate for 30 minutes before being dried and exposed to film (Chamberlain, 1979).

Other types of enhancing solutions can also be used. For autoradiography of samples containing *in vitro*-synthesized precursor proteins, we recommend boiling the gel for 5 minutes in 5% trichloroacetic acid (TCA) after electrophoresis to reduce the background radioactivity. To remove the TCA, the gel is boiled for 5 minutes in distilled water and then soaked for a further 5 minutes in 1.0 M Tris base before being placed in salicylate. Treatment with TCA is unnecessary if a purified precursor protein has been used. We routinely include 0.33% linear polyacrylamide (Polysciences, Inc., #02806; added from a 2% stock solution) in the running gel. The linear polyacrylamide increases the tensile strength of the gel without altering its separating characteristics.

In most cases, proteins undergo one or more proteolytic processing steps after import. Precursor and cleaved forms can be resolved by using the appropriate acrylamide concentration in the running gel. The radioactivity present in each band can be quantified, either by densitometry or by spectrophotometric measurement of silver grains eluted from the X-ray film (Suissa, 1983). As a standard, it is convenient to load one lane of the gel with a quantity of the precursor protein equal to 10–20% of the amount added to each import reaction. In some cases, especially with purified precursor proteins, a portion of the precursor may be lost through adsorption to tube walls, etc. This problem can be circumvented by dissolving 50 μg of mitochondria in sample buffer, heating for 2 minutes at 95°C, then adding the precursor standard and heating again.

By itself, proteolytic processing is not always an adequate measure of import. Contaminating proteases in the reaction mixture can generate "pseudomature" species. The addition of valinomycin or uncouplers (see above) should block only cleavages that are specific to the mitochondrial import reaction. However, even membrane potential-dependent cleavage does not imply that a protein has been completely imported (Schleyer and Neupert, 1985). The best method for determining the location of a labeled protein is to test its accessibility to added protease. Trypsin can be used, although proteinase K is often more effective. Proteases can be added directly to the reaction mixture after import. Typical conditions are 10–1000 μg/ml protease for 30 minutes on ice. The protease is then inhibited by adding soybean trypsin inhibitor (for trypsin) or 1 mM phenylmethylsulfonyl fluoride (PMSF; for proteinase K). In most cases, it is sufficient to sediment the membranes, remove the supernatant containing the protease, dissolve the pellet directly in SDS-containing sample buffer, and heat immediately. However, some labeled proteins may be digested even after the addition of SDS, particularly if proteinase K has been used. Proteinase K can be inactivated completely by the following procedure:

1. Add PMSF to 1 mM. Sediment the mitochondria and remove the

supernatant. Redissolve the pellet in 180 μl 0.6 M sorbitol/20 mM K$^+$-HEPES, pH 7.4/0.2 mM PMSF. Add 20 μl 50% TCA and mix.

2. Heat 5 minutes at 60°C. Place on ice 5 minutes. Spin for 5 minutes at 12,000 g in a microfuge.

3. Remove the supernatant, and resuspend the precipitate in 50 μl of SDS–PAGE sample buffer containing 40 mM Na$^+$-PIPES, pH 7.5/1 mM PMSF. If the bromophenol blue indicator dye is yellow, add 1.0-μl aliquots of Tris base until it turns blue.

4. If it is not possible to reisolate the mitochondria (e.g., if the protease treatment is done in the presence of detergent), simply add 1 mM PMSF, then add TCA to 5% and proceed to step 2.

Some precursors are not cleaved upon import, and different methods must be applied. Upon assembly of porin into the outer membrane (Ono and Tuboi, 1987) or of the adenine nucleotide translocator into the inner membrane (Pfanner and Neupert, 1987), these proteins become inextractable by 100 mM Na$_2$CO$_3$, pH 11.5 (Fujiki *et al.*, 1982). However, carbonate inextractability is not an absolute criterion for membrane integration and must be interpreted with care. In the case of the adenine nucleotide translocator, the fully assembled protein acquires the ability to bind carboxyatractyloside (Schleyer and Neupert, 1984).

VI.   Subfractionation of Mitochondria after Import

Although inaccessibility to added protease in intact mitochondria normally means that a protein has been imported, it may reside in various locations within the organelle. Cleavage by the matrix protease indicates that at least the amino-terminal portion of the precursor has crossed the inner membrane; similarly, modification by intermembrane space-localized enzymes indicates that a protein has reached that compartment. In many cases, additional information can be obtained by subfractionating the mitochondria.

A common question is whether part or all of an imported protein is located in the intermembrane space. Mitoplasts are generated simply by lowering the sorbitol concentration: the matrix swells under these conditions and the outer membrane ruptures while the inner membrane remains largely intact (Daum *et al.*, 1982). (The outer membrane will not reseal if the osmotic strength is subsequently increased.) After the import reaction, the entire mixture is diluted with 5–9 volumes of 20 mM K$^+$-HEPES, pH 7.4, for 15–30 minutes, on ice. The mitochondria can also be spun in a microfuge and resuspended in import buffer before being converted to mitoplasts. If it is necessary to eliminate the membrane potential, FCCP is the compound of choice, as valinomycin–K^+ can lead to excessive swelling of the matrix.

If a protein is soluble in the intermembrane space, it should remain in the supernatant after centrifugation of mitoplasts. It may be necessary to treat the mitoplasts with salt to reduce binding to the membranes. Both soluble and membrane-bound proteins facing the intermembrane space are digested upon protease treatment of mitoplasts. As an intact inner membrane is required for mitoplast formation by osmotic lysis, and also to maintain the membrane potential necessary for precursor import, a large fraction (90–99%) of the import-competent mitochondria can be converted to mitoplasts under the appropriate dilution conditions. In contrast, the overall efficiency of mitoplast formation (defined by the release of a soluble intermembrane space marker such as cytochrome b_2) will be somewhat lower, as all mitochondrial preparations contain organelles with damaged inner membranes. Typically 70–90% of an imported matrix marker remains protected from protease after osmotic lysis.

A protein that is soluble in the matrix is released upon sonication of the mitochondria. For large reaction volumes, a probe sonicator is convenient, whereas smaller samples can be treated in a waterbath sonicator. Several rounds of freezing and thawing may help to promote membrane disruption. Membrane fragments are separated from soluble components by centrifugation for 30 minutes at 100,000 g. The efficiency of these methods is variable and should be monitored by following appropriate marker proteins.

An alternative method for subfractionating mitochondria is titration with the detergent digitonin (Schnaitman and Greenawalt, 1968; Hartl et al., 1987). Proteins are progressively solubilized by increasing digitonin levels, with components of each of the four compartments being released in a characteristic concentration range. Although this method seems to be reliable for soluble proteins of the matrix or intermembrane space, import intermediates spanning the inner membrane may be solubilized at lower digitonin concentrations than are bona fide inner membrane markers (unpublished observations). It is not known if the same problem is encountered with octyl-polyoxyethylene, which has also been used to distinguish soluble from membrane-bound imported proteins (Vestweber and Schatz, 1988).

If it is necessary to localize a protein to the inner versus the outer membrane, it should be possible to add carrier mitochondria after the import reaction, and then use established procedures to separate the two membranes (Pon et al., 1989).

VII.   Inhibition of the Matrix-Processing Protease

Proteolytic removal of presequences by the matrix-processing protease requires divalent cations such as Co^{2+}, Zn^{2+}, or Mn^{2+} (Boehni *et al.*, 1980). This enzyme can be inhibited in intact mitochondria by a combination of EDTA and *o*-phenanthroline (Zwizinski and Neupert, 1983). Under these conditions, protease-protected full-length precursor forms are obtained. To achieve strong inhibition, mitochondria can be incubated for 5 minutes at 25°C in import buffer lacking Mg^{2+} and supplemented with 5 mM EDTA plus 1 mM *o*-phenanthroline. As the mitochondria are quite fragile in the presence of chelators, the sorbitol concentration in the mixture should be raised to 1.2 M. Mitoplasts are obtained by diluting fourfold (0.3 M final sorbitol concentration) into 20 mM K^+-HEPES, pH 7.4, containing 5 mM EDTA and 1 mM *o*-phenanthroline.

ACKNOWLEDGMENTS

Thanks to Jeff Schatz for helpful suggestions. Benjamin Glick is a Merck Sharp and Dohme Fellow of the Life Sciences Foundation.

REFERENCES

Boehni, P., Gasser, S., Leaver, C., and Schatz, G. (1980). A matrix-localized mitochondrial protease processing cytoplasmically-made precursors to mitochondrial proteins. *In* "The Organization and Expression of the Mitochondrial Genome" (A. M. Kroon and C. Saccone, eds.) pp. 432–433. Elsevier/North-Holland, Amsterdam.

Chamberlain, J. P. (1979). Fluorographic detection of radioactivity in polyacrylamide gels with the water-soluble fluor, sodium salicylate. *Anal. Biochem.* **98,** 132–135.

Chen, W.-J., and Douglas, M. G. (1987). Phosphodiester bond cleavage outside mitochondria is required for the completion of protein import into the mitochondrial matrix. *Cell (Cambridge, Mass.)* **49,** 651–658.

Daum, G., Boehni, P. C., and Schatz, G. (1982). Import of proteins into mitochondria. Cytochrome b_2 and cytochrome c peroxidase are located in the intermembrane space of yeast mitochondria. *J. Biol. Chem.* **257,** 13028–13033.

Deshaies, R. J., Koch, B. D., Werner-Washburne, M., Craig, E. A., and Schekman, R. (1988). A subfamily of stress proteins facilitates translocation of secretory and mitochondrial precursor polypeptides. *Nature (London)* **332,** 800–805.

Eilers, M., and Schatz, G. (1986). Binding of a specific ligand inhibits import of a purified precursor protein into mitochondria. *Nature (London)* **322,** 228–232.

Endo, T., and Schatz, G. (1988). Latent membrane perturbation activity of a mitochondrial precursor protein is exposed by unfolding. *EMBO J.* **4,** 1153–1158.

Fujiki, Y., Hubbard, A. L., Fowler, S., and Lazarow, P. B. (1982). Isolation of intracellular membranes by means of sodium carbonate treatment: Application to the endoplasmic reticulum. *J. Cell Biol.* **93,** 97–102.

Gasser, S. M., Ohashi, A., Daum, G., Boehni, P., Gibson, J., Reid, G., Yonetani, T., and Schatz, G. (1982). Imported mitochondrial proteins cytochrome b_2 and cytochrome c_1 are processed in two steps. *Proc. Natl. Acad. Sci. U.S.A.* **79**, 267–271.

Geli, V., and Glick, B. (1990). Mitochondrial protein import. *J. Bioenerg. Biomembr.* **22**, 725–751.

Hallermeyer, G., Zimmermann, R., and Neupert, W. (1977). Kinetic studies on the transport of cytoplasmically synthesized proteins into the mitochrondria in intact cells of *Neuprospora crassa. Eur. J. Biochem.* **81**, 523–532.

Hartl, F.-U., Ostermann, J., Guiard, B., and Neupert, W. (1987). Successive translocation into and out of the mitochondrial matrix: Targeting of proteins to the intermembrane space by a bipartite signal peptide. *Cell (Cambridge, Mass.)* **51**, 1027–1037.

Hartl, F.-U., Pfanner, N., Nicholson, D. W., and Neupert, W. (1989). Mitochondrial protein import. *Biochim. Biophys. Acta* **988**, 1–45.

Hines, V., Brandt, A., Griffiths, G., Horstmann, H., Bruetsch, H., and Schatz, G. (1990). Protein import into yeast mitochondria is accelerated by the outer membrane protein MAS70. *EMBO J.* **9**, 3191–3200.

Hurt, E. C., Pesold-Hurt, B., and Schatz, G. (1984). The amino-terminal region of an imported mitochondrial precursor polypeptide can direct cytoplasmic dihydrofolate reductase into the mitochondrial matrix. *EMBO J.* **3**, 3149–3156.

Hwang, S., and Schatz, G. (1989). Translocation of proteins across the mitochondrial inner membrane, but not into the outer membrane, requires nucleoside triphosphates in the matrix. *Proc. Natl. Acad. Sci. U.S.A.* **86**, 8432–8436.

Kaput, J., Brandriss, M. C., and Prussak-Wieckowska, T. (1989). *In vitro* import of cytochrome *c* peroxidase into the intermembrane space: Release of the processed form by intact mitochondria. *J. Cell Biol.* **109**, 101–112.

Kozlowski, M., and Zagorski, W. (1988). Stable preparation of yeast mitochondria and mitoplasts synthesizing specific polypeptides. *Anal. Biochem.* **172**, 382–391.

Krieg, P. A., and Melton, D. A. (1984). Functional messenger RNAs are produced by SP6 *in vitro* transcription of cloned cDNAs. *Nucleic Acids Res.* **12**, 7057–7070.

Maccecchini, M.-L., Rudin, Y., Blobel, G., and Schatz, G. (1979). Import of proteins into mitochondria: Precursor forms of the extramitochondrially made F_1-ATPase subunits in yeast. *Proc. Natl. Acad. Sci. U.S.A.* **76**, 343–347.

Murakami, H., Pain, D., and Blobel, G. (1988). 70-kD heat shock-related protein is one of at least two distinct cytosolic factors stimulating protein import into mitochondria. *J. Cell Biol.* **107**, 2051–2057.

Ono, H., and Tuboi, S. (1987). Integration of porin synthesized *in vitro* into outer mitochondrial membranes. *Eur. J. Biochem.* **168**, 509–514.

Ostermann, J., Horwich, A. L., Neupert, W., and Hartl, F.-U. (1989). Protein folding in mitochondria requires complex formation with hsp60 and ATP hydrolysis. *Nature (London)* **341**, 125–130.

Pelham, H. R. B., and Jackson, R. J. (1976). An efficient mRNA-dependent translation system from reticulocyte lysates. *Eur. J. Biochem.* **67**, 247–256.

Pfanner, N., and Neupert, W. (1987). Distinct steps in the import of ADP/ATP carrier into mitochondria. *J. Biol. Chem.* **262**, 7528–7536.

Pon, L., Moll, T., Vestweber, D., Marshallsay, B., and Schatz, G. (1989). Protein import into mitochondria: ATP-dependent protein translocation activity in a submitochondrial fraction enriched in membrane contact sites and specific proteins. *J. Cell Biol.* **109**, 2603–2616.

Reid, G. A., and Schatz, G. (1982). Import of proteins into mitochondria. Extramitochondrial pools and posttranslational import of mitochondrial protein precursors *in vivo. J. Biol. Chem.* **257**, 13062–13067.

Sambrook, J., Fritsch, E. F., and Maniatis, T. (1989). "Molecular Cloning, A Laboratory Manual," 2nd ed., pp. E.37-E.38. Cold Spring Harbor Lab., Cold Spring Harbor, New York.

Schleyer, M., and Neupert, W. (1984). Transport of ADP/ATP carrier into mitochondria. Precursor imported *in vitro* acquires functional properties of the mature protein. *J. Biol. Chem.* **259,** 3487–3491.

Schleyer, M., and Neupert, W. (1985). Transport of proteins into mitochondria: Translocational intermediates spanning contact sites between outer and inner membranes. *Cell (Cambridge, Mass.)* **43,** 339–350.

Schnaitman, C., and Greenawalt, J. W. (1968). Enzymatic properties of the inner and outer membranes of rat liver mitochondria. *J. Cell Biol.* **38,** 158–175.

Stueber, D., Ibrahimi, I., Cutler, D., Dobberstein, B., and Bujard, H. (1984). A novel *in vitro* transcription–translation system: Accurate and efficient synthesis of single proteins from cloned DNA sequences. *EMBO J.* **3,** 3143–3148.

Suissa, M. (1983). Spectrophotometric quantitation of silver grains eluted from autoradiograms. *Anal. Biochem.* **133,** 511–514.

Vestweber, D., and Schatz, G. (1988). A chimeric mitochondrial precursor protein with internal disulfide bridges blocks import of authentic precursors into mitochondria and allows quantitation of import sites. *J. Cell Biol.* **107,** 2037–2043.

von Jagow, G., and Klingenberg, M. (1970). Pathways of hydrogen in mitochondria of *Saccharomyces carlsbergensis. Eur. J. Biochem.* **12,** 583–592.

Yaffe, M. P. (1990). The analysis of mitochondrial function and assembly. *In* "Methods in Enzymology" (C. Guthrie and G. R. Fink, eds.), vol. **194,** pp. 627–643. Academic Press, San Diego, California.

Zwizinski, C., and Neupert, W. (1983). Precursor proteins are transported into mitochondria in the absence of proteolytic cleavage of addtional sequences. *J. Biol. Chem.* **258,** 13340–13346.

Chapter 21

Mitochondrial Inner Membrane Protease I of Saccharomyces cerevisiae

ANDRÉ SCHNEIDER

Biocenter, University of Basel
CH-4056 Basel, Switzerland

I. Introduction

A. Translocation Proteases

Transport of proteins across pro- and eukaryotic membranes is usually accompanied by proteolytic removal of targeting sequences from the transported precursor polypeptides (Verner and Schatz, 1988). This process is catalyzed by a group of enzymes referred to as translocation proteases. These include soluble and membrane-bound proteases. The best characterized member of the soluble proteases is the matrix protease of mitochondria, which removes the mitochondrial-targeting signals from precursor proteins imported into the matrix space. The enzyme from *Saccharomyces cerevisiae* consists of two subunits of 48 and 52 kDa, which have 26% sequence identity. The activity of the protease requires both subunits as well as divalent metal ions (Boehni *et al.*, 1983; Hawlitschek *et al.*, 1988; Pollock *et al.*, 1988; Yang *et al.*, 1988; Geli *et al.*, 1990). A similar enzyme appears to be present in the stroma of chloroplasts; it removes the transit sequences from preproteins targeted to the internal compartments of chloroplasts (Robinson and Ellis, 1984).

The two known prokaryotic translocation proteases are both membrane bound. Leader peptidase I cleaves the leader sequences of preproteins exported across the bacterial plasma membrane. It is an integral protein of the inner bacterial membrane, faces the periplasm, consists of a single 36-kDa subunit, and is essential for life. Its activity is stimulated by acidic phospholipids (Wolfe *et al.*, 1983; de Vrije *et al.*, 1988; Bilgin *et al.*, 1990). Prolipoprotein signal peptidase is a more specialized enzyme that cleaves only the leader sequences of modified prolipoproteins. It is a 17-kDa monomer (Dev and Ray, 1984).

The best characterized member of eukaryotic membrane-bound translocation proteases is the microsomal signal peptidase. The enzyme, an integral protein of the endoplasmic reticulum, faces the lumenal side of this membrane system. Depending on the source and the isolation conditions, it consists of five (in dogs) (Evans *et al.*, 1986) or two (in chickens) subunits (Baker and Lively, 1987). The glycosylated subunits of the dog and the chicken enzymes show partial sequence identity (Shelness *et al.*, 1988). A similar enzyme appears to be present in yeast (Boehni *et al.*, 1988; Shelness and Blobel, 1990).

Mitochondria and chloroplasts also contain membrane-bound proteases involved in the intraorganellar sorting of imported proteins. A processing protease on the inner side of the thylakoid membrane removes the sorting signal of the transit peptide of precursors transported to the thylakoid lumen (Kirwin *et al.*, 1987). Although the enzyme has not yet been purified and its subunit composition is unknown, its substrate specificity resembles that of bacterial leader peptidase and of microsomal signal peptidase (Halpin *et al.*, 1989).

B.   Inner Membrane Proteases of *Saccharomyces cerevisiae*

Cytochrome b_2 (cytb_2) and cytochrome c_1 (cytc_1) are cleaved in two successive steps during their import into mitochondria. Cytb_2 is soluble in the intermembrane space whereas cytc_1 is bound to the outer face of the inner membrane. The amino-proximal matrix-targeting signal is first removed by the matrix protease, then the second part of the presequence is cleaved off by other enzymes. The enzyme acting upon the partly cleaved cytb_2 precursor is termed "inner membrane protease I" (Schneider *et al.*, 1991).

A mutant (*pet ts2858*) that has a temperature-sensitive inner membrane protease I is temperature sensitive for the second cleavage of the cytb_2 precursor and for growth on a nonfermentable carbon source. It is also deficient in the processing of the mitochondrially encoded cytochrome oxidase subunit II precursor, a protein inserted from the matrix side across the inner membrane (Pratje *et al.*, 1983; Pratje and Guiard, 1986). Cytc_1 is processed normally, suggesting that the sorting sequence of cytc_1 is removed by yet another protease. The gene deficient in *pet ts2858* encodes a 21.4-kDa protein that is bound to the inner membrane. A 101-amino acid region of this protein shows 30% sequence identity with the second transmembrane segment and the periplasmic part of bacterial leader peptidase I (Behrens *et al.*, 1991). Mutant *pet ts2858* has been very useful in devising a specific assay for inner membrane protease I and for characterizing the enzyme.

II. *In Vitro* Assay and Solubilization of Inner Membrane Protease I

A. Principle of the Assay

As the enzyme is inactive toward *in vitro*-synthesized pre-cytb_2 or the cytb_2 intermediate generated from pre-cytb_2 by addition of purified matrix protease, the substrate is the cytb_2 intermediate in a detergent extract of mitochondria from mutant *pet ts2858*. Conversion of this intermediate to the mature form is analyzed by immunoblotting with antiserum against cytb_2. When this assay is performed with crude extracts of wild-type mitochondria as a source of enzyme, the mature cytb_2 in these wild-type mitochondria causes a high background. This background can be eliminated by using mitochondria from cytb_2-deficient yeast cells.

B. Preparation of Mitochondria Containing the cytb_2 Intermediate

The yeast strain *pet ts2858* is grown on 10% glucose at 23°C with subsequent induction for 12 hours at 37°C in a phosphate buffer containing DL-lactate (Pratje and Guiard, 1986). Mitochondria are isolated (Daum *et al.*, 1982), resuspended at a protein concentration of 6 mg/ml in 20 m*M*

Tris-Cl, pH 7.5; 0.6 M sorbitol; 10 mM N-ethylmaleiimide (NEM); 1 mM phenylmethanesulfonyl fluoride (PMSF); they are then incubated for 10 minutes on ice. NEM is included to inhibit any residual inner membrane protease I in the mutant mitochondria; it is quenched by adding dithio-threitol to 20 mM. The resulting mitochondrial suspension serves as a source of substrate; it is stable for several months if frozen in liquid nitrogen and stored in aliquots at $-70°C$.

C.   Source of Enzyme

A yeast strain carrying a disrupted cytb_2 gene is grown in 1% Bacto-yeast extract, 1% Bacto-peptone, and 2% galactose, and mitochondria are prepared as described by Daum *et al.* (1982).

D.   Preparation of Phosphatidylserine Liposomes

A volume corresponding to 10 mg of the desired phospholipid (supplied in chloroform or chloroform: methanol by Sigma Chemical Co.) is dried under a stream of nitrogen. The dried residue is suspended in 1 ml of 20 mM Tris-Cl, pH 7.5, and 0.5 mM EDTA; the suspension is sonicated five times for 1 minute each in a bath-type sonifier (Transsonic T400, ELMA Co.) containing ice water. The resulting suspension is frozen in 0.2-ml aliquots in liquid nitrogen and stored at $-70°C$. Just before use, each aliquot is thawed and sonicated again for 15 seconds.

E.   The Standard Assay

Use 20 μl of NEM-treated mitochondria from mutant *pet ts2858* (120 μg protein; containing 0.1–0.5 μg of cytb_2 intermediate; see preceding section); mix with the desired amount of enzyme (derived from cytb_2-deficient mitochondria) and adjust to a final volume 0.1 ml and final concentrations of the following components: Tris-Cl, pH 7.5, 20 mM; MgCl$_2$, 10 mM; octyl-polyoxyethylene (Fluka Co., Switzerland), 0.4% (v/v); PMSF, 1 mM. Phosphatidylserine liposomes are then added to 0.3–0.6 mg/ml (see below) and the mixture is gently agitated on a Vortex mixer, followed by a 15-second sonication in the bath type sonifier at room temperature. After incubation for 1 hour at 30°C, the reaction is stopped by adding 0.1 ml of 3 × SDS–sample buffer and heating the mixture to 95°C for 3 minutes. Finally, 70 μl of the sample is analyzed by SDS–10% PAGE and immunoblotting with antiserum against cytb_2. To obtain optimal separation of the cytb_2 intermediate from mature cytb_2, electrophoresis is continued for 1 hour after the tracking dye has electrophoresed off the gel. The immunoblot is developed with radioiodinated protein A and the bands on the fluorogram are quantified by optical scanning (Schneider *et al.*, 1991). One enzyme unit is defined as the activity cleaving 50% of the cytb_2 intermediate to mature cytb_2 under the described conditions.

F.   Comments

It is important to work at the correct detergent concentration. Although the enzyme tolerates up to 1% octyl-polyoxyethylene (see below), the assay must be performed at 0.4% octyl-polyoxyethylene, presumably because the substrate conformation is detergent sensitive. Another critical factor is concentration of phosphatidylserine; the optimal concentration is variable, probably because it is influenced by the amount of phospholipids present in the source of enzyme; however, it is generally between 0.3 and 0.6 mg/ml. Higher phosphatidylserine concentrations inhibit the enzyme.

The cleavage assay is approximately linear with respect to the amount of enzyme between 5 and 60 μg of solubilized mitochondrial protein (Fig. 1). It is also linear with time up to 60 minutes (Schneider *et al.*, 1991).

G.   Solubilization of Inner Membrane Protease I

Mitochondria from the cytb_2-deficient strain are suspended to 2 mg/ml in 20 mM Tris-Cl, pH 7.5; 1 mM PMSF; and 0.5% octyl-polyoxyethylene; centrifuge for 30 minutes at 0°C at 100,000 g. The supernatant containing

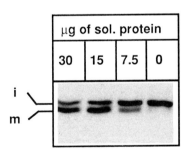

FIG. 1.   The cleavage assay. The indicated amounts of solubilized mitochondrial membrane protein were incubated under the standard assay conditions; generation of mature cytb_2 was measured as described in the text. Intermediate (i) and mature (m) forms of cytb_2 are shown.

the soluble matrix proteins is discarded and the pellet is resuspended by sonication in the bath-type sonicator in one-fifth of the orginal volume of 20 mM Tris-Cl, pH 7.5; 1 mM PMSF; and 1% octyl-polyoxyethylene. The resulting suspension is centrifuged for 90 minutes at 0°C at 100,000 g. The final supernatant usually contains more than 50% of the enzyme activity of the starting mitochondria.

III.   Properties of Inner Membrane Protease I

The enzyme consists of at least one subunit with a molecular weight of 21,400. It is the only protein known so far that shows sequence homology to bacterial leader peptidase (Behrens *et al.*, 1991) and the only defined protein-mediating intramitochondrial sorting of proteins. Overexpression of the 21.4-kDa protein does not increase the enzyme activity, suggesting that inner membrane protease I is a heterooligomer. The 21.4-kDa subunit is an integral inner membrane protein whose carboxy terminus protrudes into the intermembrane space. The enzyme requires magnesium ions and acidic phospholipid; it is inhibited by EDTA, which complexes magnesium ions, or by adriamycin, which binds to acidic phospholipids. NEM (10 mM) inhibits the enzyme by approximately 60% (Schneider *et al.*, 1991).

References

Baker, R. K., and Lively, M. O. (1987). Purification and characterization of hen oviduct microsomal signal peptidase. *Biochemistry* **26**, 8561–8567.
Behrens, M., Michaelis, G., and Pratje, E. (1991). Mitochondrial inner membrane protease I of *Saccharomyces cerevisiae* shows sequence homology to the *Escherichia coli* leader peptidase. *Mol. Gen. Genet.* (in press).
Bilgin, N., In Lee, J., Zhu, H., Dalbey, R., and von Heijne, G. (1990). Mapping of catalytically important domains in *Escherichia coli* leader peptidase. *EMBO J.* **9**, 2717–2722.
Boehni, P. C., Daum, G., and Schatz, G. (1983). Import of proteins into mitochondria: Partial purification of a matrix-located protease involved in cleavage of mitochondrial precursor polypeptides. *J. Biol. Chem.* **258**, 4937–4943.
Boehni, P. C., Deshaies, R. J., and Schekman, R. (1988). Sec11 is required for signal peptide processing and yeast cell growth. *J. Cell Biol.* **106**, 1035–1042.
Daum, G., Gasser, S., and Schatz, G. (1982). Import of proteins into mitochondria. Energy-dependent, two-step pcofessing of the intermembrane space enzyme cytochrome b_2 by isolated yeast mitochondria. *J. Biol. Chem.* **257**, 13075–13080.
Dev, I. K., and Ray, H. P. (1984). Rapid assay and purification of a unique signal peptidase that processes the prolipoprotein from *E. coli* B. *J. Biol. Chem.* **259**, 11114–11120.

de Vrije, T., de Swart, R. L., Dowhan, W., Tommassen, J., and de Kruijff, B. (1988). Phosphatidylglycerol is involved in protein translocation across *Escherichia coli* inner membranes. *Nature (London)* **334,** 173–175.

Evans, E. A., Gilmore, R., and Blobel, G. (1986). Purification of microsomal signal peptidase as a complex. *Proc. Natl. Acad. Sci. U. S. A.* **83,** 581–585.

Geli, V., Yang, M., Suda, K., Lustig, A., and Schatz, G. (1990). The *MAS*-encoded processing protease of yeast mitochondria. Overproduction and characterization of its two nonidentical subunits. *J. Biol. Chem.* **265,** 19216–19222.

Halpin, H., Elderfield, P. D., James, H. E., Zimmermann, R., Dunbar, B., and Robinson, C. (1989). The reaction specifities of the thylakoidal processing peptidase and *Escherichia coli* leader peptidase are identical. *EMBO J.* **8,** 3917–3921.

Hawlitscheck, G., Schneider, H., Schmidt, B., Tropschug, M., Hartl, F.-U., and Neupert, W. (1988). Mitochondrial protein import: Identification of processing peptidase and of PEP, a processing enhancing protein. *Cell (Cambridge, Mass.)* **53,** 795–806.

Kirwin, P. M., Elderfield, P. D., and Robinson, C. (1987). Transport of proteins into chloroplasts. Partial purification of a thylakoidal processing peptidase involved in plastocyanin biogenesis. *J. Biol. Chem.* **262,** 16386–16390.

Pollock, R. A., Hartl, F.-U., Cheng, M. Y., Ostermann, J., Horwich, A., and Neupert, W. (1988). The processing peptidase of yeast mitochondria: The two co-operating components MPP and PEP are structurally related. *EMBO J.* **7,** 3493–3500.

Pratje, E., and Guiard, B. (1986). One nuclear gene control the removal of transient presequences from two yeast proteins: One encoded by the nuclear, the other by the mitochondrial genome. *EMBO J.* **5,** 1313–1317.

Pratje, E., Mannhaupt, G., Michaelis, G., and Beyreuther, K. (1983). A nuclear mutation prevents processing of a mitochondrially encoded membrane protein in *S. cerevisiae*. *EMBO J.* **2,** 1049–1054.

Robinson, C., and Ellis, R. J. (1984). Transport of proteins into chloroplasts. *Eur. J. Biochem.* **142,** 337–342.

Schneider, A., Behrens, M., Scherer, P., Pratje, E., Michaelis, G., and Schatz, G. (1991). Inner membrane protease I, an enzyme mediating intramitochondrial protein sorting in yeast. *EMBO J.* **10,** 247–254.

Shelness, G. S., and Blobel, G. (1990). Two subunits of the canine signal peptidase complex are homologous to yeast SecII protein. *J. Biol. Chem.* **265,** 9512–9519.

Shelness, G. S., Kanwar, Y. S., and Blobel, G. (1988). cDNA-derived primary structure of the glycoprotein component of canine microsomal signal peptidase complex. *J. Biol. Chem.* **263,** 17063–17070.

Verner, K., and Schatz, G. (1988). Protein translocation across membranes. *Science* **241,** 1307–1313.

Wolfe, P. B., Zwizinski, C., and Wickner, W. (1983). Purification and characterization of leader peptidase from *Escherichia coli*. *In* "*Methods in Enzymology*" (S. Fleischer and B. Fleischer, eds.), Vol. **97,** pp. 40–46. Academic Press, New York.

Yang, M., Jensen, R. E., Yaffe, M. P., Oppliger, W., and Schatz, G. (1988). Import of proteins into yeast mitochondria: The purified matrix processing protease contains two subunits which are encoded by the nuclear *MAS1* and *MAS2* genes. *EMBO J.* **7,** 3857–3862.

Chapter 22

Purified Precursor Proteins for Studying Protein Import into Yeast Mitochondria

UTE C. KRIEG AND PHILIPP E. SCHERER

Department of Biochemistry
Biocenter, University of Basel
CH-4056 Basel, Switzerland

I. Introduction

Most mitochondrial proteins are encoded by nuclear genes and are synthesized as precursor proteins on cytosolic polysomes. These precursors then reach their final destination in one of the four mitochondrial compartments by traversing one or both mitochondrial membranes. This import can occur posttranslationally *in vivo* and *in vitro*.

Radiolabeled precursors for investigating protein import *in vitro* can be produced in a cell-free protein-synthesizing system such as a reticulocyte lysate (Maccecchini *et al.*, 1979). However, this approach has several disadvantages. First, it is difficult to assess the role in mitochondrial protein import of factors (such as heat-shock proteins) present in the protein-synthesizing extract. Second, the amount of precursor obtained in such

METHODS IN CELL BIOLOGY, VOL. 34

systems is extremely low (in the femtomole range). This makes it very difficult to characterize the precursors as well as any additional import catalysts.

These limitations are circumvented by using purified precursor proteins. Direct purification of precursors from normal cells is difficult, as the steady-state concentration of precursors *in vivo* is extremely low. Ohta and Schatz (1984) purified to homogeneity the precursor for the mitochondrial ATPase β-subunit from a *rho*⁻ yeast mutant grown in the presence of a mitochondrial uncoupling reagent. The low abundance and the instability of the precursor, however, necessitated a 10,000-fold purification. The method of choice is the purification of mitochondrial precursor proteins expressed in *Escherichia coli*. Several such attempts have been described in the literature, but precursor expression was either low (Sheffield *et al.*, 1986; Jaussi *et al.*, 1987) or yielded insoluble, denatured precursor molecules (Murakami *et al.*, 1988).

In order to overcome some of these problems, we have employed fusion proteins that consist of a soluble nonmitochondrial "passenger" protein fused to a mitochondrial matrix-targeting signal. Successful purification of such a hybrid protein depends to a large extent on the right choice of the passenger protein. Such a passenger molecule should have a conformation that is not destabilized by fusion to a mitochondrial-targeting sequence. It should also not require extensive posttranslational modifications, so that it can be isolated in its native state from a heterologous system. Finally, it should be nontoxic to both yeast and *E. coli*.

Using fusion proteins over authentic precursors affords the following advantages: (1) usually increased stability, as mentioned above; (2) if so chosen, presence of an enzymatic activity, which aids in the purification (see below) and in intracellular or intramitochondrial protein localization studies; and (3) the possibility to probe any sequence for its ability to target the passenger protein to mitochondria, thereby excluding the effect of redundant targeting information in the "mature" part of the protein (Hurt and Schatz, 1987).

The passenger protein that proved to be the most versatile in our laboratory is mouse dihydrofolate reductase (DHFR). It is a stable protein with a relative molecular mass of 21,000 Da and can be isolated from the cytosol of rodent cells. Mouse DHFR fusion proteins can be expressed at high levels in *E. coli*, remain soluble, and can be purified quite easily to homogeneity in a simple two-step procedure, yielding milligram amounts of essentially pure, functional precursor. If an appropriate targeting sequence is attached to the amino terminus, the protein is targeted to mitochondria, translocated into the matrix compartment, and its presequence is cleaved off by the matrix-located processing protease, both *in vivo* and *in vitro*.

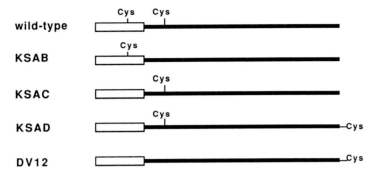

FIG. 1. Variants of the COXIV–DHFR fusion protein ("wild type"). The positions of cysteine residues in the COXIV presequence (open box) and the DHFR moiety (bar) are indicated (see text for details).

DHFR is, of course, not the only passenger protein that can be fused to mitochondrial-targeting sequences. For example, Douglas *et al.* (1984) fused β-galactosidase or metallothionine (Chen and Douglas, 1987) to parts of the β-subunit of the F_1-ATPase, and Schmitz and Lonsdale (1989) and Schmitz *et al.* (1990) have fused β-glucuronidase to targeting sequences of plant mitochondrial proteins. None of these hybrid proteins has been isolated in large amounts from *E. coli*.

Below, we describe purification procedures for a set of fusion proteins containing the first 22 amino acids of yeast cytochrome oxidase subunit IV (COXIV) fused to the amino terminus of mouse DHFR via a three-amino acid linker (Hurt *et al.*, 1984a,b). Some features of the different fusion proteins (henceforth called COXIV–DHFR) are pictured in Fig. 1.

II. Purification of Mitochondrial Precursor Proteins Expressed in *E. coli*

A. Vector/Host Systems

For the expression of the different variants of our fusion protein, we use the vector pKK 223-3 (Pharmacia, Uppsala, Sweden; cat. no. 27–4935–01). It contains the strong *trp–lac* (*tac*) promoter (de Boer *et al.*, 1983), that includes the -35 region of the *trp* promoter and the -10 region, operator, and ribosome-binding site of the *lac* $UV-5$ promoter. Because expression of mitochondrial precursors often seems to be toxic for *E. coli*, the basal expression of the precursor protein should be kept as low as possible. This can be achieved in a host overexpressing the *lac* repressor.

Such *lac* I^Q strains are, for instance, W3110 − lac I^Q L8 (Amann *et al.*, 1983) or JM105 (Pharmacia; cat. no. 27–1550–01). The promoter may be derepressed by the addition of isopropyl-β-D-thiogalactoside (IPTG).

B.   Factors Affecting the Expression Level

Expression levels are affected by the length of the induction period. However, induction times longer than 2 hours should be avoided; they do not increase the steady-state level of fusion protein and can lead to cell lysis. The amount of fusion protein that can be expressed also depends on how tightly the DHFR moiety is folded. Point mutations that destabilize the DHFR molecule decrease the expression level.

C.   Tools for Purification

The isolation of pure precursor protein (Table I) takes advantage of the fact that the folate analog methotrexate binds tightly to DHFR. Methotrexate can therefore be used as a ligand for affinity chromatography of these precursor proteins directly from cell extracts. However, point mutations that lower the affinity of DHFR for the ligand also lower the effectiveness of this purification step. The bound COXIV–DHFR can be eluted with an excess of the natural ligand, dihydrofolate. However, this eluate still contains several contaminating proteins.

In a second step, COXIV–DHFR is purified to homogeneity by anion-exchange chromatography. The protein binds to the support due to the presence of several positively charged amino acids in the COXIV presequence. All other contaminants in the eluate from the affinity column are found in the flow-through fraction. The precursor is then eluted in presence of high ionic strength.

Radiochemically (but not chemically) pure precursor protein can be isolated using a shortened purification procedure involving only anion-exchange chromatography (Table II; see also Table III).

The import competence of both unlabeled and radiolabeled precursors is monitored by the standard ATP-driven mitochondrial import assay, as described by Glick (Chapter 20).

III.   Use of Purified Precursors in the Study of Mitochondrial Protein Import

Initially, purified precursor proteins were used to establish their structural features that were required for import competence. Endo and Schatz (1988) did the first detailed structural comparison between a mitochondrial

TABLE I

PURIFICATION PROCEDURE FOR NONRADIOACTIVE COXIV–DHFR

Step	Procedure[a]
Cell growth	Inoculate 200 ml of a stationary *E. coli* preculture into 10 liters of LB-Amp. Grow cells to an OD_{600} of 0.6 (about 2 hours).
Induction	Add IPTG to 1 mM. Induce for 2 hours.
Harvesting	Harvest cells by centrifugation at 6000 g (room temperature, 10 minutes).[b]
Lysis	Resuspend cells in 100 ml of solution A. Incubate on ice for 10 minutes. Centrifuge at 6000 g (4°C, 10 minutes). Swell cells in 150 ml of ice-cold H_2O. Incubate on ice for 10 minutes. Centrifuge at 6000 g (4°C, 10 minutes). Resuspend cells in 100 ml of solution B. Homogenize with 15 strokes in a glass Dounce homogenizer.[c,d] Incubate for 35 minutes on ice. After 15 and 30 minutes, add fresh PMSF, repeat homogenization. At 30 minutes, also add octyl-POE[e] to 1%. Dilute with 250 ml solution C. Centrifuge at 17,000 g for 10 minutes. To the supernatant add 50 ml of solution D dropwise with constant stirring. Continue to stir for 45 minutes. Centrifuge at 17,000 g for 10 minutes.
Affinity chromatography	Load supernatant onto a 20-ml methotrexate–agarose column (flow rate 200 ml/hour), preequilibrated with 50 mM K-phosphate, pH 5.6. Wash with 50 ml of 50 mM K-phosphate, pH 5.6, 50 ml solution E, 100 ml solution E plus 1 M KCl, and 100 ml solution E. Elute with 50 ml solution E containing 2 mg/ml dihydrofolate followed by 50 ml of solution E.
Anion-exchange chromatography	Load eluate (about 100 ml) onto a 15 ml CM–CL Sepharose 4B column (flow rate 60 ml/hour), preequilibrated with solution E. Wash with 100 ml solution E. Elute with solution E plus 0.5 M KCl. Monitor elution by OD_{280} measurement. Pool protein peak.
Dialysis	Dialyze three times against 1 liter solution F, then two times against 1 liter solution F plus 10% glycerol.
Storage	Freeze in small aliquots, store at −70°C.[f]

[a] The compositions of media and solutions are given in Table III.

[b] Cells can be washed in H_2O, reisolated, frozen in liquid nitrogen, and stored at −70°C at this stage.

[c] We recommend the use of siliconized plastic and glassware in all subsequent purification steps and whenever purified precursors are handled; this includes the glass homogenizer used here (Braun AG, Melsungen, Germany. We use it with the tight-fitting pestle).

[d] All further purification steps are carried out at 4°C.

[e] The nonionic detergent octyl-oxypoly(ethylene) (octyl-POE; Bachem Feinchemikalien AG, Bubendorf, Switzerland; cat. no. P-1140) is used routinely in our laboratory to solubilize membranes. Another suitable detergent is Triton X-100.

[f] An average preparation will yield about 3 to 4 mg pure precursor ("wild-type" COXIV–DHFR and KSAD; Fig. 1). Frozen in liquid nitrogen and stored at −70°C, these precursors remain active for more than 1 year. However, they lose import competence when frozen and thawed repeatedly.

TABLE II

PURIFICATION PROCEDURE FOR RADIOLABELED COXIV–DHFR

Step	Procedure[a]
Cell growth	Inoculate 20 ml of LB-Amp with a single *E. coli* colony. Grow overnight. Pellet the cells at 3000 g (room temperature, 5 minutes). Resuspend in 200 ml sulfate-limiting medium.[b] Grow for 3 hours.
Starvation and induction	Pellet cells as above. Wash two times by resuspending in distilled H_2O. Resuspend in 200 ml prewarmed sulfate-free labeling medium. Incubate at 37°C for 15 minutes.
Labeling	Add 15 mCi [^{35}S]O_4^{2-}. Continue incubation for 30 minutes.
Harvesting	Harvest cells by centrifugation at 6000 g (4°C, 10 minutes).
Lysis	Resuspend cells in 2 ml of ice-cold solution A.[c] Incubate on ice 10 minutes. Centrifuge at 15,000 g 2 minutes. Wash cells once in 2 ml H_2O. Repeat centrifugation. Resuspend cells in 2 ml solution B. After homogenization and membrane solubilization in presence of 1% octyl-POE (as described in Table I), dilute the preparation with 16 ml solution G. Adding 2 ml solution H, proceed as described in Table I.
Anion-exchange chromatography	Load onto 10 ml CM–CL Sephadex 4B column (flow rate 0.5 ml/minute), preequilibrated with 70 ml solution I. Wash column with 50 ml solution I. Elute bound proteins with an 80-ml linear gradient in solution I, from 25 to 500 mM KCl (flow rate 1 ml/minute). Analyze protein profile by SDS–12% PAGE and fluorography.[d]
Dialysis and storage	Described in Table I.

[a] The compositions of media and solutions are given in Table III.

[b] This procedure can also be used in a slightly modified form for the isolation of *unlabeled* precursors, whose DHFR moiety does not bind to the methotrexate affinity column. In this case, inoculate 5 ml of a stationary preculture into 200 ml LB-Amp. Grow cells to an approximate OD_{600} of 5. Induce for 30 minutes in presence of 1 mM IPTG. Harvest and proceed as above. Analyze column fractions by SDS–12% PAGE and silver staining. This preparation will yield about 50 μg precursor of about 70% purity.

[c] Beginning with this step, we work with siliconized microcentrifuge tubes. All procedures are carried out at 4°C.

[d] We routinely collect 1.5-ml fractions and analyze 20 μl of each fraction, after precipitation with trichloroacetic acid, by SDS–12% PAGE. Exposing the fluorographs for 30 minutes is sufficient.

precursor polypeptide and the corresponding presequence-free protein. In this work they also showed that unfolded precursors disturbed the phospholipid bilayer structures of membranes. Eilers and Schatz (1986) and Eilers *et al.* (1987, 1988) used radiolabeled precursor proteins to demonstrate that the DHFR moiety must at least partly unfold, and that an

TABLE III

MEDIA AND SOLUTIONS[a]

Medium/solution	Components
LB-Amp	1% (w/v) bacto-tryptone, 0.5% (w/v) bacto-yeast extract, 0.5% (w/v) NaCl, 50 μg/ml ampicillin
Solution A	30% (w/v) sucrose, 20 mM K-phosphate, pH 8.0, 1 mM EDTA, pH 8.0
Solution B	250 mM KCl, 20 mM K-phosphate, pH 8.0, 2 mM EDTA, pH 8.0, 1 mM EGTA, pH 8.0, 10 mM dithiothreitol (DTT), 0.5 mM phenylmethylsulfonyl fluoride (PMSF),[b] 1 μg/ml soybean trypsin inhibitor (SBTI), protease inhibitor cocktail (see below), 1 mg/ml lysozyme, 0.5 mg/ml DNAse I
Solution C	50 mM K-phosphate, pH 5.6, 2 mM EDTA, pH 7.0, 1 mM EGTA, pH 8.0, 1 mM DTT,[b] 0.5 mM PMSF, 1 mg/ml SBTI, protease inhibitor cocktail
Solution D	Solution C containing 2% (w/v) protamine sulfate
Solution E	20 mM K-phosphate, pH 7.5, 1 mM EDTA, pH 7.5, 1 mM DTT
Solution F	20 mM K-phosphate, pH 7.5, 50 mM KCl, 1 mM DTT
Solution G	20 mM K-phosphate, pH 7.0, 1 mM EDTA, pH 7.0, 1 mM DTT, 1 mM PMSF, 1 mg/ml SBTI, 0.5% octyl-POE
Solution H	Solution G containing 2% (w/v) protamine sulfate
Solution I	25 mM KCl, 20 mM K-phosphate, pH 7.5, 1 mM EDTA, pH 7.5, 0.5% octyl-POE, 1 mM DTT[b]
Protease inhibitor cocktail	Use a 2000-fold dilution of the following stock mixture: 2.5 mg/ml leupeptin, 4 mg/ml antipain, 0.5 mg/ml chymostatin, 0.5 mg/ml elastinal, 10 mg/ml pepstatin, dissolve in dimethyl sulfoxide stored at $-20°$C
Sulfate-limiting medium	1 mM K-phosphate, pH 7.5, 40 μg/ml of each amino acid (except Cys and Met), 1 μg/ml thiamine, 50 μg/ml ampicillin, 0.1 mM MgSO$_4$, 10-fold diluted MES concentrate (see below)
Labeling medium	Same composition as sulfate-limiting medium, but lacking MgSO$_4$; contains 1 mM IPTG
MES concentrate	0.4 M MES-KOH,[c] pH 7.5, 40 mM Tricine-KOH, pH 7.5, 0.1 mM FeCl$_3$, 0.1 M NH$_4$Cl, 2.5 mM KCl, 0.5 M NaCl, 50 mM MgCl$_2$, 50 μM CaCl$_2$, 100-fold diluted micronutrient mix (see below)
Micronutrient mix	3.7 μg/ml NH$_4$MoO$_4$, 18.6 μg/ml boric acid, 7.2 μg/ml CoCl$_2$, 2.5 μg/ml CoSO$_4$, 15.8 μg/ml MnCl$_2$, 2.9 μg/ml ZnSO$_4$

[a] Most of our reagents are purchased from Merck Chemicals, Darmstadt, Germany; enzymes are from Boehringer, Mannheim, Germany, protease inhibitors and protamine sulfate from Sigma, St. Louis, MO, and growth media from Difco, Detroit MI.

[b] Add this and the following ingredients just prior to use.

[c] Abbreviation: MES, 2-(N-morpholino)ethanesulfonic acid.

energized inner membrane as well as ATP are required for the precursor to be translocated across both mitochondrial membranes. Vestweber and Schatz (1988 a,b,c) correlated a decreased stability of the precursor (due to the introduction of mutations) with higher import rates. Using chemically modified forms of the precursor, they were able to quantify the number of import sites per mitochondrion, probe the physical constraints imposed on carboxy-terminally coupled passenger moieties, and identify the first component of the membrane-associated import machinery (Vestweber *et al.,* 1989; see also Hines and Baker, Chapter 19, and Scherer and Krieg, Chapter 23).

In order to identify other components of the mitochondrial import machinery, we have generated a number of different precursor molecules containing cysteine residues at various sites in the protein. These residues were chosen because they can be coupled to photoreactive probes (Scherer and Krieg, Chapter 23) and other reporter groups. However, some point mutations that replace cysteine residues of wild-type DHFR by other amino acids destabilize DHFR to the extent that expression *in vivo* is lowered. Furthermore, the purification of these destabilized forms is more difficult as they do not bind to the methotrexate affinity column. A schematic representation of the constructs tested is given in Fig. 1.

The wild-type version of the COXIV–DHFR fusion protein contains two cysteine residues, one at position 18 in the presequence and the other at position 7 of the DHFR moiety. The cysteine in the presequence is not vital for targeting of the protein to mitochondria, does not affect the stability of the DHFR moiety, and can thus be replaced by serine without significantly altering the properties of the fusion protein. On the other hand, the cysteine at position 7 in DHFR is crucial for tight folding of the protein (Vestweber and Schatz, 1988a). Replacing this residue by serine (constructs KSAB and DV12) abolishes DHFR activity and methotrexate binding. All constructs lacking this cysteine can only be isolated in small amounts according to the protocol for ^{35}S-labeled precursor.

In order to combine the advantages offered by the DV12 precursor (which contains a unique carboxy-terminal chemically reactive cysteine) with the simple purification protocol for tightly folded DHFR, the KSAD version of COXIV–DHFR was created. It lacks the cysteine at position 18 of the presequence, retains the cysteine at position 7 of the DHFR moiety, and has a cysteine at the carboxy terminus of DHFR. This precursor can be isolated in large amounts and can still be selectively modified at its carboxy terminus, because the cysteine at the carboxy terminus of DHFR is much more accessible than that at position 7.

REFERENCES

Amann, E. Brosius, J., and Ptashne, M. (1983). Vectors bearing a hybrid *trp–lac* promoter useful for regulated expression of cloned genes in *E. coli. Gene* **25**, 167–178.

Chen, W.-J., and Douglas, M. G. (1987). The role of protein structure in the mitochondrial import pathway. Unfolding of mitochondrially bound precursors is required for translocation. *J. Biol. Chem.* **262**, 15605–15609.

de Boer, H. A., Comstock, L. J., and Vasser, M. (1983). The *tac* promoter: A functional hybrid derived from the *trp* and *lac* promoters. *Proc. Natl. Acad. Sci. U.S.A.* **78**, 21–25.

Douglas, M. G., Geller, B. L., and Emr, S. D. (1984). Intracellular targeting and import of an F_1-ATPase β-subunit–β-galactosidase hybrid protein into yeast mitochondria. *Proc. Natl. Acad. Sci. U.S.A.* **81**, 3983–3987.

Eilers, M., and Schatz, G. (1986). Binding of a specific ligand inhibits import of a purified precursor protein into mitochondria. *Nature (London)* **322**, 228–232.

Eilers, M., Oppliger, W., and Schatz, G. (1987). Both ATP and an energized inner membrane are required to import a purified precursor protein into mitochondria. *EMBO J.* **6**, 1073–1077.

Eilers, M., Hwang, S., and Schatz, G. (1988). Unfolding and refolding of a purified precursor protein during import into isolated mitochondria. *EMBO J.* **7**, 1139–1145.

Endo, T., and Schatz, G. (1988). Latent membrane perturbation activity of a mitochondrial precursor protein is exposed by unfolding. *EMBO J.* **7**, 1153–1158.

Hurt, E. C., and Schatz, G. (1987). A cytosolic protein contains a cryptic mitochondrial targeting signal. *Nature (London)* **325**, 499–503.

Hurt, E. C., Pesold-Hurt, B., and Schatz, G. (1984a). The amino-terminal region of an imported mitochondrial precursor polypeptide can direct cytoplasmic dihydrofolate reductase into the mitochondrial matrix. *EMBO J.* **3**, 3149–3156.

Hurt, E. C., Pesold-Hurt, B., and Schatz, G. (1984b). The cleavable prepiece of an imported mitochondrial protein is sufficient to direct cytosolic dihydrofolate reductase into the mitochondrial matrix. *FEBS Lett.* **178**, 306–310.

Jaussi, R., Behra, R., Giannattasio, S., Flura, T., and Christen, P. (1987). Expression of cDNAs encoding the precursor and the mature form of chicken mitochondrial aspartate aminotransferase in *E. coli. J. Biol. Chem.* **262**, 12434–12437.

Maccecchini, M.-L., Rudin, Y., Blobel, G., and Schatz, G. (1979). Import of proteins into mitochondria: Precursor forms of the extramitochondrially made F_1-ATPase subunits in yeast. *Proc. Natl. Acad. Sci. U.S.A.* **76**, 343–347.

Murakami, K., Amaya, Y., Takiguchi, M., Ebina, Y., and Mori, M. (1988). Reconstitution of mitochondrial protein transport with purified ornithine carbamoyltransferase precursor expressed in *E. coli. J. Biol. Chem.* **263**, 18437–18442.

Ohta, S., and Schatz, G. (1984). A purified precursor polypeptide requires a cytosolic protein fraction for import into mitochondria. *EMBO J.* **3**, 651–657.

Schmitz, U. K., and Lonsdale, D. M. (1989). A yeast mitochondrial presequence functions as a signal for targeting to plant mitochondria *in vivo. Plant Cell* **1**, 783–791.

Schmitz, U. K., Lonsdale, D. M., and Jefferson, R. A. (1990). Application of the β-glucoronidase gene fusion system to *Saccharomyces cerevisiae. Curr. Genet.* **17**, 261–264.

Sheffield, W. P., Nguyen, M., and Shore, G. C. (1986). Expression in *E. coli* of functional precursor to the rat liver mitochondrial enzyme, ornithine carbamyltransferase. Precursor import and expression *in vitro. Biochem. Biophys. Res. Commun.* **134**, 21–28.

Vestweber, D., and Schatz, G. (1988a). Point mutations destabilizing a precursor protein enhance its post-translational import into mitochondria. *EMBO J.* **7**, 1147–1151.

Vestweber, D., and Schatz, G. (1988b). A chimeric mitochondrial precursor protein with internal disulfide bridges blocks import of authentic precursors into mitochondria and allows quantitations of import sites. *J. Cell Biol.* **107,** 2037–2043.

Vestweber, D., and Schatz, G. (1988c). Mitochondria can import artificial precursor proteins containing a branched polypeptide chain or a carboxy-terminal stilbene disulfonate. *J. Cell Biol.* **107,** 2045–2049.

Vestweber, D., Brunner, J., Baker, A., and Schatz, G. (1989). A 42K outer-membrane protein is a component of the yeast mitochondrial protein import site. *Nature (London)* **341,** 205–209.

Chapter 23

Cross-Linking Reagents as Tools for Identifying Components of the Yeast Mitochondrial Protein Import Machinery

PHILIPP E. SCHERER AND UTE C. KRIEG

Department of Biochemistry
Biocenter, University of Basel
CH-4056 Basel, Switzerland

I. Introduction

Cross-linking approaches have proved useful in identifying pairs of molecules that interact specifically (affinity labeling of receptor–ligand pairs) (e.g., Hanstein, 1979, and references therein) and in characterizing the architecture of multicomponent complexes by establishing near-neighbor relationships (e.g., Capaldi *et al.,* 1979, and references therein).

METHODS IN CELL BIOLOGY, VOL. 34

Recently, cross-linking techniques have played an important role in identifying proteins that are located adjacent to polypeptides translocating across biological membranes.

In the endoplasmic reticulum (ER) system, cross-linking has been successful in identifying the signal sequence receptor (Wiedmann *et al.*, 1987), and mp39, a potential component of the translocation channel across the membrane (Krieg *et al.*, 1989); it has also been instrumental in establishing that the 54-kDa subunit of the signal recognition particle interacts directly with the signal sequence of a nascent polypeptide (Kurzchalia *et al.*, 1986; Krieg *et al.*, 1986).

Cross-linking has also been used to characterize a putative mitochondrial receptor for synthetic matrix-targeting peptides (p30) (Gillespie, 1987). As discussed in detail below, Vestweber and colleagues used photocrosslinking of a translocation-arrested precursor to identify the first membrane component located at the mitochondrial protein import site (see also Hines and Baker, Chapter 19).

Cross-linking approaches can be very powerful in identifying molecules that are located next to each other. When the cross-link forms, the two molecules may exist as a stable complex or may be trapped in a transient state. In the latter case, the experiment requires careful synchronization of the experimental system, so that the cross-linking reaction (usually a low-yield procedure) is carried out in a homogeneous sample. In any case, the formation of a covalent bond between two macromolecules establishes only their spatial proximity, and not necessarily a functional relationship. This latter point must be verified by functional tests, such as competition assays, the use of specific inhibitors, or genetic analysis of the interaction.

In principle, individual components of multimeric complexes can also be identified following coimmunoprecipitation with antibodies raised against one of the known constituents. This approach, however, requires that the complex under study remains intact during the time required for immunoprecipitation; in the case of membrane-associated complexes, the structure must also survive solubilization with detergents. Furthermore, the important epitopes recognized by the antibodies must be accessible in the nondissociated complex. Most importantly, coimmunoprecipitation does not allow one to draw conclusions about the molecular arrangement of polypeptides within a larger structure. Cross-linking, on the other hand, can overcome some of these limitations. As convalent bonds are formed, the analysis of a cross-linked product can be carried out under denaturing conditions, directly demonstrating the physical proximity of the two components involved. Depending on the chemistry of the cross-linking reagent, cross-linking may also be sensitive to the functional and/or physical state of a complex, and may thus detect possible conformational changes.

II. Cross-Linking Using Bifunctional Reagents

A. Chemical and Physical Properties of Cross-Linking Reagents

A wide variety of cross-linkers is commercially available (e.g., from Pierce Chemical Company, Rockford, IL 61105). These reagents can be subdivided into two general classes: homobifunctional cross-linkers, which carry two identical functional groups, and heterobifunctional cross-linkers, which carry two reactive groups with different specificities. These later reagents have the advantage that cross-linking can be performed in a stepwise manner. The reactive groups discussed below are most commonly used in protein cross-linking experiments.

N-Hydroxysuccinimide esters (NHS esters) react specifically with deprotonated primary amines such as the ϵ-amino group of lysine or the amino-terminal α-amino group of a protein. The reaction is best carried out at slightly alkaline pH and results in the formation of an amide bond between the cross-linking reagent and the protein. The solubility of cross-linkers in water is usually limited, but has been greatly increased by introducing a sulfonate group into the NHS moiety. This offers the possibility of using a membrane-permeant (nonpolar) and a membrane-impermeant (polar) version of a given cross-linker to probe a target for its cis or trans location with respect to a membrane barrier (see below). As NHS esters are quickly hydrolyzed, their solutions should be prepared only immediately before use.

Maleimides are widely used for the derivatization of sulfhydryl groups under slightly acidic to neutral conditions. The property to react only with a single type of functional group can be an advantage. On the other hand, this lowers the chance of obtaining a cross-link, as potential reactive groups may not be available at the proper distance and in the proper steric conformation.

Photochemical cross-linking reagents are particularly useful, as the cross-linking is performed in two independent steps. (This is in contrast to chemical reagents, which react during the same incubation with both cross-linking moieties.) In a first step, the cross-linker is reacted with a protein via a maleimide or NHS group, as described above. In a second step, the photolabile group is activated by light. Photolysis usually yields strongly electrophilic species that react fairly nonselectively with even the most unreactive groups of organic molecules, such as carbon–hydrogen bonds. Photoactivatable groups include the following reagents:

Azides are used, most commonly the aryl azides [for a comprehensive discussion of azide photochemistry, see Bayley and Staros (1984)], which can be activated at wavelengths above 300 nm. Thus, irradiation is considered nondamaging to protein samples. Illumination initially generates nitrenes, which may subsequently form secondary reactive species. These intermediates react with many functional groups found in polypeptides (this has been shown for C—H, aryl—, N—H, O—H, and S—H bonds, but others are also predicted). Due to their high reactivity, nitrenes are easily quenched by surrounding water molecules. Azides are very light sensitive and often require handling in safety light. The photoactivation step can also be performed in liquid nitrogen, which is an advantage if a biochemical reaction under investigation needs to be stopped at specific times.

Diazirines are another useful class of photoreactive reagents whose photochemistry is relatively well understood (e.g., Brunner, 1989, and references therein). Photoactivation at wavelengths around 350 nm produces highly reactive carbenes. The photoproducts generated by these carbenes are generally more stable during subsequent analysis of the cross-linking experiment than those formed by activated azides. Also, diazirines do not require handling under safety light, but can be used in day light.

B.   General Considerations

The ability to cross-link two polypeptides depends on two conditions. First, there must be two suitably located target sites on two proteins that can be attacked by an activated moiety of the cross-linker. The more reactive these groups, the higher the chances for the formation of a cross-link. Second, these two target sites must have the right distance from each other. Some flexibility is offered by cross-linkers in which the two reactive groups are separated from each other by spacer arms of different lengths. The longer the spacer, the more distant the partner proteins to be cross-linked can be.

One potential problem in probing near-neighbor relationships by cross-linking is abnormal migration of the cross-linked products on Sodium dodecyl sulfate–polyacrylamide gel electrophoresis (SDS–PAGE); these products usually migrate more slowly than predicted by the sum of the molecular masses of the individual components. This probably reflects branched structures of these species.

The relative ease of obtaining cross-linked products and the difficulty of analyzing such covalent complexes have prompted the development of cross-linkers with a cleavable spacer between the two activated moieties. Such cleavable spacers may contain disulfides that can be reduced using agents such as dithiothreitol or β-mercaptoethanol, diazo groups that can be cleaved with dithionite, or several other cleavable functions. Following cross-linking, the sample can be subjected to immunoprecipitation if antibodies against one of the components are available, and the cross-linked product present in the immunoprecipitate can then be cleaved into its components. Finally, these components can be identified by SDS–PAGE. Alternatively, the cross-linked product can be excised from an SDS–PA gel, cleaved, and analyzed on a second SDS–PA gel. Sensitivity of detection can be increased by radioiodinating the proteins contained in the gel slice (Elder *et al.*, 1977). This method allows detection of proteins in the nanogram range.

An additional advantage is offered by cleavable cross-linkers whose cleavage transfers a radioactive label to the cross-linked target protein. Such reagents include, e.g., sulfosuccinimidyl 2-(p-azidosalicylamido)-ethyl-1,3'-dithiopropionate (SASD) and N-[4-(p-azidosalicylamido)butyl]-3'(2'-pyridyldithio)propionamide (APDP); both are available from Pierce Chemical Co.

In general, there is no rule as to which cross-linker will be the one of choice for a particular problem. Success is a matter of trial and error. Still, it usually pays to consider the chemical properties of the cross-linker and the probable location of the putative partner protein(s). The following examples demonstrate the use of cross-linking reagents in detecting proteins that catalyze the import of precursor proteins into mitochondria.

C. Identification of a Complex Containing a Mitochondrial 70-kDa Stress Protein and a Partly Translocated Precursor Protein

To identify protein components that may play a role in the mitochondrial import of precursor proteins, mitochondria were first allowed to accumulate partially translocated precursor molecules. Translocation was arrested by attaching a tightly folded bovine pancreatic trypsin inhibitor (BPTI) moiety to the carboxy-terminal cysteine residue of a purified authentic precursor protein (Vestweber and Schatz, 1988). When the mitochondria were then treated with the homobifunctional, membrane-permeable cross-linker dithiobis (succinimidopropionate) (DSP), two high-molecular-mass products were formed that contained the stuck precursor and at least one additional protein (Scherer *et al.*, 1990). The cross-linked complexes were also obtained after solubilization of the membranes in Triton X-100. However, if the membrane-impermeant analog of DSP, 3,3'-dithiobis-(sulfosuccinimidyl)propionate (DTSSP), was used, the cross-links were *only* obtained *after* solubilization and not with intact mitochondria. This suggested that the target molecule was located in the matrix space. The approximate molecular mass of the cross-linked protein was assessed by iodinating mitochondria, after accumulation of the stuck precursor and cross-linking, with the membrane-permeant Bolton–Hunter reagent (Bolton and Hunter, 1973). The cross-linked complexes were immunoprecipitated with antiserum raised against the precursor protein, and then analyzed by SDS–PAGE. The bands of interest were excised from the gel, the cross-linker was cleaved by the addition of DTT, and the products were analyzed on a second SDS–PA gel. This revealed that the precursor protein had been cross-linked to a 70-kDa protein. In order to isolate this protein and to study its function, we raised antibodies against 70-kDa mitochondrial proteins and tested them for their ability to immunoprecipitate the cross-linked product. After affinity purification of the sera against mitochondrial proteins that had been separated by anion-exchange chromatography, we obtained a monospecific antibody preparation. These antibodies were used to confirm the submitochondrial location of the cross-linked 70-kDa protein using immunoelectron microscopy. This led to the unambigous identification of the 70-kDa protein as the mitochondrial 70-kDa stress protein, the gene product of the nuclear *SSC1* gene (Craig *et al.*, 1987).

III. Directed Cross-Linking Using a Photoreactive Mitochondrial Precursor Derivative

A. General Considerations

As an alternative to using cross-linking reagents that are added to the sample after the desired functional state of a given system has been achieved, one can use (photo)chemical derivatives of a molecule that is itself part of the structure under investigation. This greatly increases the efficiency and specificity of cross-linking. However, it can also cause major problems, as the derivatized macromolecule may no longer be functional. This point must be checked in all such cross-linking studies.

B. Identification of a 42-kDa Outer Membrane Protein as a Component of the Mitochondrial Protein Import Site

Vestweber *et al.* (1989) used a chimeric mitochondrial fusion protein derivatized with a trifunctional cross-linker to identify an outer membrane protein that was located in close proximity to a translocation-arrested precursor. This outer membrane protein (termed ISP42) was subsequently shown to be essential for the import of precursor molecules and for cell viability (Baker *et al.,* 1990). Each of the cross-linker's three functional groups could be activated individually. Two of these groups were sequentially reacted to link the precursor protein to BPTI. (BPTI caused the product to become stuck across the two mitochondrial membranes; see above.) First, the NHS group was used to derivatize BPTI. After removal of excess cross-linker by gel filtration, the derivatized BPTI was attached via one of its maleimide groups to the unique carboxy-terminal cysteine of the DV12 variant of the cytochrome oxidase subunit IV–dihydrofolate reductase (COXIV–DHFR) hybrid precursor (see Krieg and Scherer, Chapter 22). When the resulting COXIV–DHFR–BPTI adduct was then incubated with mitochondria, it jammed the mitochondrial protein import sites. At this stage, the third, photoreactive diazirine group of the cross-linker was activated. The photocross-linking reaction proceeded with high efficiency (almost 10% of all stuck precursor became cross-linked to the same target molecule) and resulted in a convalent complex of the precursor with a 42-kDa protein.

The success of these two studies can most likely be attributed to the design of the experimental system. The goal was to work with a synchronized population of imported precursor molecules. The "stuck precursor" approach provided the basis for examining the molecular environment of a precursor in transit across the two mitochondrial membranes from two different angles. First, the directed cross-linking (with the photoreactive group being a part of the investigated system) yielded a single target molecule; this reflects the close physical proximity of the carboxy terminus of precursor and ISP42. No other proteins were cross-linked to the stuck precursor, either because ISP42 was the only neighbor molecule, or, more likely, because of steric constraints of the cross-linker; other proteins may simply have been too far away. Second, using a small soluble chemical cross-linker in the same system resulted in the identification of one major covalent complex that involved a protein of the mitochondrial matrix. Again, no other proteins were cross-linked to the stuck precursor. This is probably explained by the fact that binding of the stuck precursor to the 70-kDa stress protein locks the position of the imported protein and allows little flexibility for cross-linking to more distant proteins. In addition, cross-link formation requires two reactive ϵ-amino groups in a defined steric conformation.

The two cross-linking techniques give completely different, although complementary, information about the environment around different domains of the same precursor. These examples document the possibilities, and the limitations, of applying cross-linking techniques to membrane systems.

REFERENCES

Baker, K. P., Schaniel, A., Vestweber, D., and Schatz, G. (1990). A yeast mitochondrial outer membrane protein essential for protein import and cell viability. *Nature (London)* **348,** 605–609.

Bayley, H., and Staros, J. V. (1984). Photoaffinity labeling and related techniques. *In* "Azides and Nitrenes: Reactivity and Utility" (E. F. V. Scriven, ed.), pp. 433–490. Academic Press, Orlando, Florida.

Bolton, A. E., and Hunter, W. M. (1973). The labelling of proteins to high specific radioactivities by conjugation to a ^{125}I-containing acylating agent. *Biochem. J.* **133,** 529–539.

Brunner, J. (1989). Photochemical labeling of apolar phase of membranes. *In* "Methods in Enzymology" (S. Fleischer and B. Fleischer, eds.), Vol. **172,** pp. 628–687. Academic Press, San Diego, California.

Capaldi, R. A., Briggs, M. M., and Smith, R. J. (1979). Cleavable bifunctional reagents for studying near neighbor relationships among mitochondrial inner membrane complexes. *In* "Methods in Enzymology" (S. Fleischer and L. Packer, eds.), Vol. **56,** pp. 630–642. Academic Press, New York.

Craig, E. A., Kramer, J., and Kosic-Smithers, J. (1987). SSC1, a member of the 70-kDa heat shock protein multigene family of *Saccharoymces cerevisiae,* is essential for growth. *Proc. Natl. Acad. Sci. U. S. A.* **84,** 4156–4160.

Elder, J. H., Pickett, R. A., II, Hampton, J., and Lerner, R. A. (1977). Radioiodination of proteins in single polyacrylamide gel slices. *J. Biol. Chem.* **252,** 6510–6515.

Gillespie, L. (1987). Identification of an outer mitochondrial membrane protein that interacts with a synthetic signal peptide. *J. Biol. Chem.* **262,** 7939–7942.

Hanstein, W. G. (1979). Photoaffinity labeling of membrane components. *In* "Methods in Enzymology" (S. Fleischer and L. Packer, eds.), Vol. **56,** pp. 653–683. Academic Press, New York.

Krieg, U. C., Walter, P., and Johnson, A. E. (1986). Photocrosslinking of the signal sequence of nascent preprolactin to the 54-kilodalton polypeptide of the signal recognition particle. *Proc. Natl. Acad. Sci. U. S. A.* **83,** 8604–8608.

Krieg, U. C., Johnson, A. E., and Walter, P. (1989). Protein translocation across the endoplasmic reticulum membrane: Identification by photocrosslinking of a 39-kD integral membrane glycoprotein as part of a putative translocation tunnel. *J. Cell Bio.* **109,** 2033–2043.

Kurzchalia, T. V., Wiedmann, M., Gishovich, A. S., Bochkarva, E. S., Bielka, H., and Rapoport, T. A. (1986). The signal sequence of nascent preprolactin interacts with the 54K polypeptide of the signal recognition particle. *Nature (London)* **320,** 634–636.

Scherer, P. E., Krieg, U. C., Hwang, S. T., Vestweber, D., and Schatz, G. (1990). A precursor protein partly translocated into yeast mitochondria is bound to a 70 kDa mitochondrial stress protein. *EMBO J.* **9,** 4315–4322.

Vestweber, D., and Schatz, G. (1988). A chimeric mitochondrial precursor protein with internal disulfide bridges blocks import of authentic precursors into mitochondria and allows quantitation of import sites. *J Cell Biol.* **107,** 2037–2043.

Vestweber, D., Brunner, J., Baker, A., and Schatz, G. (1989). A 42K outer-membrane protein is a component of the yeast mitochondrial protein import site. *Nature (London)* **341,** 205–209.

Wiedmann, M., Kurzchalia, T. V., Hartmann, E., and Rapoport, T. A., (1987). A signal sequence receptor in the endoplasmic reticulum membrane. *Nature (London)* **328,** 830–833.

INDEX

CONTENTS OF
RECENT VOLUMES

Volume 31

Vesicular Transport

Part A.

Part I. *Gaining Access to the Cytoplasm*

Volume 32

Vescular Transport

Part B.

Volume 33

Flow Cytometry